Fiber Reinforced Thermoplastic Composites: Processing/Structure/Performance Inter-relationships

Fiber Reinforced Thermoplastic Composites: Processing/Structure/Performance Inter-relationships

Abderrahmane Ayadi
Patricia Krawczak
Chung Hae Park

Basel • Beijing • Wuhan • Barcelona • Belgrade • Novi Sad • Cluj • Manchester

Abderrahmane Ayadi
Centre for Materials
and Processes
IMT Nord Europe
Institut Mines-Télécom
University of Lille
Douai
France

Patricia Krawczak
Centre for Materials
and Processes
IMT Nord Europe
Institut Mines-Télécom
University of Lille
Douai
France

Chung Hae Park
Centre for Materials
and Processes
IMT Nord Europe
Institut Mines-Télécom
University of Lille
Douai
France

Editorial Office
MDPI AG
Grosspeteranlage 5
4052 Basel, Switzerland

This is a reprint of articles from the Special Issue published online in the open access journal *Polymers* (ISSN 2073-4360) (available at: www.mdpi.com/journal/polymers/special_issues/45KNPC2392).

For citation purposes, cite each article independently as indicated on the article page online and using the guide below:

Lastname, A.A.; Lastname, B.B. Article Title. *Journal Name* **Year**, *Volume Number*, Page Range.

ISBN 978-3-7258-1866-2 (Hbk)
ISBN 978-3-7258-1865-5 (PDF)
https://doi.org/10.3390/books978-3-7258-1865-5

Cover image courtesy of Patricia Krawczak

© 2024 by the authors. Articles in this book are Open Access and distributed under the Creative Commons Attribution (CC BY) license. The book as a whole is distributed by MDPI under the terms and conditions of the Creative Commons Attribution-NonCommercial-NoDerivs (CC BY-NC-ND) license (https://creativecommons.org/licenses/by-nc-nd/4.0/).

Contents

About the Editors vii

Preface ix

Hongfu Li, Tianyu Wang, Changwei Cui, Yuxi Mu and Kangmin Niu
Low-Density and High-Performance Fiber-Reinforced PP/POE Composite Foam via Irradiation Crosslinking
Reprinted from: *Polymers* 2024, 16, 745, doi:10.3390/polym16060745 1

Yiliang Sun, Jingwen Li and Boming Zhang
Processing Method and Performance Evaluation of Flame-Retardant Corrugated Sandwich Panel
Reprinted from: *Polymers* 2024, 16, 696, doi:10.3390/polym16050696 17

Zhenhua Wang, Weili Feng, Jiachen Ban, Zheng Yang, Xiaomin Fang and Tao Ding et al.
Sisal-Fiber-Reinforced Polypropylene Flame-Retardant Composites: Preparation and Properties
Reprinted from: *Polymers* 2023, 15, 893, doi:10.3390/polym15040893 27

Janos Birtha, Eva Kobler, Christian Marschik, Klaus Straka and Georg Steinbichler
Using Heating and Cooling Presses in Combination to Optimize the Consolidation Process of Polycarbonate-Based Unidirectional Thermoplastic Composite Tapes
Reprinted from: *Polymers* 2023, 15, 4500, doi:10.3390/polym15234500 40

Jan-Christoph Zarges, André Schlink, Fabian Lins, Jörg Essinger, Stefan Sommer and Hans-Peter Heim
Influence of Different Hot Runner-Systems in the Injection Molding Process on the Structural and Mechanical Properties of Regenerated Cellulose Fiber Reinforced Polypropylene
Reprinted from: *Polymers* 2023, 15, 1924, doi:10.3390/polym15081924 61

Fabian Lins, Christian Kahl, Jan-Christoph Zarges and Hans-Peter Heim
Modification of Polyamide 66 for a Media-Tight Hybrid Composite with Aluminum
Reprinted from: *Polymers* 2023, 15, 1800, doi:10.3390/polym15071800 80

Mei-Xian Li, Hui-Lin Mo, Sung-Kwon Lee, Yu Ren, Wei Zhang and Sung-Woong Choi
Rapid Impregnating Resins for Fiber-Reinforced Composites Used in the Automobile Industry
Reprinted from: *Polymers* 2023, 15, 4192, doi:10.3390/polym15204192 93

María José Paternina Reyes, Jimy Unfried Silgado, Juan Felipe Santa Marín, Henry Alonso Colorado Lopera and Luis Armando Espitia Sanjuán
Cashew Nutshells: A Promising Filler for 3D Printing Filaments
Reprinted from: *Polymers* 2023, 15, 4347, doi:10.3390/polym15224347 120

Hyunkyung Lee, Minsu Kim, Gyungha Kim and Daeup Kim
Effect of the Chemical Properties of Silane Coupling Agents on Interfacial Bonding Strength with Thermoplastics in the Resizing of Recycled Carbon Fibers
Reprinted from: *Polymers* 2023, 15, 4273, doi:10.3390/polym15214273 134

Sofie Verstraete, Bart Buffel, Dharmjeet Madhav, Stijn Debruyne and Frederik Desplentere
Short Flax Fibres and Shives as Reinforcements in Bio Composites: A Numerical and Experimental Study on the Mechanical Properties
Reprinted from: *Polymers* 2023, 15, 2239, doi:10.3390/polym15102239 148

Antoine Runacher, Mohammad-Javad Kazemzadeh-Parsi, Daniele Di Lorenzo, Victor Champaney, Nicolas Hascoet and Amine Ammar et al.
Describing and Modeling Rough Composites Surfaces by Using Topological Data Analysis and Fractional Brownian Motion
Reprinted from: *Polymers* **2023**, *15*, 1449, doi:10.3390/polym15061449 . . . . **167**

Sujith Sidlipura, Abderrahmane Ayadi and Mylène Lagardère Deléglise
Assessing Intra-Bundle Impregnation in Partially Impregnated Glass Fiber-Reinforced Polypropylene Composites Using a 2D Extended-Field and Multimodal Imaging Approach
Reprinted from: *Polymers* **2024**, *16*, 2171, doi:10.3390/polym16152171 . . . . **182**

About the Editors

Abderrahmane Ayadi

Abderrahmane Ayadi completed his engineering preparatory classes in Maths and Physics in 2009 at the Engineering Preparatory Institute of Sfax (Tunisia). He earned an Engineering degree in Materials Engineering from the National Engineering School of Sfax (Tunisia) in 2012, following the completion of his end-of-studies project at the Department of Mechanical Engineering at Penn State University (USA). In 2016, he received his Ph.D. in Mechanics, Energetics, and Materials from the University of Lille 1, France. After two years of post-doctoral research at IMT Nord Europe (2016–2018), he joined the Centre for Materials and Processes (CERI-MP) as an associate professor in October 2018. Dr. Ayadi's research activities focus on experimental and numerical mechanics applied to manufacturing processes of thermoplastics and thermoplastic matrix composites. His expertise includes injection moulding, thermocompression, and thermoforming processes, with a particular emphasis on image-based characterization of process-induced microstructures, in situ mechanical testing, and in operando instrumentation using imaging techniques. He has co-authored over 20 publications in peer-reviewed journals and conferences, referenced in Scopus. Dr. Ayadi actively participates in scientific events organized by the European Scientific Association for Material Forming, the French Mechanics Society, and the European Mechanics Society, contributing significantly to the academic community.

Patricia Krawczak

Patricia Krawczak graduated with an engineering degree (1989) from Mines Douai, France, then with a M.Sc. (1990) and a Ph.D. in organic and macromolecular chemistry (1993), and finally a diploma of habilitation in physics (1999) from the University of Lille, France. She is currently a full professor at IMT Nord Europe, a top-level graduate school of engineering and a research center of the Institut Mines-Telecom (IMT), part of the number one public group of engineering and management graduate schools in France. After having managed IMT Nord Europe's research and education department Polymers and Composites Technology and Mechanical Engineering (2000 to 2019), she has been in charge of the strategic program Technology Platforms of the Institut Mines Telecom since 2019. She has gained deep experience in the field of plastics and composites engineering through numerous collaborations with industry and academia, participated as a principal investigator in numerous national and European industry-oriented projects, and supervised a number of master's and Ph.D. students. She has co-authored over 400 journal or conference publications and book chapters and is ranked among the world's top 2 percent of scientists (according to Stanford University–Elsevier data updates). Her research interests cover advanced processing–manufacturing technologies and the physics and mechanics of polymers and polymer composites. Additionally, she is a member of the Executive Management Board of the French Society of Plastics Engineers (SFIP) and a member of the Scientific and Strategic Advisory Board of the French Technical Center for Plastics and Composites (IPC). She is also a Knight of the National Order of Academic Palms (decoration for valuable services to the universities in education and teaching) and a Knight of the National Order of the Legion of Honor (highest French order of merit, decoration for public service and professional activity with eminent merits).

Chung Hae Park

Chung-Hae Park graduated with B.Sc. and M.Sc. degrees (1996 and 1998) from Seoul National University (Korea), a dual Ph.D. degree (2003) in both mechanical and materials engineering and mechanical and aerospace engineering from Mines Saint-Etienne (France) and Seoul National University (Korea), and a diploma of habilitation in Mechanics (2011) from University Le Havre (France). After two experiences in both industry and academia—first as a senior research engineer (2003–2005) at LG Chemical's Technology Centre in Korea and then as a faculty member (2006–2013) at University Le Havre in France—he has been appointed as a full professor at IMT Nord Europe (France) since 2013. He has been the deputy director (in charge of innovation) of the Centre for Materials and Processes of this top-level engineering graduate school since 2021. As a principal investigator and coordinator in numerous national and European industry-oriented projects, he has supervised a number of master's and Ph.D. students. He has co-authored more than 150 journal or conference publications and book chapters and has four patents. His expertise areas cover high-performance structural composite materials (thermoset or thermoplastic matrices and carbon, glass, or vegetal fibres) and processing and manufacturing issues (liquid composite moulding, resin transfer moulding, SMC, compression, and injection moulding), with a focus on numerical modelling and simulation.

Preface

This special issue presents a diverse collection of research articles, categorized as follows:

1. Composite Material Development

- Li et al. (2024): developed low-density, high-performance PP/POE composite foam via irradiation crosslinking, improving mechanical properties and thermal shrinkage resistance.
- Sun et al. (2024): focused on flame-retardant thermoplastic corrugated sandwich panels, highlighting flame-retardant additives' potential in expanding application ranges.
- Wang et al. (2023): investigated sisal-fiber-reinforced polypropylene composites, enhancing fire resistance in natural-fiber-reinforced materials.

2. Process Optimization and Applications

- Birtha et al. (2023): optimized the consolidation of unidirectional fiber-reinforced thermoplastic tapes, enhancing bonding strength and reducing voids.
- Zarges et al. (2023): studied the impact of hot runner systems on regenerated-cellulose-fiber-reinforced polypropylene.
- Lins et al. (2023): modified polyamide 66 for media-tight hybrid composites with aluminum, targeting automotive uses.
- Mei-Xian Li et al. (2023): investigated rapid impregnating resins for automotive composites, focusing on optimizing resin impregnation.

3. Sustainability and Environmental Impact

- Paternina Reyes et al. (2023): used cashew nut shells as a filler for 3D printing filaments, promoting sustainability by utilizing agricultural waste.
- Lee et al. (2023): examined silane coupling agents in improving bonding strength with thermoplastics in recycled carbon fibers.
- Verstraete et al. (2023): used flax stems as reinforcements in polylactic acid bio-composites, offering an eco-friendly alternative.

4. Advanced Modeling Techniques

- Runacher et al. (2023): modeled rough composite surfaces using topological data analysis and fractional Brownian motion, providing insights into surface roughness.

5. Advanced Characterization Techniques

- Sidlipura et al. (2024): utilized 2D extended-field and multimodal imaging to analyze porosity and surface area fractions in partially impregnated glass-fiber-reinforced polypropylene composites, showcasing the approach's potential at the meso-scale.

This collection of articles underscores the diversity of research directions and the potential of fiber-reinforced thermoplastic composites in various applications, driven by the ongoing pursuit of sustainability, performance optimization, and material innovation.

Abderrahmane Ayadi, Patricia Krawczak, and Chung Hae Park
Editors

Article

Low-Density and High-Performance Fiber-Reinforced PP/POE Composite Foam via Irradiation Crosslinking

Hongfu Li *, Tianyu Wang, Changwei Cui, Yuxi Mu and Kangmin Niu *

School of Materials Science and Engineering, University of Science and Technology Beijing, Beijing 100083, China
* Correspondence: lihongfu@ustb.edu.cn (H.L.); niukm@ustb.edu.cn (K.N.)

Abstract: This study addresses the challenge of achieving foam with a high expansion ratio and poor mechanical properties, caused by the low melt viscosity of semi-crystalline polypropylene (PP). We systematically employ a modification approach involving blending PP with polyolefin elastomers (POE), irradiation crosslinking, and fiber reinforcement to prepare fiber-reinforced crosslinked PP/POE composite foam. Through optimization and characterization of material composition and processing conditions, the obtained fiber-reinforced crosslinked PP/POE composite foam exhibits both low density and high performance. Specifically, at a crosslinking degree of 12%, the expansion ratio reaches 16 times its original value, and a foam density of 0.057 g/cm^3 is reduced by 36% compared to the non-crosslinked PP/POE system with a density of 0.089 g/cm^3. The density of the short-carbon-fiber-reinforced crosslinked sCF/PP/POE composite foam is comparable to that of the crosslinked PP/POE system, but the tensile strength reaches 0.69 MPa, representing a 200% increase over the crosslinked PP/POE system and a 41% increase over the non-crosslinked PP/POE system. Simultaneously, it exhibits excellent impact strength, tear resistance, and low heat shrinkage. Irradiation crosslinking is beneficial for enhancing the melt strength and resistance to high temperature thermal shrinkage of PP/POE foam, while fiber reinforcement contributes significantly to improving mechanical properties. These achieve a good complementary effect in low-density and high-performance PP foam modification.

Keywords: polypropylene; melt strength; thermoplastic foam; irradiation crosslinking; fiber composite; mechanical property

Citation: Li, H.; Wang, T.; Cui, C.; Mu, Y.; Niu, K. Low-Density and High-Performance Fiber-Reinforced PP/POE Composite Foam via Irradiation Crosslinking. *Polymers* 2024, *16*, 745. https://doi.org/10.3390/polym16060745

Academic Editors: Patricia Krawczak, Chung Hae Park and Abderrahmane Ayadi

Received: 26 January 2024
Revised: 28 February 2024
Accepted: 4 March 2024
Published: 8 March 2024

Copyright: © 2024 by the authors. Licensee MDPI, Basel, Switzerland. This article is an open access article distributed under the terms and conditions of the Creative Commons Attribution (CC BY) license (https:// creativecommons.org/licenses/by/ 4.0/).

1. Introduction

Due to the ease of synthesis, excellent comprehensive performance, and low cost of polypropylene (PP), it has become one of the most widely used plastics globally, and is extensively applied in various aspects of our daily lives [1]. For instance, PP can achieve lightweight design and high specific strength structural application through foaming processes, while possessing characteristics such as thermal insulation. Particularly driven by the increasing demand in recent years, there is a significant need for PP in emerging fields, like sandwich structures of automotive structures, interiors, food packaging, and other fields [1–3]. Enhancing the foaming expansion ratio and mechanical properties of PP foam to reduce the usage of chemically synthesized materials represents a crucial direction in the development of polymer foam applications under the impetus of a low carbon footprint.

However, due to the uniform chain structure and high crystallinity of PP, its low melt strength near foaming temperature makes it challenging to support the stable growth of high expansion ratio bubbles, leading to the occurrence of bubble rupture. For instance, Lee [4] investigated the relationship between the extrusion foaming ratio and processing temperature of PP. It was found that, during the foaming process, if the temperature is too high, the melt strength of PP significantly decreases, resulting in severe gas escape and potential damage to the bubble structure. On the other hand, if the temperature is too low, the melt strength becomes excessively high, hindering the growth of bubbles. The optimal foaming temperature for the system was identified as 190 °C, and it was emphasized that the foaming temperature needs to consider the impact of the system and foaming agent type. Burt [5] evaluated the foaming temperature window for PP and found a narrow temperature range of only 4 °C suitable for PP foaming. Therefore, to achieve PP foaming materials with a high expansion ratio, enhancing the melt strength of PP during the foaming process and expanding the operational temperature window are crucial technical challenges.

To enhance the melt strength of PP, various approaches such as blending with thermoplastic elastomers [6–8], introducing side-chain branching to PP [9,10], and crosslinking [11–14] can be employed. For instance, Wang and Gong [8,15] investigated blending modifications of PP and polyolefin elastomers (POE), revealing that POE exhibits higher melt strength, good flowability, lower melting temperature, compatibility with PP, and ease of processing. Additionally, incorporating POE can reduce the crystallinity of PP, which is advantageous for minimizing the sensitivity of melt strength to temperature changes, and enhancing the impact resistance of PP. Han [16] utilized a two-step method involving a foaming agent, crosslinking agent blending modification, and irradiation crosslinking to improve the melt strength of PP. Subsequently, foaming of the crosslinked PP was achieved using compression molding, resulting in PP foam with uniform pore distribution, small pore size, and excellent mechanical properties. Notably, physical crosslinking methods such as irradiation [17–19] are more environmentally friendly compared to chemical crosslinking methods. They effectively avoid the use of volatile peroxide initiators and chemical crosslinking agents like benzoyl peroxide (BPO), dicumyl peroxide (DCP), and triallyl isocyanurate (TAIC), aligning with development needs for a low carbon footprint and low VOC emissions [13].

However, the high foaming ratio results in a significant reduction in material density, and it is challenging to simultaneously maintain excellent mechanical properties. Well-dispersed particles like clay [20–22], talc [23,24], nano-$CaCO_3$ [25], hollow molecular-sieve particles [26], CNT [27], and graphene [28] are generally used to enhance PP cell structures and mechanical properties. However, the reinforcing effect of particles is often limited. Fiber reinforcement has proven to be an effective method for enhancing the performance of polymer materials [29,30] and concrete [31,32]. In recent years, the concept of fiber reinforced polymer composites has gradually been applied to the research on the reinforcement and modification of high-performance polymer foams. Relevant results indicate that fiber parameters such as strength, length, and content can influence the thermal and mechanical properties of polymers, as well as their foaming behavior [33–36]. Sebaey [37,38] investigated the energy absorption behavior of polyurethane (PU) foam-filled carbon-fiber-reinforced polymer composite tubes under impact. Chuang [39] developed carbon-fiber- and glass-fiber-reinforced foamed PU composite multifunctional protective boards suitable for diversified environments. Kumar [40] emphasized the reinforcement of rigid polyurethane foam (RPUF) via the addition of glass fibers for diverse engineering applications. Mechanical properties were found to be improved with addition of GF content due to the increased foam density and decreased cell size. Kuranchie [41] reviewed the effect of natural fiber reinforcement on PU composite foams. Shen [42] carried out foaming analysis of polystyrene-carbon nanofiber nanocomposite. However, for PP polymer, research is primarily concentrated on fiber-reinforced non-foamed PP composites [43–45], with limited reports on attempts to obtain fiber-reinforced PP composite foam. Wang [46] attempted to enhance PP fine-celled foaming with the incorporation of short carbon fibers via melt compounding in a twin-screw extruder, followed by injection molding and foaming with supercritical CO_2. The most uniform foam size distribution and cell morphology were achieved with 25 wt.% carbon fibers. Nonetheless, a large number of unfoamed regions could still be observed. Furthermore, a low foaming rate, indicated by a high foam density of 0.58 g/cm^3, was achieved due to the increased melt viscoelasticity at high fiber content, which inhibited the foaming process. Cai [35] predicted the elastic moduli of the foamed glass-fiber-reinforced PP composite foam through experiments and by constructing a multilayer RVE model. Rachtanapun [47] investigated wood fiber reinforced PP/HDPE blend composites to determine the effects of processing condition, blend composition, and wood fiber content on the void fraction and cell morphology of the materials without involving the study of mechanical properties. Therefore, it is evident that there are still challenges in achieving a high foaming expansion ratio while maintaining high strength for fiber-reinforced PP.

Based on this, we simultaneously employed a blend modification approach using POE thermoplastic elastomer and physical irradiation crosslinking to enhance the melt strength of PP, aiming to produce a higher expansion ratio and low-density PP foam. Simultaneously, we utilized short glass fibers (sGF) and short carbon fibers (sCF) to enhance the mechanical properties of the PP foam. Through optimization of material composition, process conditions, and characterization assessments, we systematically investigated the influence of thermoplastic elastomers, crosslinked network structures, fiber types, and content on the foaming, thermal stability, and mechanical performance of PP, ultimately achieving the fabrication of low-density and high-performance fiber-reinforced PP/POE composite foam.

2. Materials and Methods

2.1. Materials

Isotactic polypropylene (PP), a brand of S131, with a molecular weight (M_w) of 420,804 and a melt flow index of 1.3 g/10 min was supplied by Sumitomo Chemical Co., Ltd., Tokyo, Japan. Polyolefin elastomer (POE), a brand of 6202, exhibiting a Shore hardness of 64 A and a melt flow index of 9.1 g/10 min was obtained from ExxonMobil Corporation, Spring, TX, USA. The foaming agent ADC 271, primarily composed of azodicarbonamide, was sourced from Baerlocher GmbH, Unterschleißheim, Germany. Stearic acid was utilized as a lubricant during processing. The antioxidant 1076G was obtained from BASF. The nucleating agent TMA-3, identified as an α-type nucleating agent for improving the cell distribution of foaming materials, was procured from Anhui Xiangyun Rubber Plastic Co., Ltd., Hefei, China. Triallyl isocyanurate (TMPTMA), a brand of TMPTMA-P, was used as a crosslinking agent and was provided by Anhui Xiangyun Chemical Co., Ltd., Hefei, China. Short carbon fibers (sCF), specifically T300 with a length of 2 mm, were procured from Toray. Short glass fibers (sGF) with a length of 3 mm were supplied by Shanghai Lishuo Composite Materials Co., Ltd., Shanghai, China.

2.2. Preparation of PP/POE Blend Sheets Ready to Foam

In this study, a two-step compression molding foaming process was employed to prepare PP foam samples. This method requires low flowability of the polypropylene melt, making it particularly suitable for convenient operations when modifying the foaming system through blending, incorporating fibers, or introducing other inorganic fillers [48]. As shown in Figure 1, initially, according to the proportions listed in Table 1, a banbury mixer (XH-401CE-160, Guangdong Xihua Machinery Co., Ltd., Dongguan, China) was utilized to mix PP, POE, stearic acid, antioxidant, and either sCF or sGF at 160 °C for 5 min at medium speeds and high speeds, respectively, resulting in the preparation of a fiber-reinforced PP/POE premix. Subsequently, the foaming agent, nucleating agent, and crosslinking agent were added, followed by low-speed mixing for 5 min and high-speed mixing for an additional 5 min. The mixed material was then placed on a flat vulcanization bed, degassed at 170 °C for 1 min, and held under high pressure for 90 s to mold the pre-mixture into PP/POE sheets ready to foam.

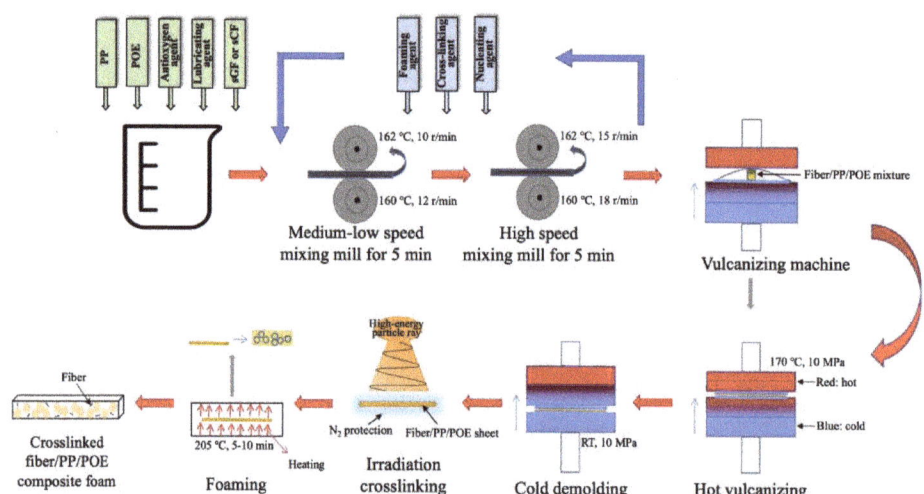

Figure 1. Diagram for the preparation of PP/POE and fiber-reinforced PP/POE blend sheet, irradiation crosslinking, and foaming processes.

Table 1. The composition of PP/POE blend and fiber-reinforced PP/POE composite foams.

Material	PP	POE	Nucleating Agent	Lubricating Agent	Antioxygen Agent	Foaming Agent	Crosslinking Agent	Reinforced Fibers
Brand	S131	6202	TMA-3	Stearic acid	1076G	ADC	TMPTMA	sCF or sGF
Content (g)	60	40	0.4	1.0	0.5	15	0–16	0–20

2.3. Radiation Crosslinking and Foaming of PP/POE Blend Sheet

The premixed samples of PP/POE or fiber-reinforced PP/POE sheets were placed in plastic bags. After purging air and introducing nitrogen, the bags were sealed. An electron accelerator (GJ-2, Sichuan Atomic Energy Research Institute) was used to induce active radicals and crosslinking in PP by adjusting the irradiation dose settings to 2/4/8/10/15/20 kGy. The irradiated sheets were then heat treated at 120 °C for 2 h to eliminate residual free radicals generated during the irradiation process. Subsequently, the heat treated premix sheets were placed in a coating machine for foaming, with a foaming temperature of 200–220 °C and a dwell time of 5–7 min, resulting in the final production of crosslinked PP/POE blended foamed sheets.

2.4. Characterizations

The density of PP premix sheets and foams was measured using a solid density analyzer (KW-300Y). The foaming expansion ratio V_f was determined by Formula (1):

$$V_f = \frac{\rho_u}{\rho_f} \quad (1)$$

where V_f represents the foaming expansion ratio, ρ_u is the density of the unfoamed sample in g/cm^3, and ρ_f is the density of the foamed sheet in g/cm^3.

The degree of crosslinking was assessed through gel content in PP. Approximately 0.2 g of the crosslinked and unfoamed sample was weighed and recorded as W. After the sample was wrapped in a copper mesh with a 200 mesh and dried at 80 °C for 30 min, the dried mass was recorded as W_1. Following this, the sample was placed in 140 °C xylene and stirred reflux for 6 h. Finally, the PP mass, after removing the copper mesh and drying at 80 °C for 1 h, was recorded as W_2. The gel content G was calculated using Formula (2):

$$G = \left(1 - \frac{W_1 - W_2}{W}\right) \times 100\% \qquad (2)$$

high temperature thermal shrinkage performance involved placing 20 mm × 20 mm × 4 mm samples in a 160 °C oven for 10 min. After the samples cooled to room temperature and were allowed to stand for 2 h, the changes in thickness before (H_u) and after (H_ρ) thermal treatment were measured, and the thermal stability parameter α was calculated according to Formula (3):

$$\alpha = \frac{H_u - H_\rho}{H_\rho} \qquad (3)$$

the dynamic rheological properties were tested using a rotational rheometer (MCR302). A parallel plate mold with a diameter of 35 mm and a 1 mm sample gap was used. The frequency was set from 0.1 to 100 rad/s, and the linear viscoelastic range strain was 2%. The storage modulus, loss modulus, viscosity, and tan δ of the polymer matrix were tested at 200 °C in a N_2 environment.

Tensile strength of the samples followed ASTM D638 standard, with sample dimensions of 200 mm × 15 mm × 4 mm and a stretching rate of 10 mm/min. Impact strength was evaluated using a simple beam impact tester (XJ-300) for specimens without notches, measuring 75 mm × 15 mm × 4 mm, with an impact speed of 2.9 m/s. Tear strength testing followed ASTM D 3574 standard, with a sample size of 150 mm × 20 mm × 4 mm, containing a triangular notch in the middle. The crosshead speed of grip separation was 500 mm/min.

Microscopic foam pore morphology was observed using a scanning electron microscope (SEM). Image J software v1.54 was employed for pore size and distribution statistics.

3. Results and Discussion

3.1. Irradiation Crosslinking Process Optimization of PP/POE Foam

3.1.1. Crosslinking Degree

The mechanism of chemical structure crosslinking induced by irradiation in PP is illustrated in Figure 2. Under the influence of irradiation, high-energy particles excite PP to generate tertiary carbon radicals. In the absence of a crosslinking agent in the system, the generated tertiary carbon radicals tend to undergo β-scission reactions [49], leading to the main chain breakage of PP. This results in a decrease in molecular weight, and further decomposition into small molecules. The melt strength of these small molecules is lower, making it less favorable for foaming. However, when a crosslinking agent with double bonds and multifunctional groups is present in the PP system, its carboxyl functional group can effectively inhibit β-scission reactions under irradiation. The carboxyl functional group binds with tertiary carbon radicals, forming more stable radicals. These stable radicals can undergo coupling termination reactions with the remaining tertiary carbon radicals, forming a crosslinked structure or network structure. The addition of a crosslinking agent and irradiation dosage are the two main influencing parameters affecting the degree of crosslinking in PP [50]. The degree of crosslinking is generally characterized by the gel content of the material, where a higher gel content corresponds to a higher degree of crosslinking. In this study, Trimethylolpropane trimethacrylate (TMPTMA) was chosen as the crosslinking agent, primarily controlling the crosslinking reaction through its double bond structure, as depicted in Figure 2.

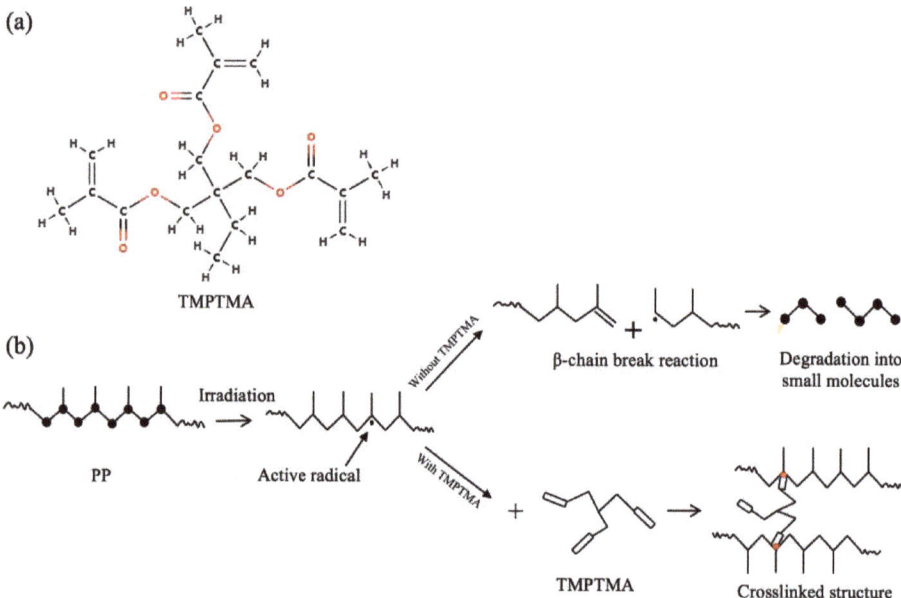

Figure 2. (**a**) Molecular structure of the crosslinking agent TMPTMA, (**b**) mechanism of PP chain scission and crosslinking induced by irradiation.

Figure 3a illustrates the relationship between the crosslinking degree and the amount of crosslinking agent at an irradiation dosage of 10 kGy. It is evident that, with a crosslinking agent content below 8 phr, the crosslinking degree increases with an increasing crosslinking agent. However, beyond 8 phr, the trend becomes more gradual. This is attributed to the excess tertiary carbon radicals generated in the system due to irradiation. As the content of the crosslinking agent increases, the crosslinking degree shows a positive correlation. However, the limited production of tertiary carbon radicals at 10 kGy can result in an insufficient supply of radicals to match the increased amount of crosslinking agent and facilitate additional crosslinking. At this point, the crosslinking degree reaches a peak, entering a plateau phase.

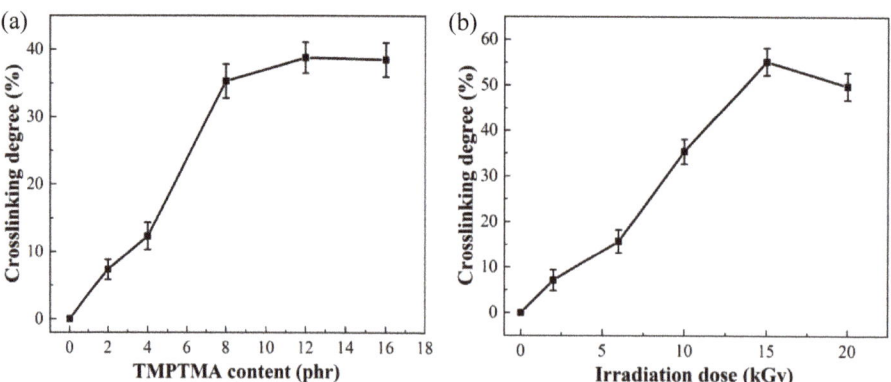

Figure 3. The change in crosslinking degree with variations in (**a**) crosslinking agent content and (**b**) irradiation dosage. (phr: an abbreviation for "Parts per Hundred". In this study, it specifically means "with PP/POE as 100 parts, and the proportion of other materials relative to PP/POE in terms of parts").

Figure 3b illustrates the relationship between crosslinking degree and irradiation dosage at a crosslinking agent content of 8 phr. It is evident that an initial increase in irradiation dosage significantly enhances the crosslinking degree. At an irradiation dosage of 15 kGy, the crosslinking degree reaches its maximum at 55%. Subsequently, with further increases in irradiation dosage, the crosslinking degree shows a declining trend. The observed decrease may be attributed to the accumulation of tertiary carbon radicals in the system with increasing irradiation dosage. However, since the crosslinking agent content in the system is fixed, the number of tertiary carbon radicals available for forming a crosslinked system with the crosslinking agent is limited. Excessive radicals can trigger the aforementioned β-scission reactions, causing PP to degrade into numerous small molecules and resulting in a decrease in gel content.

It can be concluded that the amount of the crosslinking agent and the irradiation dose jointly determine the PP/POE crosslinking degree. Achieving a high crosslinking degree requires a high irradiation dose and an adequate amount of crosslinking agent corresponding to the irradiation dose. Additionally, under the experimental conditions of this study, the crosslinking degree of the crosslinked PP/POE blend system can be regulated within a relatively broad range of 0–60%. Therefore, by adjusting the ratio of irradiation dose to crosslinking agent content, the crosslinking degree, melt strength, and flowability of the PP/POE blend can be controlled. This, in turn, allows for the control of the foaming behavior, mechanical properties, and thermal stability of PP foams.

3.1.2. Foaming and Rheological Behaviors

The crosslinking degree has a certain impact on the melt strength and flowability of the system during foaming, thereby influencing the foaming effect of the foam material. We employ different PP/POE premix sheets with varying crosslinking degrees and compare their lowest achieved foaming density to explore the influence of the crosslinking degree on the foaming effect. Figure 4 shows the relationship between the density and the crosslinking degree of PP/POE foam. It is evident that with an increase in the crosslinking degree, the density of the foamed material significantly decreases. Particularly, when the crosslinking degree reaches 12%, the density reaches its lowest point, with a foaming density of only 0.057 g/cm^3 and a corresponding foaming ratio of 16. The density is much lower than that of non-crosslinked PP/POE blend and that reported in the literatures [13,25,46]. As the crosslinking degree continues to increase to 35%, the foaming density shows a noticeable increase and then maintains a constant plateau up to 55%, with a density of 0.076 g/cm^3. However, the density is still lower than that of the non-crosslinked material, which has a density of 0.089 g/cm^3. It is evident that the foaming density does not exhibit a purely positive correlation with the crosslinking degree. This is because, with an increase in the crosslinking degree within the system, the melt strength of the foaming system continues to improve. However, when the crosslinking degree is too high, such as 55%, the flowability of the melt is significantly restricted due to high viscosity, leading to insufficient expansion and growth of the bubbles, affecting both the size and quantity of the generated pores. Ultimately, this results in a decrease in the foaming ratio. Han reported that the melt viscosity drops remarkably above 2 kGy, decreasing to below the melt viscosity of the virgin PP [16]. Overall, a crosslinking degree of 12% demonstrates the best foaming effect for the system, achieving a foaming ratio of 16.

The impact of increasing crosslinking degree on the enhancement of melt strength in the PP foaming system and its influence on foam morphology can be elucidated through the rheological properties presented in Figure 5. Figure 5a illustrates the relationship between the storage modulus of different crosslinking degrees in the crosslinked PP/POE blend foaming matrix and shear rate. It can be observed that the storage modulus increases with the increment of shear rate, and the crosslinked PP/POE blended system with a higher crosslinking degree exhibits a larger storage modulus. Since the storage modulus at low shear viscosity reflects the elasticity of the melt, it is evident that the elasticity of the melt is significantly enhanced after the introduction of crosslinking modification, contributing to the growth of bubbles.

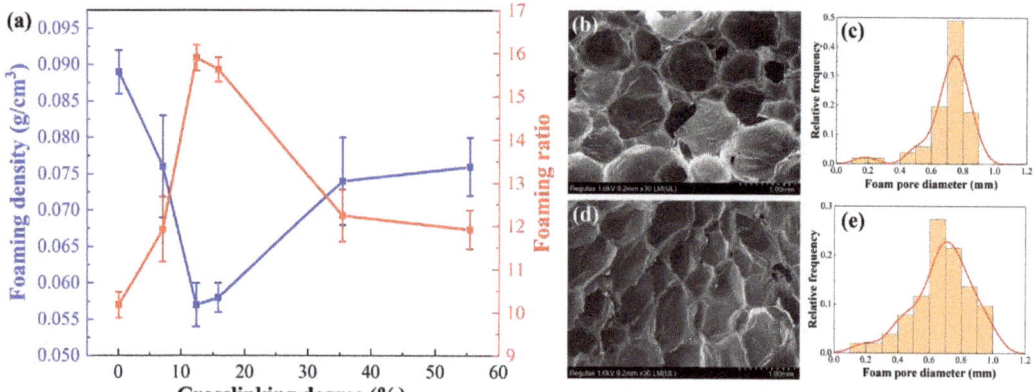

Figure 4. (**a**) The influence of crosslinking degree on foaming density and foaming expansion ratio, as well as the cellular morphology and pore size distribution at crosslinking degrees of (**b**,**c**) 12% and (**d**,**e**) 55%.

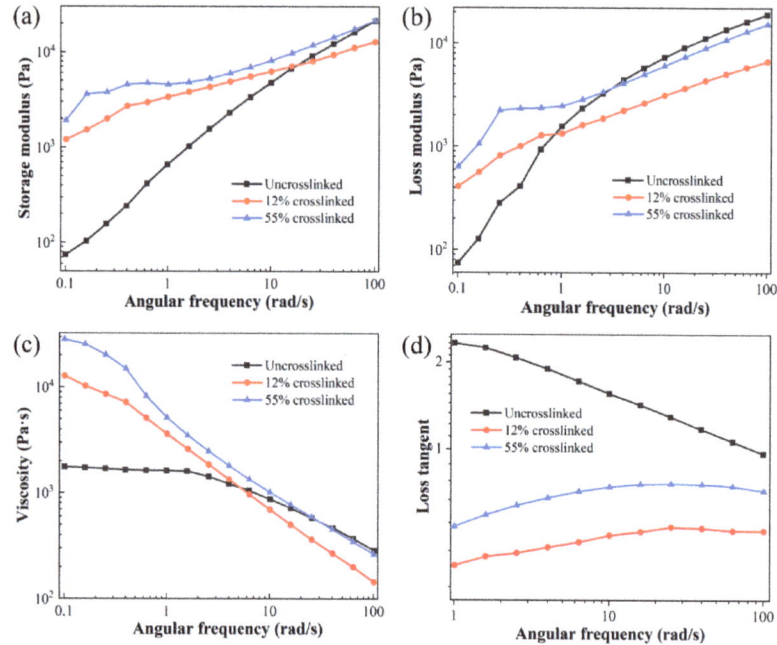

Figure 5. The impact of crosslinking degree on the melt properties of (**a**) storage modulus, (**b**) loss modulus, (**c**) viscosity, and (**d**) loss factor.

Figure 5b depicts the relationship between the loss modulus of the melt and shear rate. At low shear rates, the crosslinked PP/POE blended system has a higher loss modulus compared to the non-crosslinked blended system, which may be attributed to the formation of a crosslinked network in the crosslinked system, limiting the movement of the melt. Figure 5c shows that the viscosity of the PP/POE blended system exhibits a typical shear-thinning phenomenon due to the gradual orientation of the POE and PP polymer chains during oscillatory shear. At low shear rates, both the storage modulus and viscosity of the crosslinked system increase significantly when compared to the non-crosslinked system. As the shear rate increases, they begin to become indistinguishable. However, due to the constraints imposed by the crosslinked structure, the viscosity of the crosslinked PP/POE blended system is higher than that of the non-crosslinked system.

Furthermore, by examining the loss factor, the ratio of the loss modulus to the storage modulus, it is evident that the loss factor of the crosslinked system is significantly smaller than that of the non-crosslinked system (Figure 5d). Simultaneously, the loss factor of the crosslinked system is less than one, indicating that the elastic modulus in the crosslinked system is significantly higher than the loss modulus, and the melt exhibits better elasticity. This implies that the crosslinked PP/POE system can to some extent prevent the merging and rupture of bubbles. Additionally, it can be observed that when the crosslinking degree reaches 55%, the loss factor of the system slightly increases. This also explains the reduction in pore size and uneven distribution at a 55% crosslinking degree: the increased viscosity in the viscoelasticity leads to poor melt flowability, restricting the flow of the melt and causing insufficient growth dynamics of pore size. Therefore, from the perspective of rheological behavior, crosslinking modification shows a significant improvement in the foaming effect of the system. However, there is a peak value, and an excessively high crosslinking degree is unfavorable for the growth of bubbles.

3.1.3. Thermal and Mechanical Properties

Thermal Shrinkage Performance

Figure 6 illustrates the dimensional changes of foam materials with different crosslinking degrees after high temperature treatment at 160 °C for 10 min. It is evident that with the increase in crosslinking degree, the thermal shrinkage rate of the material at high temperatures significantly decreases. On one hand, compared to the non-crosslinked system, the foam material thickness in the crosslinked PP/POE blend system has increased, reaching a maximum value of 11.1 mm at a crosslinking degree of 12% and indicating a higher foaming ratio at this point. On the other hand, the thermal shrinkage rate of the crosslinked system has notably decreased, especially at a crosslinking degree of 55% where the shrinkage rate is only 13.4%, and the thickness after shrinkage remains at 8.4 mm, indicating good thermal stability. Thus, it is observed that an increase in the crosslinking degree significantly enhances the thermal stability of the material in the crosslinked system. Since foam materials often undergo thermal processing during fabrication, the thermal stability of the material is crucial. In the crosslinked system, this requirement is essentially met, and a higher crosslinking degree evidently contributes to improved thermal stability without compromising foaming efficiency.

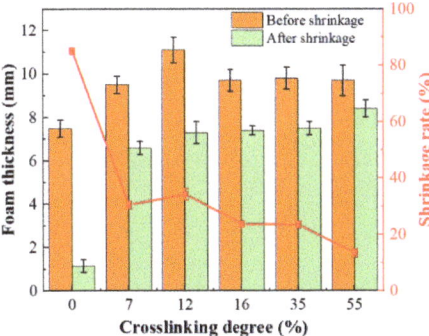

Figure 6. The thermal shrinkage rate of PP/POE foams as a function of crosslinking degree.

Mechanical Properties

The influence of the crosslinking degree on the tensile strength, tear strength, and impact toughness of the PP/POE blend system is shown in Figure 7. From Figure 7, which represents the change in tensile performance of the unfoamed PP/POE blend system, it is observed that the tensile strength of the crosslinked system is not significantly improved when compared to the non-crosslinked system. However, for the foamed system, an increase in crosslinking degree leads to a decrease followed by an increase in the tensile strength of foamed materials. At a crosslinking degree of 12%, the system exhibits the lowest tensile strength of 0.23 MPa, representing a 53% reduction compared to the non-crosslinked system. When the crosslinking degree reaches 55%, the tensile strength returns to 0.51 MPa. Thus, it can be inferred that the crosslinking degree itself has minimal impact on the tensile strength of unfoamed sheets. Instead, the tensile strength of foamed material is primarily determined by the foaming effect, i.e., foaming ratio and pore structure. Considering the SEM morphology in Figure 4, at a slight crosslinking degree of 12%, the pore size increases, foaming ratio increases, and pore density decreases, resulting in a loss of mechanical strength in the final foamed material. However, when the crosslinking degree is 55%, the foaming ratio decreases and the pore size is minimized, leading to an improvement in the tensile strength of the material. This phenomenon is also observed in tear strength and impact strength, exhibiting similar curves to the tensile strength variation, i.e., a decrease followed by an increase in strength. At a crosslinking degree of 12%, the tear strength and impact toughness are at their lowest, measuring 2.0 kN/m and 1.1 kJ/m^2, respectively. As the crosslinking degree increases to 55%, tear strength and impact toughness rebound to 3.2 kN/m and 1.9 kJ/m^2. This phenomenon is similar to the variation in tensile strength, influenced by the pore structure and pore size of the foamed material.

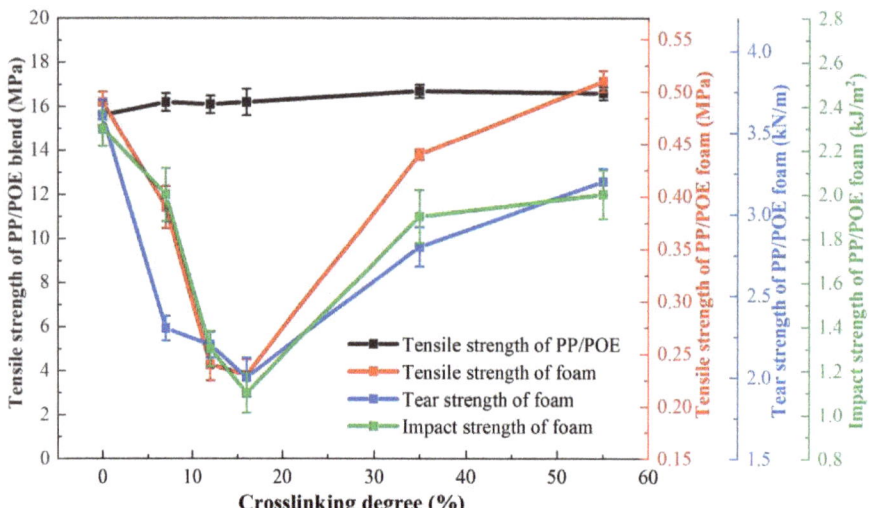

Figure 7. The tensile strength, tear strength, and impact strength of the PP/POE blend matrix and foam as a function of crosslinking degree.

3.2. Fiber-Reinforced Crosslinked PP/POE Composite Foam

From the above results of irradiation crosslinking in the PP/POE blend matrix, it can be observed that mild crosslinking effectively enhances the melt strength of PP, significantly increases the foaming expansion ratio, reduces foam density, and improves high temperature thermal stability. However, it comes with a significant decrease in mechanical properties. Therefore, we further carried out fiber reinforcement modification on the aforementioned 12% crosslinked PP/POE blend system, corresponding to a 4 kGy irradiation dose in the experiment, to compensate for the adverse effects on mechanical properties caused by the high foaming expansion ratio.

3.2.1. Impact of Fibers on Foaming Behavior

The density and cell morphology of fiber-reinforced irradiation-crosslinked PP/POE composite foams are shown in Figure 8. When the fiber content is less than 10 phr, there is no significant change in the foam density, only an increase of 10%. As the fiber content continues to increase beyond 15 phr, the density of fiber-reinforced crosslinked PP/POE composite foams sharply increases, and the foaming expansion ratio significantly decreases. Specifically, when the addition is 25 phr, the densities of sCF/PP/POE and sGF/PP/POE composite foams are 0.114 g/cm^3 and 0.097 g/cm^3, respectively, which have increased by 100% and 70%, respectively, compared to the PP/POE polymer foam without fiber addition at 0.057 g/cm^3. This indicates that the foaming behavior of the crosslinked PP/POE blend system is significantly adversely affected at a high fiber content.

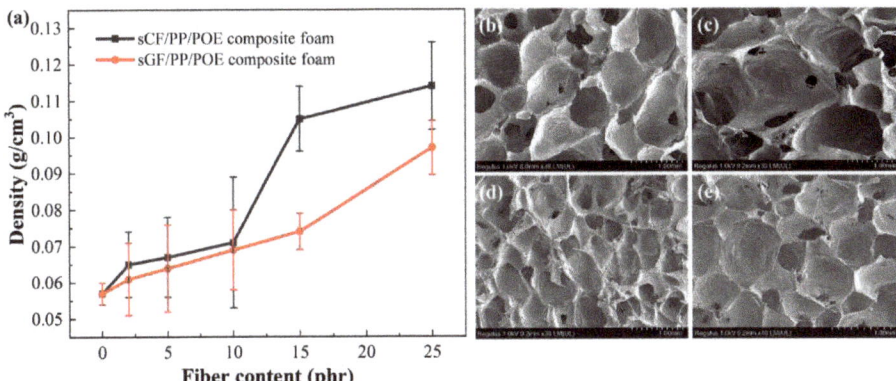

Figure 8. Effects of fiber content on (**a**) the PP/POE composite foam density, and pore cell morphology of (**b**,**c**) 10 phr- and 25 phr-reinforced sCF/PP/POE foam, and (**d**,**e**) 10 phr- and 25 phr-reinforced sGF/PP/POE foam.

By examining the SEM cross-sectional morphology in Figure 8b–e of foams from 10 phr and 25 phr sCF/PP/POE and sGF/PP/POE composite materials, it can be observed that with 10 phr fiber addition, the cell morphology of both materials is relatively clear, and the cell structure is relatively uniform. However, compared to the system without fiber in Figure 4, the foaming structure of fiber-reinforced PP/POE composite materials is still somewhat damaged, showing some irregularly shaped cells and increased cell openings. With the addition of 25 phr fibers, the cell morphology of fiber-reinforced PP/POE composite materials is significantly affected, especially in the case of sCF/PP/POE composite foam, where the foam structure is severely disrupted with irregular cell walls and collapsing and merging phenomena occurring between cells.

3.2.2. Impact of Fibers on Thermal and Mechanical Properties of Foam

Thermal Shrinkage Performance

As shown in Figure 9, the addition of sCF is conducive to a slight reduction in the thermal shrinkage of PP/POE material, while sGF has almost no effect. This may be attributed to the negative thermal expansion coefficient of CF, which acts as a counterforce against certain degrees of thermal shrinkage. However, these changes are not significant. A possible reason is that although the addition of fibers is beneficial for increasing the thermal stability of thermoplastic polymers [51,52], in the case of foam structure, the shrinkage behavior is predominantly influenced by the continuous network structure dominated by polymers. Therefore, the contribution of fibers with a dispersed and low faction is limited. In addition, the introduction of fibers deteriorates the overall integrity of the foam and increases the pore size, leading to an increased probability of pore collapse after heating, offsetting the positive contribution of fibers. In the end, the thermal shrinkage stability is not significantly enhanced. However, despite the improvement in material viscosity and thermal shrinkage to some extent with the addition of carbon fibers in sCF/PP/POE, a further comparison with Figure 6 reveals that its contribution is far less significant than the enhanced thermal shrinkage stability brought about by irradiation crosslinking.

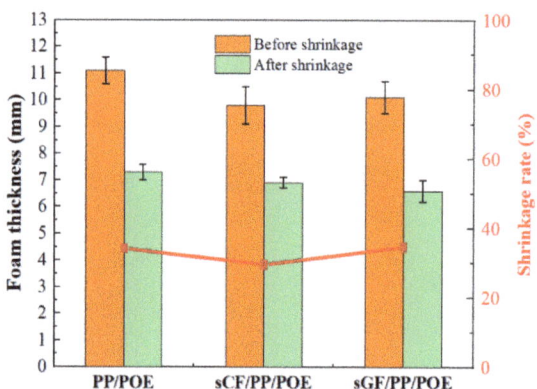

Figure 9. Impact of fibers on thermal shrinkage of PP/POE foam.

Mechanical Properties

Figure 10a illustrates the relationship between tensile strength and fiber content for sCF/PP/POE and sGF/PP/POE composite foamed materials. It can be observed that the addition of sCF and sGF significantly enhances the tensile strength of PP/POE composite foamed materials, with both reaching their maximum values at 10 phr. However, the reinforcing effect of sCF is more pronounced than that of sGF. Specifically, at a sCF content of 10 phr, the tensile strength of the sCF/PP/POE composite foamed material increases to 0.69 MPa. This represents a 200% increase compared to the 0.23 MPa of the 12% crosslinked PP/POE system and a 41% increase compared to the 0.49 MPa of the non-crosslinked PP/POE system. This improvement is attributed to the uniform distribution of fibers within the pore walls in the crosslinked sCF/PP/POE composite foam system, which significantly enhances the polymer matrix. When the foamed material is loaded, fiber fracture and pull-out from the matrix can effectively absorb energy, increase the load required for material failure, and thereby enhance the mechanical properties of the foamed system (Figure 10b,c). For sCF, its individual fiber strength is higher than that of sGF, so the increase in pore wall strength is more pronounced after its addition. However, with a further increase in sCF content to 15 phr, a sharp decrease in tensile strength is observed for the sCF/PP/POE composite foamed material. At an sCF content of 25 phr, the tensile strength is only 0.29 MPa, approaching the mechanical strength of the system without added fibers. As shown in the SEM cross-sectional pore morphology in Figure 8, the addition of sCF at high content severely disrupts the internal pore structure, leading to significant loss of mechanical strength in the pores and ultimately resulting in a decrease in the overall mechanical strength of the material. This is because high fiber content often leads to increased matrix viscosity [46], hindering foaming, and it also introduces a large number of stress concentration points, making the cell walls prone to rupture and unable to form perfectly connected cell walls for load bearing. According to the classical theory $F = \sigma A$, a large number of defects in the cell walls mean that the effective load-bearing area A decreases. Under the constant intrinsic strength σ of the material, the load-bearing capacity F decreases.

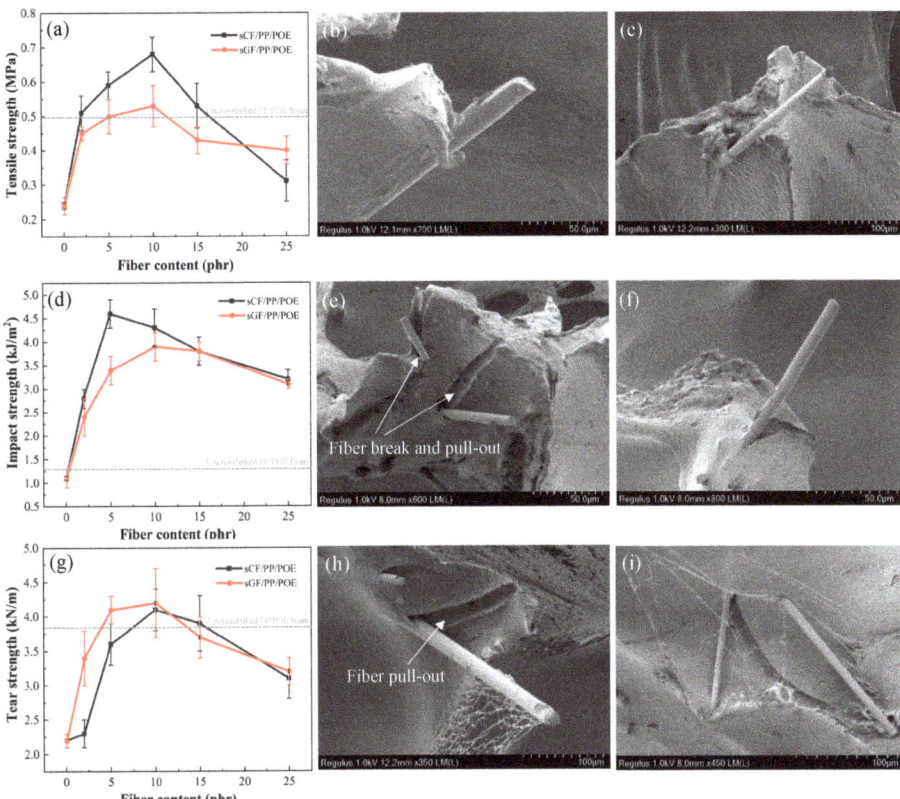

Figure 10. Effect of fiber content on the tensile behaviors of (**a**) tensile strength, SEM failure cross-section of crosslinked, (**b**) sGF/PP/POE composite foam, and (**c**) sCF/PP/POE composite foam; on the impact behaviors of (**d**) impact strength, SEM failure cross-section of crosslinked, (**e**) sGF/PP/POE composite foam, and (**f**) sCF/PP/POE composite foam; and on the tear behaviors of (**g**) tear strength, SEM failure cross-section of crosslinked, (**h**) sGF/PP/POE composite foam, and (**i**) sCF/PP/POE composite foam.

Similarly, with the increase in fiber content, both sCF/PP/POE and sGF/PP/POE composite foams exhibit a similar pattern of initially increasing and then decreasing impact strength (Figure 10d). However, there is a significant overall improvement in performance across the entire addition range. For sCF, the impact strength reaches the maximum value of 4.5 kJ/m^2 at a content of 5 phr. This represents a 246% increase compared to the 1.3 kJ/m^2 of the 12% crosslinked PP/POE system and a 96% increase compared to the 2.3 kJ/m^2 of the non-crosslinked PP/POE system. As for sGF, its impact strength reaches the maximum value of 3.6 kJ/m^2 at a content of 10 phr. Similarly to the tensile process, the impact fracture failure process is accompanied by fiber fracture and pull-out, so the reinforcing effect of sCF is better than that of sGF.

Figure 10g shows the variation curve of the tear strength of sCF/PP/POE and sGF/PP/POE composite foamed materials with fiber content. It can be observed that the tear strength of PP/POE composite foamed materials is significantly improved with the addition of sCF and sGF, and both start to decrease after the addition exceeds 10 phr. When comparing the two fibers, the enhancement of tear strength by sGF is more pronounced, reaching a maximum tear strength of 4.2 kN/m. There is no obvious fiber fracture from the fractured surface morphology after tearing, and the main failure mode is the debonding at the fiber/matrix interface and fiber pull-out. Due to the better interface compatibility of sGF with the matrix, compared to the unmatched epoxy sizing of sCF, sGF shows a more pronounced enhancement in tear strength.

In summary, simultaneously combining the results from this section with those of Section 3.1, crosslinking modification has a significant enhancing effect on the melt strength and high temperature thermal stability of PP/POE foam, with little impact on mechanical properties. Conversely, fibers have a significant impact on improving mechanical properties, with little effect on melt strength and high temperature thermal stability. In the fiber-reinforced crosslinked PP/POE blend foam system, the two mechanisms achieve a good complementary effect.

4. Conclusions

This study addresses the challenge of achieving high foaming ratios and high performance of polypropylene (PP) foam, caused by its low melt strength. A systematic exploration of material composition, irradiation crosslinking, and fiber reinforcement was conducted. Ultimately, a fiber-reinforced PP/POE composite foam with both low-density and high-performance characteristics was obtained. The main conclusions are as follows:

(1) The crosslinking degree of the PP/POE blend can be effectively controlled within the range of 0–60% by adjusting the irradiation dose and crosslinking agent content. The crosslinking degree is positively correlated with the irradiation dose and crosslinking agent content, while it exhibits an increase and then decrease trend with the foaming ratio. At a crosslinking degree of 12%, the foaming ratio of the crosslinked PP/POE blend can reach 16, representing a 36% reduction in density and a 60% increase in foaming ratio compared to the non-crosslinked PP/POE blend.

(2) Mild crosslinking at 12% effectively enhances the melt strength of PP, significantly improves the foaming ratio, reduces foam density, and enhances resistance to high temperature thermal shrinkage. However, it leads to a 53% decrease in mechanical properties. Through further fiber reinforcement, the mechanical strength of the crosslinked PP/POE composite material is significantly improved. Specifically, the foam density of 10 phr carbon-fiber-reinforced crosslinked sCF/PP/POE composite material is comparable to that of the 12% crosslinked PP/POE system, but the tensile strength reaches 0.69 MPa, increasing by 200% compared to the 12% crosslinked PP/POE system and by 41% compared to the non-crosslinked PP/POE system. It also exhibits a 96% increase in impact strength and an 11% increase in tear resistance when compared to the non-crosslinked PP/POE system.

(3) In summary, irradiation crosslinking is beneficial for enhancing the melt strength and resistance to high temperature thermal shrinkage of PP/POE foam, while fiber reinforcement contributes significantly to improving mechanical properties. These achieve a good complementary effect in low-density and high-performance PP foam modification.

Author Contributions: Conceptualization, H.L. and T.W.; methodology, T.W. and C.C.; software, H.L., T.W. and Y.M.; validation, H.L. and T.W.; formal analysis, H.L., T.W. and C.C.; investigation, H.L., T.W. and Y.M.; resources, H.L. and K.N.; data curation, H.L., T.W. and C.C.; writing—original draft preparation, H.L. and T.W.; writing—review and editing, H.L.; visualization, H.L., T.W. and C.C.; supervision, K.N.; project administration, K.N.; funding acquisition, K.N. All authors have read and agreed to the published version of the manuscript.

Funding: This research was funded by CGT Changshu Co., Ltd. grant number 2020-0981 and Hebei Construction Group Co., Ltd. grant number 2023-1439.

Institutional Review Board Statement: Not applicable.

Data Availability Statement: Data are contained within the article.

Acknowledgments: The authors would like to acknowledge CGT Changshu Co., Ltd. and Hebei Construction Group Co., Ltd. for providing the research background and experimental conditions.

Conflicts of Interest: The authors declare no conflicts of interest.

References

1. Hossain, M.T.; Shahid, M.A.; Mahmud, N.; Habib, A.; Rana, M.M.; Khan, S.A.; Hossain, M.D. Research and application of polypropylene: A review. *Discov. Nano* **2024**, *19*, 2. [CrossRef]
2. Sadiku, R.; Ibrahim, D.; Agboola, O.; Owonubi, S.J.; Fasiku, V.O.; Kupolati, W.K.; Jamiru, T.; Eze, A.A.; Adekomaya, O.S.; Varaprasad, K. Automotive components composed of polyolefins. In *Polyolefin Fibres*; Elsevier: Amsterdam, The Netherlands, 2017; pp. 449–496.
3. Patil, A.; Patel, A.; Purohit, R. An overview of polymeric materials for automotive applications. *Mater. Today Proc.* **2017**, *4*, 3807–3815. [CrossRef]
4. Lee, P.C.; Kaewmesri, W.; Wang, J.; Park, C.B.; Pumchusak, J.; Folland, R.; Praller, A. Effect of die geometry on foaming behaviors of high-melt-strength polypropylene with CO_2. *J. Appl. Polym.* **2008**, *109*, 3122–3132. [CrossRef]
5. Burt, J.G. The elements of expansion of thermoplastics Part II. *J. Cell. Plast.* **1978**, *14*, 341–345. [CrossRef]
6. Tang, L.; Zhai, W.; Zheng, W. Autoclave preparation of expanded polypropylene/poly (lactic acid) blend bead foams with a batch foaming process. *J. Cell. Plast.* **2011**, *47*, 429–446. [CrossRef]
7. Antunes, M.; Velasco, J.I.; Haurie, L. Characterization of highly filled magnesium hydroxide-polypropylene composite foams. *J. Cell. Plast.* **2011**, *47*, 17–30. [CrossRef]
8. Gong, W.; Fu, H.; Zhang, C.; Ban, D.; Yin, X.; He, Y.; He, L.; Pei, X. Study on foaming quality and impact property of foamed polypropylene composites. *Polymers* **2018**, *10*, 1375. [CrossRef] [PubMed]
9. Wang, L.; Wan, D.; Zhang, Z.; Liu, F.; Xing, H.; Wang, Y.; Tang, T. Synthesis and structure–property relationships of polypropylene-g-poly (ethylene-co-1-butene) graft copolymers with well-defined long chain branched molecular structures. *Macromolecules* **2011**, *44*, 4167–4179. [CrossRef]
10. Zhai, W.; Wang, H.; Yu, J.; Dong, J.Y.; He, J. Foaming behavior of isotactic polypropylene in supercritical CO_2 influenced by phase morphology via chain grafting. *Polymer* **2008**, *49*, 3146–3156. [CrossRef]
11. Chen, P.; Zhao, L.; Gao, X.; Xu, Z.; Liu, Z.; Hu, D. Engineering of polybutylene succinate with long-chain branching toward high foamability and degradation. *Polym. Degrad. Stab.* **2021**, *194*, 109745. [CrossRef]
12. Wang, L.; Ishihara, S.; Ando, M.; Minato, A.; Hikima, Y.; Ohshima, M. Fabrication of high expansion microcellular injection-molded polypropylene foams by adding long-chain branches. *Ind. Eng. Chem. Res.* **2016**, *55*, 11970–11982. [CrossRef]
13. Liu, H.; Chuai, C.; Iqbal, M.; Wang, H.; Kalsoom, B.B.; Khattak, M.; Qasim Khattak, M. Improving foam ability of polypropylene by crosslinking. *J. Appl. Polym.* **2011**, *122*, 973–980. [CrossRef]
14. Yang, C.; Xing, Z.; Zhao, Q.; Wang, M.; Wu, G. A strategy for the preparation of closed-cell and crosslinked polypropylene foam by supercritical CO_2 foaming. *J. Appl. Polym.* **2018**, *135*, 45809. [CrossRef]
15. Wang, S.; Wang, K.; Pang, Y.; Li, Y.; Wu, F.; Wang, S.; Zheng, W. Open-cell polypropylene/polyolefin elastomer blend foams fabricated for reusable oil-sorption materials. *J. Appl. Polym.* **2016**, *133*, 43812. [CrossRef]
16. Han, D.H.; Jang, J.H.; Kim, H.Y.; Kim, B.N.; Shin, B.Y. Manufacturing and foaming of high melt viscosity of polypropylene by using electron beam radiation technology. *Polym. Eng. Sci.* **2006**, *46*, 431–437. [CrossRef]
17. Feng, J.M.; Wang, W.K.; Yang, W.; Xie, B.H.; Yang, M.B. Structure and properties of radiation cross-linked polypropylene foam. *Polym. Plast. Technol. Eng.* **2011**, *50*, 1027–1034. [CrossRef]
18. Yang, C.; Xing, Z.; Zhang, M.; Zhao, Q.; Wang, M.; Wu, G. Supercritical CO_2 foaming of radiation crosslinked polypropylene/high-density polyethylene blend: Cell structure and tensile property. *Radiat. Phys. Chem.* **2017**, *141*, 276–283. [CrossRef]
19. Cheng, S.; Phillips, E.; Parks, L. Processability improvement of polyolefins through radiation-induced branching. *Radiat. Phys. Chem.* **2010**, *79*, 329–334. [CrossRef]
20. Taki, K.; Yanagimoto, T.; Funami, E.; Okamoto, M.; Ohshima, M. Visual observation of CO_2 foaming of polypropylene-clay nanocomposites. *Polym. Eng. Sci.* **2004**, *44*, 1004–1011. [CrossRef]
21. Jiang, X.L.; Bao, J.B.; Liu, T.; Zhao, L.; Xu, Z.M.; Yuan, W.K. Microcellular foaming of polypropylene/clay nanocomposites with supercritical carbon dioxide. *J. Cell. Plast.* **2009**, *45*, 515–538. [CrossRef]
22. Mohebbi, A.; Mighri, F.; Ajji, A.; Rodrigue, D. Current issues and challenges in polypropylene foaming: A review. *Cell. Polym.* **2015**, *34*, 299–338. [CrossRef]
23. Wang, L.; Wu, Y.K.; Ai, F.F.; Fan, J.; Xia, Z.P.; Liu, Y. Hierarchical porous polyamide 6 by solution foaming: Synthesis, characterization and properties. *Polymers* **2018**, *10*, 1310. [CrossRef]
24. Llewelyn, G.; Rees, A.; Griffiths, C.A.; Jacobi, M. A novel hybrid foaming method for low-pressure microcellular foam production of unfilled and talc-filled copolymer polypropylenes. *Polymers* **2019**, *11*, 1896. [CrossRef]
25. Mao, H.; He, B.; Guo, W.; Hua, L.; Yang, Q. Effects of nano-$CaCO_3$ content on the crystallization, mechanical properties, and cell structure of PP nanocomposites in microcellular injection molding. *Polymers* **2018**, *10*, 1160. [CrossRef]
26. Yang, C.; Wang, M.; Xing, Z.; Zhao, Q.; Wang, M.; Wu, G. A new promising nucleating agent for polymer foaming: Effects of hollow molecular-sieve particles on polypropylene supercritical CO_2 microcellular foaming. *RSC Adv.* **2018**, *8*, 20061–20067. [CrossRef]
27. Albooyeh, A. The effect of addition of Multiwall Carbon Nanotubes on the vibration properties of Short Glass Fiber reinforced polypropylene and polypropylene foam composites. *Polym. Test.* **2019**, *74*, 86–98. [CrossRef]
28. Ellingham, T.; Duddleston, L.; Turng, L.S. Sub-critical gas-assisted processing using CO_2 foaming to enhance the exfoliation of graphene in polypropylene+ graphene nanocomposites. *Polymer* **2017**, *117*, 132–139. [CrossRef]

29. Yao, S.S.; Jin, F.L.; Rhee, K.Y.; Hui, D.; Park, S.J. Recent advances in carbon-fiber-reinforced thermoplastic composites: A review. *Compos. Eng.* **2018**, *142*, 241–250. [CrossRef]
30. Dixit, N.; Jain, P.K. 3D printed carbon fiber reinforced thermoplastic composites: A review. *Mater. Today Proc.* **2021**, *43*, 678–681. [CrossRef]
31. Tayeh, B.A.; Akeed, M.H.; Qaidi, S.; Bakar, B.A. Ultra-high-performance concrete: Impacts of steel fibre shape and content on flowability, compressive strength and modulus of rupture. *Case Stud. Constr. Mater.* **2022**, *17*, e01615. [CrossRef]
32. Althoey, F.; Hakeem, I.Y.; Hosen, M.A.; Qaidi, S.; Isleem, H.F.; Hadidi, H.; Shahapurkar, K.; Ahmad, J.; Ali, E. Behavior of concrete reinforced with date palm fibers. *Materials* **2022**, *15*, 7923. [CrossRef]
33. Zhang, H.; Rizvi, G.; Park, C. Development of an extrusion system for producing fine-celled HDPE/wood-fiber composite foams using CO_2 as a blowing agent. *Adv. Polym. Technol. J. Polym. Process. Inst.* **2004**, *23*, 263–276. [CrossRef]
34. Yang, J.; Li, P. Characterization of short glass fiber reinforced polypropylene foam composites with the effect of compatibilizers: A comparison. *J. Reinf. Plast. Compos.* **2015**, *34*, 534–546. [CrossRef]
35. Cai, L.; Zhao, K.; Huang, X.; Ye, J.; Zhao, Y. Effective elastic modulus of foamed long glass fiber reinforced polypropylene: Experiment and computation. *Proc. Inst. Mech. Eng. J. Mater. Des. Appl.* **2021**, *235*, 202–215. [CrossRef]
36. Yang, C.; Wang, G.; Zhao, J.; Zhao, G.; Zhang, A. Lightweight and strong glass fiber reinforced polypropylene composite foams achieved by mold-opening microcellular injection molding. *J. Mater. Res. Technol.* **2021**, *14*, 2920–2931. [CrossRef]
37. Abedi, M.M.; Nedoushan, R.J.; Yu, W.-R. Enhanced compressive and energy absorption properties of braided lattice and polyurethane foam hybrid composites. *Int. J. Mech. Sci.* **2021**, *207*, 106627. [CrossRef]
38. Sebaey, T.A.; Rajak, D.K.; Mehboob, H. Internally stiffened foam-filled carbon fiber reinforced composite tubes under impact loading for energy absorption applications. *Compos. Struct.* **2021**, *255*, 112910. [CrossRef]
39. Chuang, Y.C.; Li, T.T.; Huang, C.H.; Huang, C.L.; Lou, C.W.; Chen, Y.S.; Lin, J.-H. Protective rigid fiber-reinforced polyurethane foam composite boards: Sound absorption, drop-weight impact and mechanical properties. *Fibers Polym.* **2016**, *17*, 2116–2123. [CrossRef]
40. Kumar, M.; Kaur, R. Glass fiber reinforced rigid polyurethane foam: Synthesis and characterization. *e-Polymers* **2017**, *17*, 517–521. [CrossRef]
41. Kuranchie, C.; Yaya, A.; Bensah, Y.D. The effect of natural fibre reinforcement on polyurethane composite foams—A review. *Sci. Afr.* **2021**, *11*, e00722. [CrossRef]
42. Shen, J.; Zeng, C.; Lee, L.J. Synthesis of polystyrene—Carbon nanofibers nanocomposite foams. *Polymer* **2005**, *46*, 5218–5224. [CrossRef]
43. Yunus, R.B.; Zahari, N.; Salleh, M.; Ibrahim, N.A. Mechanical properties of carbon fiber-reinforced polypropylene composites. *Key Eng. Mater.* **2011**, *471*, 652–657. [CrossRef]
44. Li, J.; Cai, C.L. Friction and wear properties of carbon fiber reinforced polypropylene composites. *Adv. Mater. Res.* **2011**, *284*, 2380–2383. [CrossRef]
45. Rezaei, F.; Yunus, R.; Ibrahim, N. Effect of fiber length on thermomechanical properties of short carbon fiber reinforced polypropylene composites. *Mater. Des.* **2009**, *30*, 260–263. [CrossRef]
46. Wang, C.; Ying, S.; Xiao, Z. Preparation of short carbon fiber/polypropylene fine-celled foams in supercritical CO_2. *J. Cell. Plast.* **2013**, *49*, 65–82. [CrossRef]
47. Rachtanapun, P.; Selke, S.; Matuana, L. Microcellular foam of polymer blends of HDPE/PP and their composites with wood fiber. *J. Appl. Polym.* **2003**, *88*, 2842–2850. [CrossRef]
48. Saiz-Arroyo, C.; de Saja, J.A.; Velasco, J.I.; Rodríguez-Pérez, M.Á. Moulded polypropylene foams produced using chemical or physical blowing agents: Structure—Properties relationship. *J. Mater. Sci.* **2012**, *47*, 5680–5692. [CrossRef]
49. Auhl, D.; Stange, J.; Münstedt, H.; Krause, B.; Voigt, D.; Lederer, A.; Lappan, U.; Lunkwitz, K. Long-chain branched polypropylenes by electron beam irradiation and their rheological properties. *Macromolecules* **2004**, *37*, 9465–9472. [CrossRef]
50. Schulze, D.; Trinkle, S.; Mülhaupt, R.; Friedrich, C. Rheological evidence of modifications of polypropylene by β-irradiation. *Rheol. Acta* **2003**, *42*, 251–258. [CrossRef]
51. Cho, J.; Lee, S.K.; Eem, S.H.; Jang, J.G.; Yang, B. Enhanced mechanical and thermal properties of carbon fiber-reinforced thermoplastic polyketone composites. *Compos. Appl. Sci. Manuf.* **2019**, *126*, 105599. [CrossRef]
52. Gavali, V.C.; Kubade, P.R.; Kulkarni, H.B. Mechanical and thermo-mechanical properties of carbon fiber reinforced thermoplastic composite fabricated using fused deposition modeling method. *Mater. Today Proc.* **2020**, *22*, 1786–1795. [CrossRef]

Disclaimer/Publisher's Note: The statements, opinions and data contained in all publications are solely those of the individual author(s) and contributor(s) and not of MDPI and/or the editor(s). MDPI and/or the editor(s) disclaim responsibility for any injury to people or property resulting from any ideas, methods, instructions or products referred to in the content.

Article

Processing Method and Performance Evaluation of Flame-Retardant Corrugated Sandwich Panel

Yiliang Sun [1], Jingwen Li [1,*] and Boming Zhang [1,2]

[1] School of Materials Science and Engineering, Beihang University, Beijing 100191, China; ylsun@buaa.edu.cn (Y.S.)
[2] Institute of Advanced Materials, Shandong Institutes of Industrial Technology, Jinan 250102, China
* Correspondence: ljw666@buaa.edu.cn

Abstract: In this study, in order to expand the engineering application range of thermoplastic corrugated sheets, flame-retardant thermoplastic corrugated sheets were prepared by the thermoplastic molding method. Based on our previous research results, we prepared flame-retardant prepreg tapes with the flame retardant addition accounting for 15%, 20%, and 25% of the resin matrix. Then, we prepared flame-retardant thermoplastic corrugated sandwich panels with corresponding flame retardant addition amounts. The limiting oxygen index test, vertical combustion test, cone calorimetry test, and mechanical property test were carried out on each group of samples and control group samples. The results showed that when the flame retardant was added at 25%, the flame retardant level could reach the V0 level. Compared with the control group, the flexural strength and flexural modulus decreased by 2.6%, 14.1%, and 19.9% and 7.3%, 16.1%, and 21.9%, respectively. When the amount of flame retardant was 15%, 20%, and 25%, respectively, the total heat release decreased by 16.3%, 23.5%, and 34.1%, and the maximum heat release rate decreased by 12.5%, 32.4%, and 37.4%, respectively.

Keywords: flame retardant; corrugated sandwich panel; thermoplastic composites; continuous fiber

Citation: Sun, Y.; Li, J.; Zhang, B. Processing Method and Performance Evaluation of Flame-Retardant Corrugated Sandwich Panel. *Polymers* 2024, 16, 696. https://doi.org/10.3390/polym16050696

Academic Editors: Long-Cheng Tang, Abderrahmane Ayadi, Patricia Krawczak and Chung Hae Park

Received: 27 October 2023
Revised: 23 February 2024
Accepted: 26 February 2024
Published: 4 March 2024

Copyright: © 2024 by the authors. Licensee MDPI, Basel, Switzerland. This article is an open access article distributed under the terms and conditions of the Creative Commons Attribution (CC BY) license (https://creativecommons.org/licenses/by/4.0/).

1. Introduction

Compared with traditional thermosetting composites [1–3], continuous-fiber-reinforced thermoplastic composites are recyclable, in line with the concepts of low carbon and environmental protection, and also afford a fast molding speed and good impact toughness. Continuous-fiber-reinforced thermoplastic prepregs have no storage conditions. Therefore, the research and applications of thermoplastic composites have attracted increasing attention in recent years [4].

In the thermoplastic resin matrix, polypropylene has the advantages of a low molding temperature, low price, good chemical stability, and excellent comprehensive performance [5]. The application fields and market for continuous fiber-reinforced polypropylene composites are also gradually expanding. The flammable characteristics of polypropylene have led to its limitations in many application scenarios; thus, modifying polypropylene with flame retardants has important practical significance for expanding the application range of polypropylene and its composites. Research on flame retardants for polypropylene composites is relatively comprehensive and mature. However, according to a literature review, research on continuous fiber reinforcement needs to be further expanded and deepened. The flame retardancy of continuous fiber-reinforced polypropylene shares some similarities with that of the polypropylene matrix, and there are also many differences that have not received special attention in previous research [6]. We conducted extensive research on the flame retardancy of continuous glass fiber (CGF)-reinforced polypropylene. Building on previous research [7], we further investigated the structural applications of continuous-fiber-reinforced flame-retardant polypropylene. The corrugated sandwich panel is a representative structure [8–12]. Corrugated structures are simple to prepare, have strong bearing capacities, and are widely used in aerospace, marine vessels, packaging, and other fields [13–16]. Compared with other structural forms, such as the honeycomb sandwich structure, the corrugated structure is characterized by the fact that the available overall space of the corrugated structure is larger than that of the honeycomb structure. This feature can facilitate the subsequent use of corrugated sandwich panels, such as filling the corrugated space with thermal insulation to increase thermal insulation performance and also using its space to place other components. It can be said that the corrugated sandwich structure can provide more space that can be used, which is also a feature and advantage compared with other sandwich core structures. The feature can provide designers and engineers with more design possibilities. For example, when the corrugated sandwich panel structure is applied to the new energy vehicle industry, the battery part can be put into the corrugated space to realize the integration of structure and function.

At present, based on a literature survey, there is no relevant research on continuous fiber-reinforced polypropylene flame-retardant corrugated sandwich panels [17,18]. However, few studies have paid attention to the flame retardancy of continuous fiber thermoplastic corrugated sandwich panels, and whether the flame retardant performance can meet the corresponding requirements is the key to whether it can meet the needs of use scenarios. Therefore, this study focuses on further exploring the preparation and performance of a flame-retardant corrugated plate on the basis of previous research on continuous fiber-reinforced polypropylene flame-retardant prepregs.

In this study, we prepare a continuous fiber-reinforced flame-retardant polypropylene prepreg tape and further prepare a flame-retardant prepreg tape on flame-retardant corrugated panels. The process possibility of preparing corrugated core materials by thermoplastic molding was verified. The effect of the addition of flame retardant on the mechanical properties of composite materials was compared by the bending mechanical properties test. The flame retardant performance of corrugated panels is evaluated. The amount of flame retardant added to meet the different flame-retardant performance levels was determined.

2. Materials and Methods

2.1. Raw Materials and Test Equipment

Polypropylene (Bx3900), plastic granules with a melt flow index of 100.0 g/10 min (230 °C, 2.16 kg), was obtained from SK (Corporation: Seoul, Republic of Korea). Polypropylene (MF650X), plastic granules with a melt flow index of 1200.0 g/10 min (230 °C, 2.16 kg), was obtained from LyondellBasell (Corporation: Rotterdam, The Netherlands). Maleic anhydride-grafted polypropylene (MAPP), MF650X, plastic granules with a melt flow index of 20.0 g/10 min (230 °C, 2.16 kg), was obtained from Exxon Mobil Co., Ltd. (Brussels, Belgium). The intumescent flame retardant(IFR) was purchased from Xinxiu Chemical Co., Ltd. (Xinxiang, China). The Continuous fiberglass yarn (4305S), pre-treated by sizing agent and dedicated to the Polypropylene resin, was obtained from Chongqing Polycomp International Corporation (Chongqing, China).

In order to make the material and equipment information used in the experiment easily accessible and readable, it has been presented in the form of a table. Information on the materials used in the experiment is listed in Table 1. The information on test equipment is listed in Table 2.

Table 1. Raw materials used in the experiment.

Raw Materials	Provider	Product Grade
Polypropylene	SK	Bx3900
Polypropylene	LyondellBasell	MF650X
MAPP	Exxon Mobil	Exxelor PO 1020
Continuous fiberglass yarn	Chongqing International Co.	4305s
IFR	Xinxiu Chemical	IFR-PP-1

Table 2. Specifications of devices used in the experiment.

Device	Manufacturer	Device Model
Prepreg belt production M	Designed and assembled in laboratory	BUAA-2019
Molding machine	DiDa Machinery Manufacturing Factory (Chengdu, China)	Y35-100T
Limiting oxygen index tester	Jiangning Analytical Instrument Co., Ltd. (Nanjing, China)	JF-3
Combustion grade tester	Yaoke (Shanghai, China)	YK-Y0142
Cone calorimeter	VOUCH (Shanghai, China)	6810

2.2. Preparation of Flame-Retardant Prepreg Tape

We first prepared flame-retardant-modified polypropylene by melt-blending and then used a self-designed and assembled prepreg tape production line to prepare the flame-retardant prepreg tape with the fiber, as described in our previous work. In the preparation of the continuous glass fiber-reinforced flame-retardant prepreg tape, the key process parameters in the prepreg belt process were adjusted to obtain a flame-retardant prepreg belt with excellent performance for the subsequent related experiments.

2.3. Preparation of Flame-Retardant Corrugated Sandwich Panels

After preparation, the flame-retardant prepreg was cut to a certain size and used to form the corrugated plate panel and core material. The preparation of the flame-retardant corrugated plate is detailed below.

2.3.1. Preparation of Upper and Lower Panels of Corrugated Plates by Molding Method

The skin of the corrugated plate was prepared by the molding method. The prepared flame-retardant prepreg was cut and spliced into a rectangular shape with dimensions of 320 mm × 120 mm and laid in the middle of two layers of aluminum plate according to the designed laying order, then heated on the heating plate for 5 min. The resin in the prepreg in the heating plate was in a completely molten state, and the aluminum plate was immediately placed in a molding machine for molding. The molding pressure was set to 5 MPa, the temperature of the molding machine was set to 80 °C, and the heat preservation and pressure holding time was 5 min.

2.3.2. Preparation of Core Material for Corrugated Plate by Roller Pressing Method

The corrugated core material was prepared using a roller-pressing method. The flame-retardant prepreg tape was wound around the unwinding reel. The unwinding device is composed of an unwinding shaft, drive motor, and other parts. The motor drives the unwinding shaft and rotates to realize the continuous motion of the prepreg belt. The prepreg belt was heated by the heating device. The temperature of the heating device was set to 250 °C. The resin in the prepreg belt in a molten state was passed through the roll mold, and the corrugated plate core material structure was formed. The overall molding process is shown in Figure 1.

2.3.3. Hot Melt Method of Bonding Upper and Lower Panels and Corrugated Core Materials

Based on the structural characteristics of the corrugated core materials, combined with the advantages of the thermoplastic composite materials for secondary processing, a bonding device for connecting the corrugated core materials with the upper and lower panels was designed. The bonding device is shown in Figure 1, where the prepared corrugated core material was placed in the middle of the upper and lower panels. The outside of the material (the upper and lower aluminum plates) can be heated, and the heat-shaping rod is placed inside the core material. The heating temperature of the heating rod was set to 220 °C, and the heating temperature of the upper and lower heating aluminum plates was set to 220 °C. After the core material and the heating position of the upper and lower panels were molten, pressure was applied through the heating rod to complete the bonding molding of the corrugated core material and the panel.

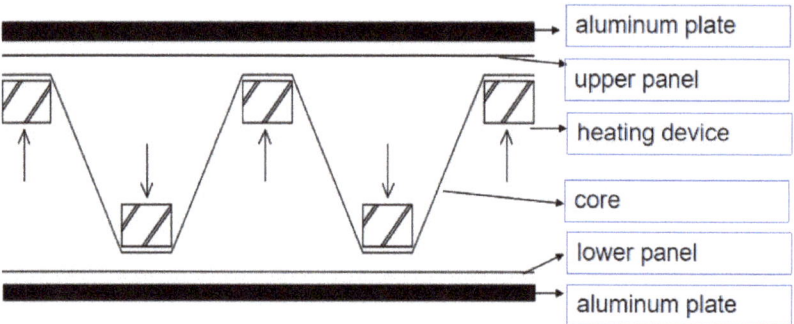

Figure 1. Bonding between the corrugated core material and the panels.

Figure 2 visually shows the overall preparation process from polypropylene pellets to flame-retardant corrugated sandwich panels.

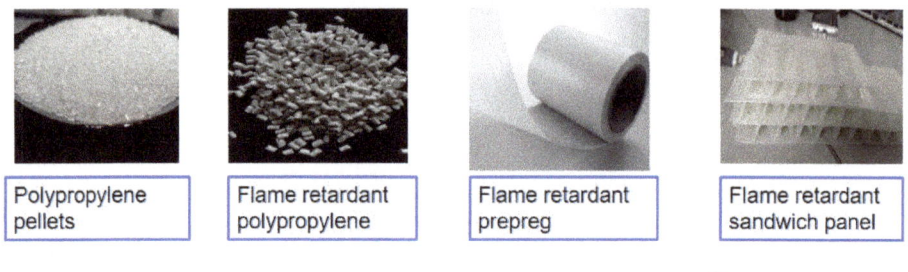

Figure 2. Schematic diagram of the experimental process.

2.3.4. Water Cutting Method for Cutting Specimens

After preparing the flame-retardant corrugated plate, the specimen was cut by water cutting to avoid melting at the edge of the cutting material caused by the heat generated during the process of cutting the thermoplastic composites.

2.4. Performance Test

2.4.1. Limiting Oxygen Index Test

The limiting oxygen index (LOI) method, also known as the LOI or critical oxygen index method, was proposed by Fenimore and Martin in 1966 to evaluate the combustion performance of plastics and textile materials. The oxygen index test is a good indicator of the combustion performance of a material. The combustion performance can be quantitatively evaluated with numerical results to a certain extent. The advantages are: the test is simple and the experimental cost is relatively low. This method has been widely used in evaluating the combustion characteristics and fire performance of materials.

The limiting oxygen index is defined as the minimum oxygen concentration at which the tested sample can maintain combustion under specified experimental conditions, that is, the lowest volume percentage of oxygen in the combustion environment of a gas-oxygen-nitrogen mixture during the experiment, expressed as follows:

$$\text{LOI} = \frac{[O_2]}{[O_2] + [N_2]} \times 100\%,$$

where $[O_2]$ and $[N_2]$ are the volumetric flow rates of oxygen and nitrogen, respectively.

2.4.2. Vertical Combustion Test

Plastic combustion grade tests are divided into horizontal and vertical combustion tests and are commonly used to evaluate the combustion performance of materials. The UL94 vertical combustion test method divides the difficulty of material combustion into V-0, V-1, and V-2 grades. The specific test methods and grade judgment standards were referenced from the test standard ASTM D3801-20a protocol (Standard Test Method for Measuring the Comparative Burning Characteristics of Solid Plastics in a Vertical Position). Vertical burning tests were conducted using a vertical burning test instrument (YK-Y0142) (Yaoke, Nanjing, China) with dimensions of 130 mm× 13 mm × 3.0 mm.

2.4.3. Cone Calorimetry Test

A cone calorimeter is used to determine the amount of heat released during combustion by measuring the oxygen consumed by the material during combustion, which in turn determines the rate of heat release of the material during the combustion test. Cone calorimetry tests (CCTs) were conducted using a cone calorimeter (Fire Testing Technology, Leeds, UK) with a heat flux of 50 kW/m^2 according to the ISO 5660 standard. Each specimen measured 100 mm × 100 mm × 12 mm.

2.4.4. Mechanical Properties Test

The mechanical property tests were performed on Changchun Kexin WDW-100 Universal Mechanical Testing Machine. According to ASTM D7264 standard, three-point bending property tests were carried out on each specimen at a beam moving speed of 2 mm/min to obtain the bending strength and bending modulus of the composites.

3. Results and Discussion

3.1. Limiting Oxygen Index and Combustion Rate Test

The results of the limiting oxygen index test are shown in Figure 3. The limiting oxygen index of the blank control group without flame retardant addition is 20.5. When 15, 20, and 25% flame retardant were each added, the limiting oxygen index was 25.2, 31.9, and 34.3, corresponding to an increase of 22.9, 55.6, and 67.3%, respectively. The test results indicate that as the amount of added flame retardant increased, the limiting oxygen index increased. The limiting oxygen index was largest when the content of the flame retardant was increased from 15% to 20%, moving from 22.9% to 55.6%; after the added amount exceeded 20%, the limiting oxygen index increased to a lesser extent. When the amount of added flame retardant was increased from 20% to 25%, the limiting oxygen index increased from 55.6% to 67.3%. This phenomenon can be explained by considering the combustion mechanism. With a flame-retardant content of 15%, the expanded carbon layer formed after the combustion of the material is relatively fluffy and not dense; thus, the flame-retardant effect is not good. With the addition of 20% flame retardant, the expanded carbon layer formed by the sample during combustion is more compact, the flame-retardant effect is further improved, and the data show that the limiting oxygen index is greatly improved. When the flame-retardant content is further increased, the density of the expanded carbon layer formed after the combustion of the sample is limited. Thus, the limiting oxygen index increased to a lesser extent.

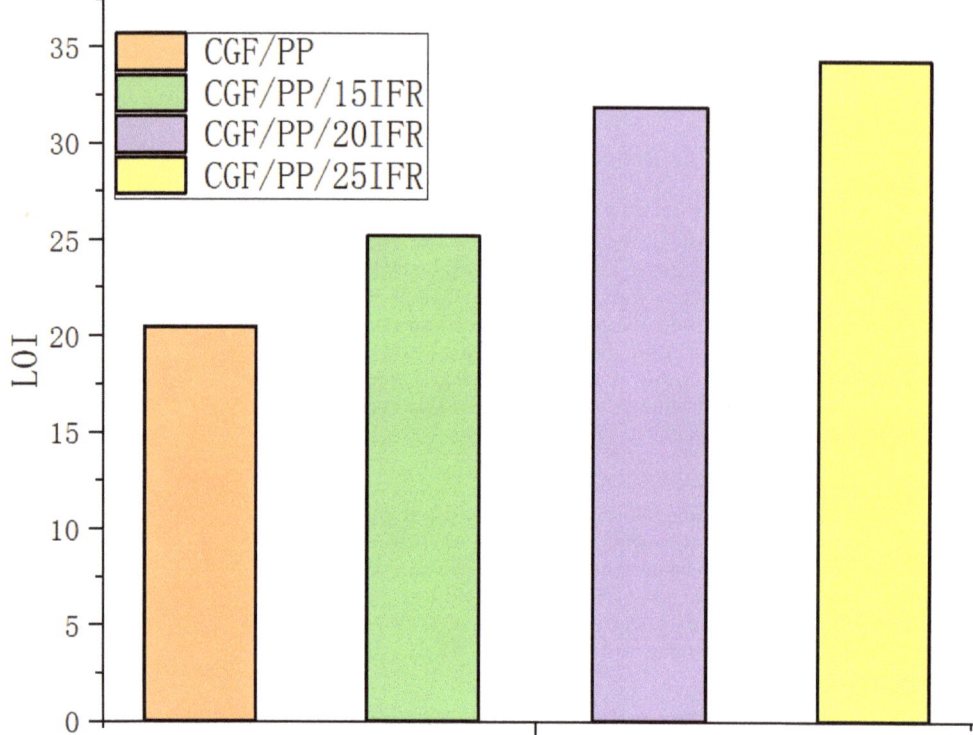

Figure 3. Comparison of limiting oxygen indices of each group of specimens.

The results of the combustion grade test are shown in Table 3. When no flame retardant was added, because of the continuous fiber used as the reinforcement, the glass fiber did not burn during the combustion process, and the specimen did not exhibit droplet behavior during combustion, which is different from the combustion of the polypropylene matrix spline. With an increase in the amount of flame retardant to 25%, the flame retardant level of the sample reached V0. This flame retardant level can meet the needs of most use scenarios. Notably, when 15% flame retardant was added, the LOI was greatly improved compared to that of the blank control group, but the vertical combustion grade still did not reach the lowest flame retardant level. Thus, for use in a scenario where the flame retardant level requirement reaches V1, the addition of 20% flame retardant should be selected, not only to reduce the amount of flame retardant and the cost of materials but also to meet the material requirements.

Table 3. Vertical combustion grade of each group of specimens.

Sample	UL-94 Rating	Dripping
CGF/PP	No rating	NO
CGF/PP/15IFR	No rating	NO
CGF/PP/20IFR	V1	NO
CGF/PP/25IFR	V0	NO

3.2. Cone Calorimetry Test

The heat release rate of the corrugated plates with different amounts of added flame retardant is shown in Figure 4. For the corrugated plates without flame retardant addition, the heat of combustion quickly reached a peak; with increasing flame retardant addition, the maximum heat release rate decreased significantly compared with when no flame retardant was added. When 15%, 20%, and 25% flame retardant were each added, the maximum heat release rate decreased by 12.5%, 32.4%, and 37.4%, respectively. When the amount of flame retardant was increased from 15 to 25%, the effect on the reduction of the exothermic peak was significant. However, beyond 20% addition, the flame retardant had a limited effect on the reduction of the heat release peak. The total heat release was 121.8 when no flame retardant was added. When 15%, 20%, and 25% flame retardant were each added, the total heat release was 101.9, 93.1, and 80.3, corresponding to a decrease of 16.3%, 23.5%, and 34.1%, respectively. The total heat release reduction data demonstrate that the effect of increasing the flame retardant content on the total heat release and heat release peak was different. When the flame retardant content was increased from 20% to 25%, the total heat release was significantly reduced from 23.5% to 34.1%, which can also be reflected in the heat release rate diagram. For several groups of specimens, the heat release curve quickly peaks after ignition, after which the heat release rate decreases rapidly, leading to a second heat release peak.

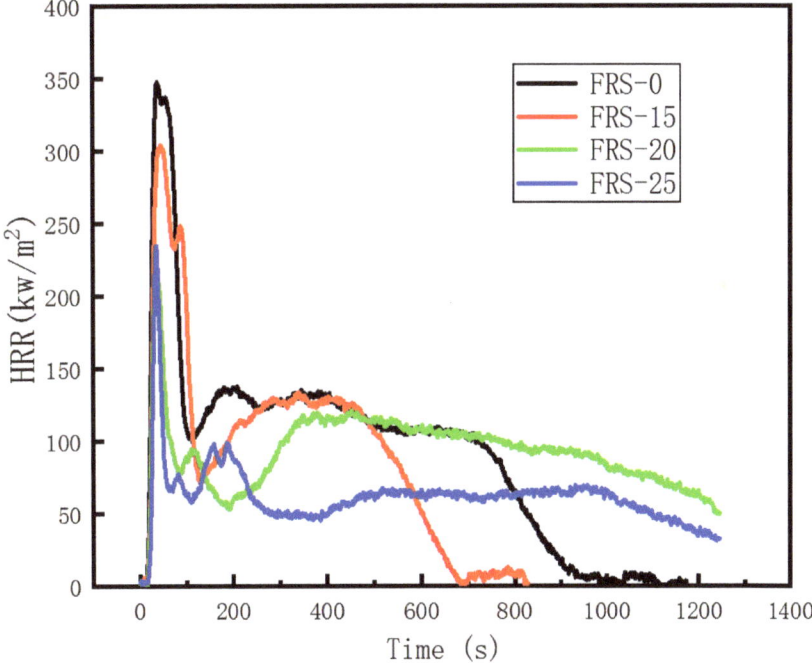

Figure 4. Comparison of combustion heat release rate of each group of flame-retardant corrugated plates.

The fire growth index (FGI) and the fire performance index (FPI) are commonly used to quantify and compare the size of the fire hazard. The FGI is the ratio of the heat release peak to the peak time; the larger the index, the faster the fire grows. The fire performance index is defined as the ratio of the ignition time to the heat release peak; the longer the ignition time, the smaller the heat release peak, and the larger the FPI value, reflecting the better flame retardant performance of the material. Many studies have shown that the FPI has a certain correlation with the development time of a fire in a closed space. The larger the FPI, the longer the boom. The boom time is an important parameter for designing fire escape methods. The FGI and the FPI of each group of corrugated plates are calculated and summarized in Table 4.

Table 4. Comparison of relevant data of various groups of flame-retardant corrugated plates.

Sample Name	FRS-0	FRS-15	FRS-20	FRS-25
Weight (g)	46.3	44.7	47.6	46.9
Heat release peak (kw/m^2)	347.7	304.3	234.7	217.5
Total heat release (J)	121.8	101.9	93.1	80.3
Time to heat release peak (s)	29	34	24	26
Time to Ignition (s)	10	12	11	13
FPI	0.0287	0.0394	0.0469	0.0598
FGI	11.99	8.95	9.78	8.36

The mechanical performance test is reflected through the bending strength test, and the bending test can better reflect the influence of flame retardants on the mechanical properties of flame-retardant prepreg. Figure 5 is the bending strength-strain curve, and Figure 6 is the line chart of the bending strength and bending modulus of various groups. Overall, the addition of flame retardants will result in a decrease in material bending strength, with the bending strength of the blank control group at 313.5 MPa and the bending modulus at 13.7 GPa. The bending strengths at 15%, 20%, and 25% flame retardant addition are 305.3, 269.2, and 251.1 MPa, respectively, representing a decrease of 2.6%, 14.1%, and 19.9% compared to the control group. The bending modulus at 15%, 20%, and 25% flame retardant addition are 12.7, 11.5, and 10.7 GPa, representing a decrease of 7.3%, 16.1%, and 21.9% compared to the control group's modulus of 13.7 GPa. The addition of flame retardants will affect the mechanical properties of the material. When the flame retardant content is 15%, the decrease in the bending strength of the material is relatively small, but when the flame retardant content is further increased, the mechanical properties of the material decrease significantly. This is because when the amount of flame retardant added is too large, it will affect the adhesive strength between the matrix and the fibers at the micro level, resulting in a significant decrease in mechanical properties at the macro level.

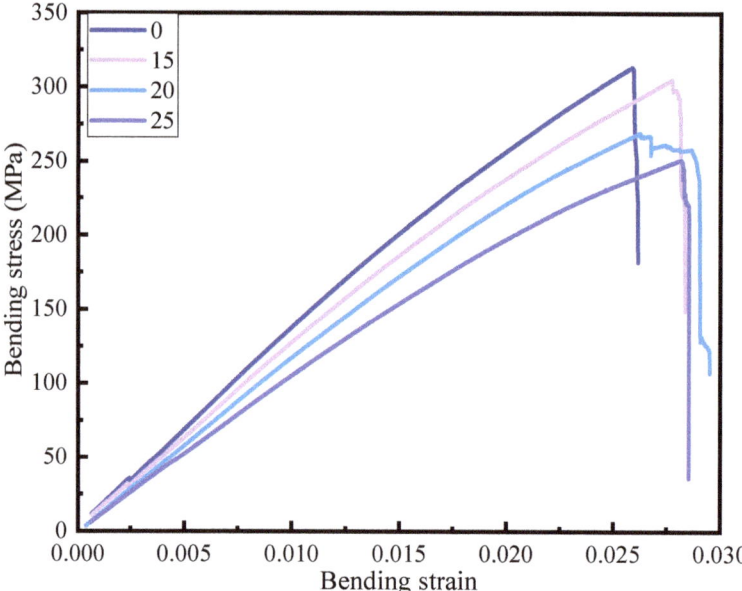

Figure 5. Bending stress-strain curves for each group of samples.

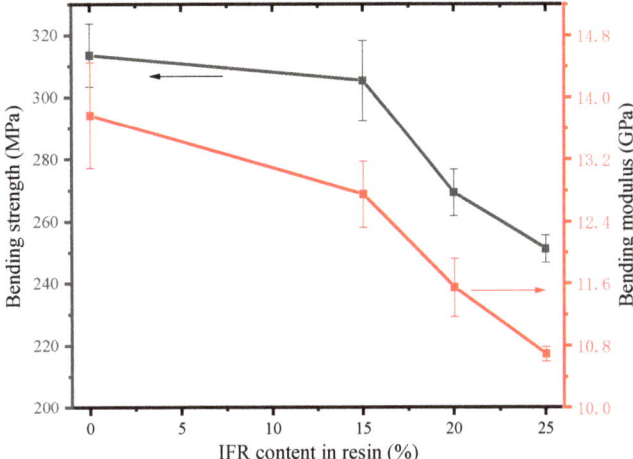

Figure 6. Comparison of flexural strength and flexural modulus of each group of samples.

From Figure 5, it can be seen that when the flame retardant content is 15%, the decrease in bending strength is not very significant, while the fracture toughness increases. When the flame retardant content is 20%, the bending strain reaches 0.025, and the bending strength of the material slightly decreases, then maintains the bending strength, and eventually, as the bending strain increases, the material fails and fractures. When the flame retardant content reaches 25%, the strain at which the material fails is comparable to that of 15% addition. By comparing the above sets of data, it can be concluded that the addition of flame retardants improves the material's bending fracture toughness. The increase in toughness shows a trend of first increasing and then decreasing with the increase in the amount of flame retardants added.

4. Conclusions

1. The hot pressing scheme for preparing corrugated plates was explored. The skin and core material of the corrugated plate were prepared by the hot pressing method. The skin and core material were welded together by the hot melt method, which not only ensures the quality of the forming process but also has high production efficiency. The production and preparation process can be further optimized and improved in subsequent research.
2. The amount of flame retardant added to the corrugated plate can be determined according to the needs of different use scenarios to achieve a balance of the flame retardant performance, molding process, and cost-effectiveness of the material. The flame retardant efficiency is the highest with the addition of 20% flame retardant. This can meet the primary flame retardant demand while reducing the amount of flame retardant added and can significantly reduce the cost of materials, thereby ensuring the economic viability of the flame-retardant corrugated plates.
3. The use of flame retardant will have a certain impact on the mechanical properties of the material, but by optimizing the molding process and selecting the appropriate amount of flame retardant, the mechanical properties can meet the requirements of the corresponding grade when the flame retardant performance meets the requirements of the corresponding rates.
4. The flame retardant performance of the corrugated plate can be further designed because the skin and core material of the corrugated plate are formed separately. The amount of flame retardant added to the core material and the skin can be different to achieve a balance between flame retardant properties, mechanical properties, and the economic cost of materials.

Author Contributions: Data curation, Writing—original draft, Writing—review and editing, Y.S.; Methodology, Project administration, Resources, Supervision, J.L.; review, Project administration, Supervision B.Z. All authors have read and agreed to the published version of the manuscript.

Funding: This work was supported by Fundamental Research Funding (No. 514010104-302) and the National Natural Science Foundation of China (Grant No. 11872086).

Institutional Review Board Statement: Not applicable.

Data Availability Statement: Data are contained within the article.

Conflicts of Interest: The authors declare no conflict of interest.

References

1. Utekar, S.; Suriya, V.K.; More, N.; Rao, A. Comprehensive study of recycling of thermosetting polymer composites—Driving force, challenges and methods. *Compos. Part B Eng.* **2021**, *207*, 108596. [CrossRef]
2. Wang, B.; Fan, S.; Chen, J.; Yang, W.; Liu, W.; Li, Y. A review on prediction and control of curing process-induced deformation of continuous fiber-reinforced thermosetting composite structures. *Compos. Part A Appl. Sci. Manuf.* **2023**, *165*, 107321. [CrossRef]
3. Mahshid, R.; Isfahani, M.N.; Heidari-Rarani, M.; Mirkhalaf, M. Recent advances in development of additively manufactured thermosets and fiber reinforced thermosetting composites: Technologies, materials, and mechanical properties. *Compos. Part A Appl. Sci. Manuf.* **2023**, *171*, 107584. [CrossRef]
4. Ozturk, F.; Cobanoglu, M.; Ece, R.E. Recent advancements in thermoplastic composite materials in aerospace industry. *J. Thermoplast. Compos. Mater.* **2023**. [CrossRef]
5. Zhao, W.; Kumar Kundu, C.; Li, Z.; Li, X.; Zhang, Z. Flame retardant treatments for polypropylene: Strategies and recent advances. *Compos. Part A Appl. Sci. Manuf.* **2021**, *145*, 106382. [CrossRef]
6. Hopmann, C.; Wilms, E.; Beste, C.; Schneider, D.; Fischer, K.; Stender, S. Investigation of the influence of melt-impregnation parameters on the morphology of thermoplastic UD-tapes and a method for quantifying the same. *J. Thermoplast. Compos. Mater.* **2019**, *34*, 1299–1312. [CrossRef]
7. Sun, Y.; Li, J.; Li, H. Flame Retardancy Performance of Continuous Glass-Fiber-Reinforced Polypropylene Halogen-Free Flame-Retardant Prepreg. *Coatings* **2022**, *12*, 976. [CrossRef]
8. Chen, H.; Wang, J.; Ni, A.; Ding, A.; Sun, Z.; Han, X. Effect of novel intumescent flame retardant on mechanical and flame retardant properties of continuous glass fibre reinforced polypropylene composites. *Compos. Struct.* **2018**, *203*, 894–902. [CrossRef]
9. Guo, H.; Yuan, H.; Zhang, J.; Ruan, D. Review of sandwich structures under impact loadings: Experimental, numerical and theoretical analysis. *Thin-Walled Struct.* **2024**, *196*, 111541. [CrossRef]
10. Cheon, Y.-J.; Kim, H.-G. An equivalent plate model for corrugated-core sandwich panels. *J. Mech. Sci. Technol.* **2015**, *29*, 1217–1223. [CrossRef]
11. Rejab, M.R.M.; Cantwell, W.J. The mechanical behaviour of corrugated-core sandwich panels. *Compos. Part B Eng.* **2013**, *47*, 267–277. [CrossRef]
12. Valdevit, L.; Wei, Z.; Mercer, C.; Zok, F.W.; Evans, A.G. Structural performance of near-optimal sandwich panels with corrugated cores. *Int. J. Solids Struct.* **2006**, *43*, 4888–4905. [CrossRef]
13. Dayyani, I.; Shaw, A.D.; Saavedra Flores, E.I.; Friswell, M.I. The mechanics of composite corrugated structures: A review with applications in morphing aircraft. *Compos. Struct.* **2015**, *133*, 358–380. [CrossRef]
14. Birman, V.; Kardomateas, G.A. Review of current trends in research and applications of sandwich structures. *Compos. Part B Eng.* **2018**, *142*, 221–240. [CrossRef]
15. Li, W.; Zhang, Z.; Jiang, Z.; Zhu, M.; Zhang, J.; Huang, H.; Liang, J. Comprehensive performance of multifunctional lightweight composite reinforced with integrated preform for thermal protection system exposed to extreme environment. *Aerosp. Sci. Technol.* **2022**, *126*, 107647. [CrossRef]
16. Lurie, S.A.; Solyaev, Y.O.; Volkov-Bogorodskiy, D.B.; Bouznik, V.M.; Koshurina, A.A. Design of the corrugated-core sandwich panel for the arctic rescue vehicle. *Compos. Struct.* **2017**, *160*, 1007–1019. [CrossRef]
17. Hörold, A.; Schartel, B.; Trappe, V.; Korzen, M.; Bünker, J. Fire stability of glass-fibre sandwich panels: The influence of core materials and flame retardants. *Compos. Struct.* **2017**, *160*, 1310–1318. [CrossRef]
18. Wi, S.; Yang, S.; Yun, B.Y.; Kang, Y.; Kim, S. Fire retardant performance, toxicity and combustion characteristics, and numerical evaluation of core materials for sandwich panels. *Environ. Pollut.* **2022**, *312*, 120067. [CrossRef] [PubMed]

Disclaimer/Publisher's Note: The statements, opinions and data contained in all publications are solely those of the individual author(s) and contributor(s) and not of MDPI and/or the editor(s). MDPI and/or the editor(s) disclaim responsibility for any injury to people or property resulting from any ideas, methods, instructions or products referred to in the content.

Article

Sisal-Fiber-Reinforced Polypropylene Flame-Retardant Composites: Preparation and Properties

Zhenhua Wang [1], Weili Feng [1], Jiachen Ban [2], Zheng Yang [1], Xiaomin Fang [1], Tao Ding [1], Baoying Liu [1,*] and Junwei Zhao [1,*]

[1] College of Chemistry and Chemical Engineering, Henan University, Kaifeng 475004, China
[2] Institute of Science and Technology, Minsheng College, Henan University, Kaifeng 475004, China
* Correspondence: liubaoying666@163.com (B.L.); zhaojunwei@henu.edu.cn (J.Z.)

Abstract: Natural-fiber-reinforced polypropylene (PP) composites with a series of advantages including light weight, chemical durability, renewable resources, low in cost, etc., are being widely used in many fields such as the automotive industry, packaging, and construction. However, the flammability of plant fiber and the PP matrix restricts the application range, security, and use of these composites. Therefore, it is of great significance to study the flame retardants of such composites. In this paper, sisal-fiber-reinforced polypropylene (PP/SF) flame-retardant composites were prepared using the two-step melt blending method. The flame retardant used was an intumescent flame retardant (IFR) composed of silane-coated ammonium polyphosphate (Si-APP) and pentaerythritol (PER). The influence of different blending processes on the flammability and mechanical properties of the composites was analyzed. The findings suggested that PP/SF flame-retardant composites prepared via different blending processes showed different flame-retardant properties. The (PP/SF)/IFR composite prepared by PP/SF secondary blending with IFR showed excellent flame-retardant performance, with a limited oxygen index of about 28.3% and passing the UL-94 V-0 rating (3.2 mm) in the vertical combustion test. Compared with the (PP/IFR)/SF composite prepared by a matrix primarily blended with IFR and then secondly blended with SF, the peak heat release rate (pk HRR) and total heat release (THR) of the (PP/SF)/IFR composite decreased by 11.3% and 13.7%, respectively. In contrast, the tensile strength of the (PP/SF)/IFR system was 5.3% lower than that of the (PP/IFR)/SF system; however, the overall mechanical (tensile, flexural, and notched impact) properties of the composites prepared using three different mixing processes were similar.

Keywords: polypropylene; sisal fiber; flame retardant; preparation; mechanical properties

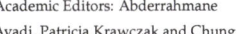

Citation: Wang, Z.; Feng, W.; Ban, J.; Yang, Z.; Fang, X.; Ding, T.; Liu, B.; Zhao, J. Sisal-Fiber-Reinforced Polypropylene Flame-Retardant Composites: Preparation and Properties. *Polymers* 2023, 15, 893. https://doi.org/10.3390/polym15040893

Academic Editors: Abderrahmane Ayadi, Patricia Krawczak and Chung Hae Park

Received: 30 December 2022
Revised: 3 February 2023
Accepted: 6 February 2023
Published: 10 February 2023

Copyright: © 2023 by the authors. Licensee MDPI, Basel, Switzerland. This article is an open access article distributed under the terms and conditions of the Creative Commons Attribution (CC BY) license (https://creativecommons.org/licenses/by/4.0/).

1. Introduction

In order to target the global warming threat, the world is committed to controlling peak carbon emissions in the hope of becoming carbon neutral in the near future. Manufacturers need greater awareness of their production chains and need to develop new materials with reusable or renewable natural resources. Regarding this, natural-plant-fiber-reinforced thermosetting and thermoplastic composite materials are rapid developments in basic research and industrial applications as they are created to be light, renewable, chemically resistant, low-cost, easy to manufacture, completely or partially recyclable, and biodegradable [1–3]. Sisal fiber (SF)-reinforced polypropylene (PP) composite (PP/SF) materials have been widely used in automotive applications because of the advantages they impart in low cost and density, as well as satisfactory mechanical properties coupled with processing [4–6]. However, the inherent flammability of the PP matrix and SF has limited the applications of their composites for many semi-structural components that require strict fire safety [5–7]. In this regard, it is of great importance to study the fire performance of natural-fiber-reinforced polymeric composites.

In recent years, flame retardant research on PP has mainly been based on the addition of the intumescent flame retardant (IFR) system composed of ammonium polyphosphate (APP) and pentaerythritol (PER), so as to further optimize the flame-retardant effect. The main research has focused on the following: (1) The microencapsulation of APP to improve the poor compatibility between APP and substrate, low flame retardant efficiency, and portability of APP [8,9]. (2) The search for an efficient carbon source to improve the carbonization [10,11]. (3) The adoption of a new acid source to replace APP [12–14]. (4) The compounds of different flame retardants [15–19]. (5) The development of one-component flame retardants [20–22]. The flame-retardant modification of natural fibers is mainly achieved by the flame retardant post-treatment of fibers or fabrics [23–27]. To address the flammability of the PP/SF composite, flame retardant is recommended for the composite to constrain and delay the spread of fire after ignition. Among the kinds of flame retardants, the current popular flame retardant system applied to PP/SF is the compound system composed of phosphorous flame retardants, among which the more classic combustible formula is APP as the main flame retardant and PER or metal hydroxide as the synergist [5–7,28,29].

Processing technology has a great influence on the physical and mechanical properties of SF-reinforced polymeric composites due to the fiber impregnation and mixing, final length and diameter, and interface adhesion being closely related to the processing variables [30–32]. The existing research has mainly focused on how to control the production quality of these kinds of composite materials in different manufacturing technologies, such as extrusion, compression molding, resin transfer molding, etc. [33,34]. However, studies on the effect of processing technology on the flame retardancy and mechanical properties of the PP/SF composite is still scarce, and there is still a gap in further research on the effect of processing technology on the flame-retardant mechanism of PP/SF composite during combustion. In light of this, flame retardant PP/SF composites were prepared with an intumescent flame retardant system consisting of silane-coated ammonium polyphosphate (Si-APP) and PER in this work, and the influence of the blending process on the flame retardancy and mechanical properties of PP/SF composites was studied; the corresponding influencing mechanism was also analyzed.

2. Materials and Methods

2.1. Materials

Homopolymerized polypropylene pellets (PP, T03, density 0.9 g/cm^3, melt flow index 3.0 g/10 min, melting point temperature 168.5 °C) were purchased from Dongguan Juzhengyuan Technology Co., Ltd., Dongguan, China. Sisal fiber (SF) roving yarn, comprising a bundle of elementary fibers without surface chemical treatment, was provided by Guangxi Longzhou Strong Hemp Industry Co., Ltd., Chongzuo, China. Ammonium polyphosphate with silane structure (Si-APP, TF-201W, phosphorus content $\geq$31%, nitrogen content $\geq$14, average degree of polymerization $\geq$1000) was purchased from the company of Shifang Taifeng new flame retardant, Shifang, China [35]. Pentaerythritol (PER) was supplied by Tianjin Kemiou, Chemical Reagent Co., Ltd., Tianjin, China.

2.2. Sample Preparation

2.2.1. Experimental Technological Process

The two-step melt blending method was used for the preparation of PP/SF flame-retardant composites. SF and IFR were mixed with a PP matrix in a different order, as shown in Figure 1. After each melt blending step, the extruded strips were cut into pellets and dried for subsequent processing. In order to better distinguish each sample, the abbreviations of the composite materials prepared under different mixing processes were simply named as (PP/IFR)/SF, (PP/SF)/IFR, and PP/IFR/SF, where (PP/IFR)/SF indicates that the matrix and flame retardant were firstly melt blended, and then remixed with SF (as shown in Figure 1a). (PP/SF)/IFR indicates that the matrix and SF were firstly melt blended, and then remixed with IFR (as shown in Figure 1b). PP/IFR/SF indicates that PP, IFR, and SF were firstly melt blended, followed by a second blend (as shown in Figure 1c).

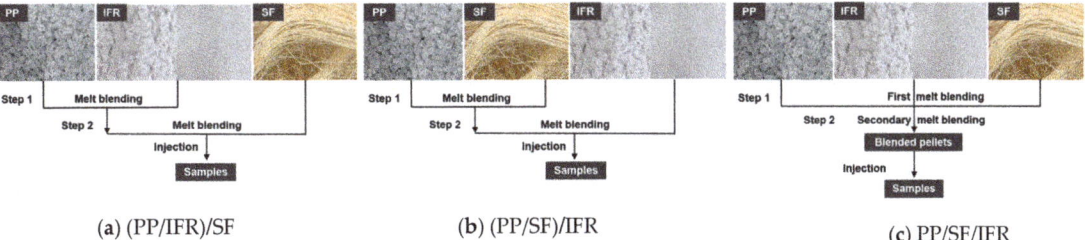

Figure 1. Flow chart of sample preparation of PP/SF flame-retardant composites.

2.2.2. Preparation of PP/SF Flame-Retardant Composites

Sisal fibers were cleaned with distilled water to ensure the removal of any surface impurities, and then dried at 80 °C in an oven for 12 h. The dried sisal fibers were soaked in sodium hydroxide solution with a weight percentage of 10 wt% for 4 h and washed with distilled water until neutrality; we then put them in the oven at 80 °C for 12 h to dry.

The PP pellets, Si-APP, and PER were dried at 80 °C in an oven for 8 h to remove any hygroscopic moisture. According to the previous research experience of our research group [36], all of the raw materials were configured according to the formula in Table 1 for melt blending. The auxiliaries here refer to the processing auxiliaries including compatibilizers and antioxidants. The equipment used for blending was a corotating twin screw extruder (AK22, Nanjing Keya Chemical Complete Equipment Company, Jiangsu, China), the extrusion temperature was from 178 °C to 180 °C, and the screw speed was 80 rpm. The standard samples for flammability and mechanical tests were injection molded from 178 °C to 180 °C using a UN90A2 injection machine (Guangdong Yizumi Precision Machinery Co., Ltd., Guangdong, China). The injection pressure and pressure holding time were 55 bar and 10 s, respectively. The prepared samples were annealed in a drying oven at 80 °C for 4 h to be tested.

Table 1. Formulation and UL-94 rating, as well as limiting oxygen index of original PP and PP/SF composites.

Designation	PP (wt%)	PER (wt%)	Si-APP (wt%)	SF (wt%)	Auxiliaries (wt%)	UL-94 (3.2 mm)	LOI (%)
PP	95	0	0	0	5	NR [a]	18.4
(PP/IFR)/SF	45	7.5	22.5	20	5	NR	24.7
(PP/SF)/IFR	45	7.5	22.5	20	5	V-0 [b]	28.3
PP/IFR/SF	45	7.5	22.5	20	5	NR	25.7

[a] No rating. [b] A single sample could be extinguished within 10 s after double fire ignition, and the total flame burning time of five samples was not more than 50 s.

2.3. Characterization

The combustion behaviors of samples were conducted using a cone calorimeter (Fire Testing Technology Ltd., East Grinstead, UK) under an external heat flux of 50 kW/m^2 in accordance with ISO 5660-1. The specimen size used was 100 × 100 × 6 mm^3.

A fire oxygen index apparatus (TTech-GBT2406-1, Testech Testing Instrument Technology Co., Ltd., Suzhou, China) was used for the limiting oxygen index (LOI) tests, according to the GB/T 2406.2-2009 standard [35]. The tests were conducted using rectangular samples (80 mm × 10 mm × 4 mm). The LOI value of the composite was the average of the minimum oxygen concentration, which supports ten replicate samples which retained combustion.

The Underwriters Laboratories-94 (UL-94) vertical burning tests were conducted using a vertical burning tester (TTech-GBT2408 [35], Testech Testing Instrument Technology Co., Ltd., Suzhou, China). The tests were performed according to the GB/T 2408-2008 standard with samples of size 125 mm × 12.5 mm × 3.2 mm. The final UL-94 rating of the composite was judged by the total flame burning time of five samples.

Tensile and flexural tests were performed on a universal testing machine (TCS-2000, GOTECH Testing Machines Inc., Qingdao, China), along with a ZBC-8400 impact testing machine (MTS Industrial System Co., Ltd., Hong Kong, China) for the notched Izod impact. All of the tests were performed at room temperature. The tensile test was conducted and evaluated according to the GB/T 1040-2006 standard [35], with measurements conducted with a crosshead speed of 50 mm/min. Flexural tests (according to the GB/T9341-2008 standard [35]) were performed at the rate of 2 mm/min with a span distance of 64 mm. The notched Izod impact behaviors of the composites were measured according to the national GB/T 1043-2008 standard [35]. All of the mechanical test results given represented an average value of at least five tests.

The thermal stability of the composites was analyzed via thermogravimetric tests (TGA) conducted using a TGA/SDTA851e Thermogravimetric Analyzer (Mettler-Toledo International Co., Ltd., Zurich, Switzerland). The measurements were conducted from room temperature to 800 °C with a heating rate of 10 °C/min under a protective atmosphere of nitrogen.

The fracture surface after the notched impact test and charred residue obtained from the cone test of the PP/SF composites were observed using a field emission scanning electron microscope (JSM-7610F, Electronics Corporation, Tokyo, Japan). Furthermore, a laser Raman spectrometer (Renishaw inVia, Gloucestershire, UK) with a laser excitation source of 780 nm was used for the detection of the internal structure of residual carbon after vertical burning tests.

3. Results and Discussion

3.1. Flammability of PP/SF Flame-Retardant Composites

The properties of polymer composites are closely related to their morphology and structure. Different processing technology will affect the internal structure of polymer products, and then affect the performance of products. Therefore, the effects of different feeding processes on the mechanical properties and flame retardancy of PP/SF flame-retardant composites were investigated. The results of the LOI test and vertical burning test of pure PP and PP/SF composites are presented in Table 1. The (PP/IFR)/SF and PP/IFR/SF samples had a high burning rate and could not reach any UL-94 rating. Pure PP is very flammable and its limiting oxygen index value is only 18.4%. The addition of sisal fiber and flame retardant can effectively improve the flammability of PP, which showed that the LOI value increased by at least 37.2%. Compared with (PP/IFR)/SF, the LOI value of PP/IFR/SF composites increased to 25.7%. The difference in LOI values may be caused by the different dispersion of flame retardants at the matrix and fiber interface. The (PP/SF)/IFR composite reached a UL-94 V-0 rating, showing excellent fire resistance, and a remarkably high LOI value of 28.3% was obtained, which increased by 14.6% compared with the (PP/IFR)/SF system.

When the sample is burned under the thermal radiation of the conical electric heater, the flame will consume a certain concentration of oxygen in the air and release a certain calorific value of combustion. The combustion behavior of PP and PP/SF composites was evaluated using a conical calorimeter. The experimental principle was to calculate and measure the heat release rate, mass loss rate, and other parameters in the combustion process according to the amount of oxygen consumed by materials during combustion, and then analyze and judge the combustion performance of the material. The results of the cone calorimeter experiment are listed in Table 2. Figure 2 shows the curves of the heat release rate (HRR) and the total heat release (THR) versus time. In general, compared with pure PP, the addition of flame retardants shortened the time to ignition (TTI) of the PP/SF composites, with the peak heat release rate (pk HRR) decreasing and the total smoke production (TSP) increasing, which indicated that the combustion process of the PP/SF composites was suppressed and incomplete. The pk HRR and THR of the PP/SF flame-retardant composites obtained in different blending sequences were different. The pk HRR of (PP/SF)/IFR and PP/IFR/SF were 11.3% and 9.3%, lower than that of (PP/IFR)/SF.

Compared with (PP/IFR)/SF, the THR value of (PP/SF)/IFR decreased to 122.2 MJ/m² and reduced by 13.7%. These results showed that the (PP/SF)/IFR composite exhibited better fire resistance, which was consistent with the LOI and vertical combustion test results.

Table 2. Cone calorimetric combustion experimental results of pure PP and PP/SF flame-retardant composites.

Designation	PP	(PP/IFR)/SF	(PP/SF)/IFR	PP/IFR/SF
TTI (s)	36	23	29	28
pk HRR (kW/m²)	646.2	362.2	321.3	328.6
THR (MJ/m²)	135.3	141.6	122.2	134.9
mean EHC (MJ/kg)	25.2	22.3	21.3	22.0
SEA (m²/kg)	299.3	451.0	477.3	461.6
Av CO (kg/kg)	0.02	0.08	0.09	0.07
Av CO₂ (kg/kg)	1.4	1.2	1.1	1.1
TSP (m²)	14.3	25.3	24.1	24.9

TTI: Time to ignition; pk HRR: Peak heat release rate; THR: Total heat release; mean EHC: Mean effective heat of combustion; SEA: Specific extinction area; Av CO: Average carbon monoxide yield; Av CO_2: Average carbon dioxide yield; TSP: Total smoke production.

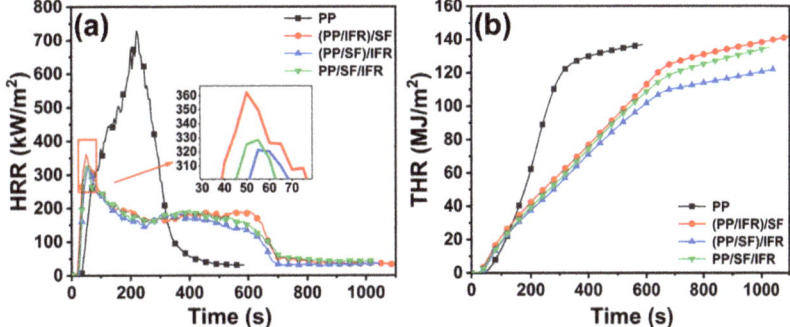

Figure 2. HRR (**a**) and THR (**b**) curves of pure PP and PP/SF flame-retardant composites.

Multiple heat release peaks appeared in the release rate curves of composites under different processes. According to the analysis of the mechanism of flame retardants during combustion [35], Si-APP can play its flame retardant role via both the condense phase and the gas phase. Si-APP was decomposed by heat during combustion and produced active free radicals such as PO· and PO_2· into the gas phase, which can trap active radicals such as H·, O· and HO· in the flame reaction, so as to inhibit the combustion reaction. In the condensed phase, Si-APP was dehydrated and decomposed by heating, and the first exothermic peak was produced (as shown in Figure 2a). As the combustion process went on, the phosphoric acid substances generated via Si-APP thermal decomposition promoted the dehydration of the matrix and SF into carbon residues; in this way, a dense charred layer was formed onto the underlying substrate surface to isolate the internal structure from air and heat, further inhibiting the combustion process. The residual charred layer and the protected matrix further decomposed to form a more stable charred layer structure, resulting in the emergence of a second exothermic peak, as seen in Figure 2a.

According to the above research results, although the total amount of flame retardant added to the matrix was the same, different processing sequences lead to different flame-retardant effects. The composite, with PP and SF blended first, and then mixed with flame retardants, shows better flame retardancy.

3.2. Morphology of Char Residues

The quality of residual carbon is closely related to the flame retardant efficiency. Figures 3 and 4 show the digital photos of the residual carbon layer of composite material after the cone calorimeter test and SEM photos, respectively, from which the carbon formation of the flame retardant system after combustion can be directly reflected. As can be seen from Figure 3, although the carbon layer of the composites obtained by different blending processes was relatively complete, cracks were not absent (as indicated by the arrow). There were many cracks on the charred layer of the (PP/IFR)/SF composite. The incomplete and fragmented charred layer provides channels for combustible gas, heat, and oxygen to enter the underlying matrix, resulting in flame retardant failure.

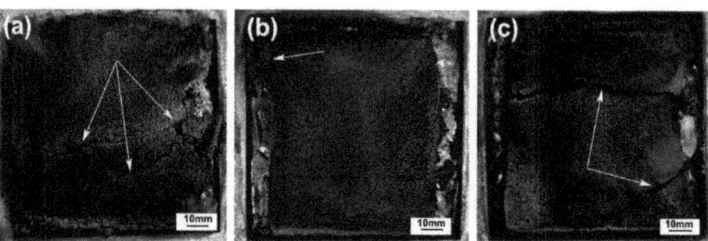

Figure 3. Digital photographs of charred residue of PP/SF composites after cone calorimeter test: (**a**) (PP/IFR)/SF; (**b**) (PP/SF)/IFR; (**c**) PP/IFR/SF.

Figure 4. SEM images of charred residue of PP/SF composites after cone calorimeter test. (**a**) (PP/IFR)/SF; (**b**) (PP/SF)/IFR; (**c**) PP/IFR/SF.

In contrast, the charred layer of PP/IFR/SF and (PP/SF)/IFR composites was relatively thick and dense. In particular, the charred layer formed by (PP/SF)/IFR was thick and continuous, which could effectively play the role of heat and oxygen isolation, so as to endow the composite with excellent flame-retardant performance.

The internal structure of the carbon layer was further analyzed via SEM (as shown in Figure 4). The (PP/IFR)/SF composite formed a fluffy and porous charred layer after combustion, and only part of the surface of sisal fibers was covered by the charred residue (as shown in Figure 4a). There were numerous holes and cracks along the interface between fibers and the matrix (as indicated by the ellipse), which provided channels for heat and the gases that support combustion. The charred layer produced by the PP/IFR/SF composite was relatively dense, and the sisal fibers were mostly coated by the charred layer (as shown in Figure 4c), which endowed the PP/IFR/SF composite with a better flame-retardant effect. The charred layer of the (PP/SF)/IFR composites was relatively compact, with fewer cavities and good quality (as indicated by the arrow in Figure 4b). The dense charred layer can prevent the contact between volatile combustible gas and external oxygen.

The quality of the carbon layer is not only related to its external morphology, but also to its internal graphitization degree. Raman spectroscopy was used to evaluate the graphitization degree of charred residue formed during combustion after the cone calorimeter test; the corresponding results are shown in Figure 5. The D-peak and G-peak are two characteristic peaks of the carbon atomic crystal near 1300 cm^{-1} and 1580 cm^{-1}, representing a lattice defect of the carbon atom and the in-plane stretching vibration of the sp^2 hybridization of the C atom, respectively [37,38]. The ratio of the integrated intensities of D and G bands (I_D/I_G) is used to estimate the graphitization degree of the charred residue [39]. In general, the smaller the I_D/I_G value, the fewer defects of the carbon atoms in the formed structure, and the higher the degree of graphitization of the charred layer, indicating a better quality of the combustion residue of the material [35,40]. A good stable carbon residue can act as a barrier to isolate the combustible medium from air at high temperatures [38,39]. The I_D/I_G of (PP/IFR)/SF, (PP/SF)/IFR, and PP/IFR/SF was 0.69, 0.61, and 0.63, respectively. The I_D/I_G of (PP/SF)/IFR composites was the lowest, which indicated that the degree of graphitization of the carbon residue was the highest.

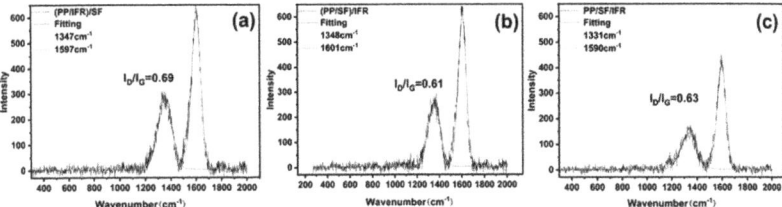

Figure 5. Raman spectra of the char residue of PP/SF flame—retardant composites. (**a**) (PP/IFR)/SF; (**b**) (PP/SF)/IFR; (**c**) PP/IFR/SF.

3.3. Thermal Stability of PP/SF Composites

The fire resistance of materials is closely related to the thermal stability of materials, which can be obtained using thermogravimetric analysis. Figure 6 and Table 3 are the TG and DTG results of PP, SF, and PP/SF flame-retardant composites. PP was thermally stable, with an initial thermal decomposition temperature ($T_{-5\%}$) as high as 409.5 °C; nonetheless, there was little residue after thermal decomposition at 800 °C. The sisal fibers were thermal-sensitive when the temperature was above 200 °C [41], and the $T_{-5\%}$ of sisal fiber was only 269.1 °C; however, the sisal fiber did have a high residue of 18.3% at 800 °C, which means that sisal fiber can be used as a carbon forming agent to enhance the flame retardation of the condensed phase of PP [2]. The introduction of flame retardants further reduces the thermal stability of composites, which is related to the action mechanism of flame retardants. In the case of fire, the thermal decomposition of flame retardants occurs before the substrate, so as to produce the corresponding chemical substances to inhibit the combustion process of the matrix and play the role of flame retardation. The thermal degradation trend of the three kinds of PP/SF flame-retardant composites was consistent. There were slight differences between the initial decomposition temperatures of (PP/SF)/IFR, (PP/IFR)/SF, and PP/IFR/SF composites, where $T_{-5\%}$ of the (PP/SF)/IFR composite was the lowest, which was 245.6 °C. However, the temperatures of 50% thermal weight loss ($T_{-50\%}$) and the maximum weight loss (T_{max}) of composites obtained under different processing sequences were similar. Additionally, the char yield of the (PP/SF)/IFR composite at 800 °C was 23.9%, slightly higher than the other two materials.

Table 3. Thermal analysis data of PP/SF flame-retardant composites.

Designation	$T_{-5\%}$/°C	$T_{-50\%}$/°C	T_{max}/°C	Residue/%
PP	409.5	455.3	492.5	0.8
SF	269.1	363.6	361.8	18.3
(PP/IFR)/SF	249.0	480.4	479.0	22.5
(PP/SF)/IFR	245.6	480.2	480.1	23.9
PP/IFR/SF	250.6	480.0	480.4	22.5

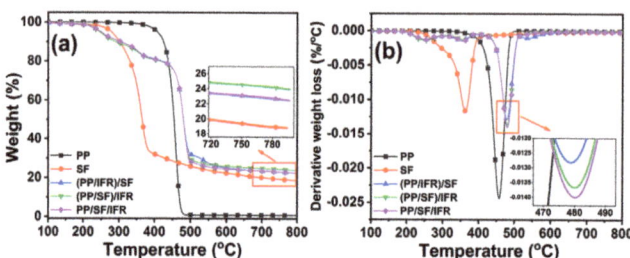

Figure 6. TGA (**a**) and DTG (**b**) of PP, SF, and PP/SF flame−retardant composites.

3.4. Flame-Retardant Mechanism

The flame-retardant mechanism that may be involved in the combustion process of the PP/SF flame-retardant composites is shown in Figure 7. During the combustion process, Si-APP catalyzed the carbon formation of sisal fiber and the PP matrix, along with the release of some volatiles from the decomposition of the substrate. In the (PP/IFR)/SF composites, the matrix was first blended with flame retardants. In this process, the flame retardants were coated by the matrix, and then blended with fiber in the secondary extrusion process, and most of the flame retardants were separated from the fibers by the matrix; less were attached to the surface of the fibers. This resulted in poor carbonization of the fiber surface during combustion, as shown in Figure 4a. The bare fiber interface provided a channel for the transport of oxygen and combustibles during the combustion process, resulting in a poor flame-retardant effect of the composites. In addition, the fibers in the (PP/IFR)/SF composite underwent only one extrusion process and had poor dispersion in the matrix; this resulted in an uneven charred layer formed during combustion and obvious cracks on the surface of the carbon layer, as shown in Figure 3a, which led to the deterioration of the flame-retardant performance of the composite. In the (PP/SF)/IFR composites, PP was blended with SF first, the fiber was wrapped by PP, and then the flame retardants were wrapped in the outer layer of PP. During the combustion process, the flame retardant decomposed in advance, promoting the carbonization of the internal matrix and forming a dense carbon layer on the fiber surface, as shown in Figures 3b and 4b. The charred layer with the expansion structure formed during combustion can effectively protect the base from thermal decomposition, endowing the composite with good flame-retardant performance. Polypropylene, fiber, and flame retardants in the PP/IFR/SF composite were randomly dispersed, and the flame-retardant effect was relatively poor.

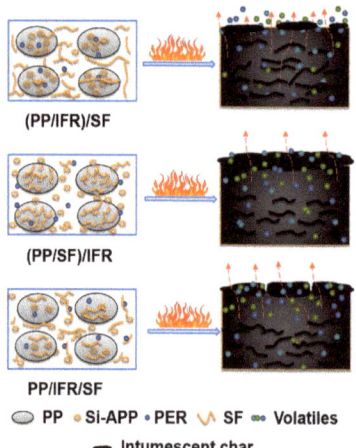

Figure 7. Possible flame-retardant mechanism of PP/SF flame-retardant composites.

3.5. Mechanical Properties

The effects of different mixing sequences on the mechanical properties of the PP/SF flame-retardant composites are shown in Figure 8. In general, the addition of fiber and flame retardants improved the tensile and flexural properties of PP, while they deteriorated the notched impact strength of PP, which was due to the introduction of fibers and flame-retardant particles increasing the defects of the system, leading to higher stress concentrations and therefore the failure of the material at a lower stress.

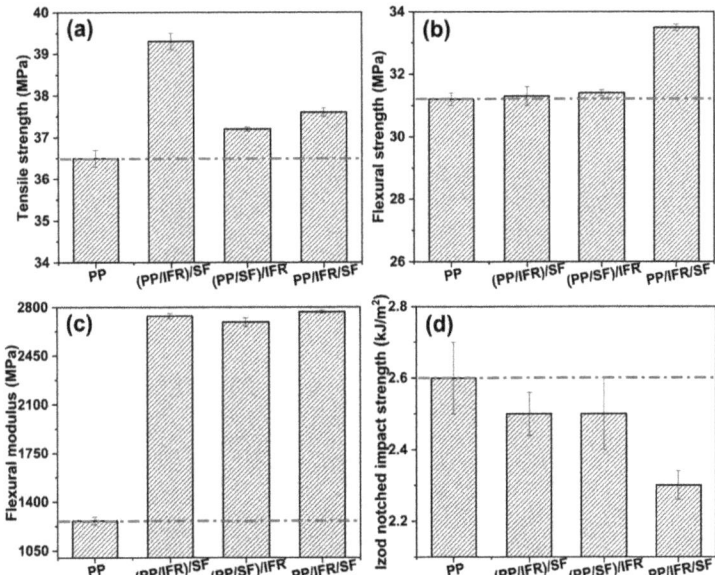

Figure 8. Mechanical properties of pure PP and PP/SF flame-retardant composites. (**a**) Tensile strength, (**b**) Flexural strength, (**c**) Flexural modulus, (**d**) Izod notched impact strength.

The flexural strength and modulus (Figure 8) of the composites showed an increase with the incorporation of sisal fibers and flame retardants in the pure PP. Flexural strength represents the load-carrying capacity of the specimen, which is related to the distribution of the fillers in the composite and the interfacial adhesion between fillers and the matrix. Both the flame retardants and fibers in the PP/IFR/SF composite underwent secondary blending and were well dispersed in the matrix, endowing PP/IFR/SF a better bending strength than that of (PP/SF)/IFR and (PP/IFR)/SF. The flexural moduli of the obtained composites were of the same order of magnitude with pure PP. As is well known, various mechanisms such as shearing, tension, and compression take place simultaneously during the bending test [42]. The failure characteristics of composites were completely changed as a result of the addition of fibers and flame-retardant particles. Moreover, the modulus of the added natural fiber was comparatively lower than that of the inorganic fibers such as glass fiber and carbon fiber; therefore, a combination of influencing factors leading to the composite modulus is not an order of magnitude higher than that of pure PP. Similar results have been observed by some research groups [5,42,43].

Compared with the (PP/IFR)/SF and PP/IFR/SF systems, the overall mechanical properties of the (PP/SF)/IFR composites were slightly worse. The tensile strengths of the PP/IFR/SF and (PP/SF)/IFR composites were relatively low, especially that of the (PP/SF)/IFR composites, which was 37.2 MPa. In general, in a certain size range, the longer the fiber length in the fiber-reinforced products, the more damage work the fiber consumes through its own fracture in the process of material failure, and thus the higher the mechanical strength of the composite material [44,45]. Therefore, the (PP/IFR)/SF composite exhibits the best mechanical properties. However, after secondary extrusion shearing, the length of the fibers in the PP/IFR/SF and (PP/SF)/IFR composites was obviously damaged, which led to the reinforcement effect of the fiber being obviously weaker than that of the (PP/IFR)/SF composites.

The interfacial bonding and fiber dispersion within the fiber-reinforced polymeric composites can be evaluated by the morphology of the impact section. The SEM micrographs of the impact section of the PP/SF composites are shown in Figure 9. As can be seen from the figure, the fibers of composite materials obtained by different blending processes had different degrees of pulling out when the samples were destroyed by external forces. As shown in the position indicated by the ellipse in Figure 9a, fiber pull-out in (PP/IFR)/SF was obvious, leaving many pulled-out marks and holes. There was little resin matrix residue on the surface of the pulled-out fiber. However, there was no obvious peeling phenomenon between the fiber and matrix in Figure 9b and c, which indicated that the interface bonding between the fiber and the matrix in (PP/SF)/IFR and PP/IFR/SF was better than that of (PP/IFR)/SF. This also showed that the secondary blending of the fiber and matrix was conducive to the interfacial bonding between the fiber and matrix. Moreover, it can be seen that the length of the fiber pulled out in the (PP/SF)/IFR sample was shorter than that of the PP/IFR/SF sample, indicating that the interfacial bonding between the fiber and the matrix of (PP/SF)/IFR was better than that of PP/IFR/SF, showing a slightly higher notched impact strength of 2.5 kJ/m^2.

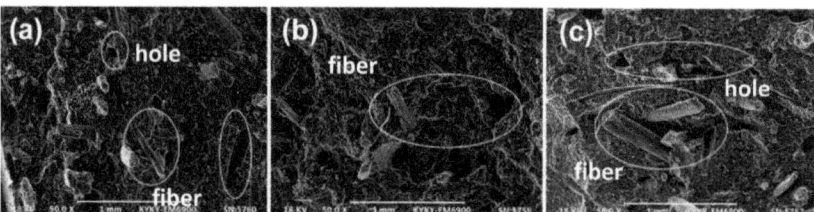

Figure 9. SEM micrographs of PP/SF flame-retardant composites. (**a**) (PP/IFR)/SF; (**b**) (PP/SF)/IFR; (**c**) PP/IFR/SF.

4. Conclusions

Sisal-fiber-reinforced polypropylene flame-retardant composites were prepared via twin-screw extrusion and the effects of different blending procedures on the fire retardancy and mechanical performance of the composites were investigated. In the case of the same raw material formula, the (PP/SF)/IFR composite prepared by PP, first blended with SF, and then with flame retardants for secondary blending, showed an excellent flame-retardant performance, with a limited oxygen index of 28.3% and reaching a UL-94 V-0 rating. The tensile strength of the (PP/SF)/IFR system was 5.3% lower than that of the (PP/IFR)/SF system; however, the comprehensive mechanical properties of (PP/SF)/IFR were not much different from those of the composites prepared via the other two mixing processes. It is clear from the present study that the blending procedure has a great influence on the properties of the composites, especially the flame-retardant properties. A useful composite with good fire resistance and strength could be successfully developed through the process of matrix blending with fiber first, and then secondary blending with flame retardants. These good properties allow sisal-fiber-reinforced polypropylene flame-retardant composites to exhibit great application prospects in construction and the automotive industry.

However, adding large amounts of flame retardants is still the main means to realize the flame retardancy of existing natural-fiber-reinforced polymeric composite products, which will undoubtedly lead to the deterioration of the mechanical properties of the composites, especially the notched impact properties as shown in this study. In order to balance the flame retardancy and mechanical properties of nature fiber-reinforced polymeric composites, some key technical problems such as the compatibility between the matrix and fillers, the multiple interfacial processing technologies between fiber, matrix, and flame retardants, and the interfacial heat conduction problem need to be solved. From a practical and commercial point of view, the majority of scientific research workers and enterprises still need to commit to developing halogen-free, low-smoke, low-toxicity flame retardants and new methods of flame retardant modification with high efficiency and environmental protection, simplifying the natural fiber flame retardancy treatment process, and ultimately realizing the development of high efficiency halogen-free flame retardants and high-performance natural-fiber-reinforced polymeric flame-retardant composites.

Author Contributions: Conceptualization, B.L., X.F., T.D. and J.Z.; methodology, B.L. and W.F.; validation, B.L. and J.Z.; formal analysis, B.L., W.F. and X.F.; investigation, Z.W., W.F. and J.B.; resources, B.L., X.F., T.D. and J.Z.; data curation, Z.W., W.F., J.B. and Z.Y.; writing—original draft preparation, Z.W. and W.F.; writing—review and editing, B.L. and Z.W.; visualization, Z.Y.; supervision, T.D. and J.Z.; project administration, X.F.; funding acquisition, B.L. and X.F. All authors have read and agreed to the published version of the manuscript.

Funding: Natural Science Foundation of China (51703051) and Key scientific research project of colleges and universities in Henan province (21A430005).

Institutional Review Board Statement: Not applicable.

Data Availability Statement: Available data can be obtained from the corresponding author upon request.

Conflicts of Interest: The authors declare no conflict of interest.

References

1. Madyaratri, E.W.; Ridho, M.R.; Aristri, M.A.; Lubis, M.A.R.; Iswanto, A.H.; Nawawi, D.S.; Antov, P.; Kristak, L.; Majlingová, A.; Fatriasari, W. Recent advances in the development of fire-resistant biocomposites—A review. *Polymers* **2022**, *14*, 362. [CrossRef]
2. Mukhopadhyay, S.; Srikanta, R. Effect of ageing of sisal fibres on properties of sisal-polypropylene composites. *Polym. Degrad. Stab.* **2008**, *93*, 2048–2051. [CrossRef]
3. Sun, Z.; Wu, M. Effects of sol-gel modification on the interfacial and mechanical properties of sisal fiber reinforced polypropylene composites. *Ind. Crops Prod.* **2019**, *137*, 89–97. [CrossRef]
4. Kozlowski, R.; Wladyka-Przybylak, M. Flammability and fire resistance of composites reinforced by natural fibers. *Polym. Adv. Technol.* **2008**, *19*, 446–453. [CrossRef]
5. Jeencham, R.; Suppakarn, N.; Jarukumjorn, K. Effect of flame retardants on flame retardant, mechanical, and thermal properties of sisal fiber/polypropylene composites. *Compos. Part B Eng.* **2014**, *56*, 249–253. [CrossRef]
6. Suppakarn, N.; Jarukumjorn, K. Mechanical properties and flammability of sisal/PP composites: Effect of flame retardant type and content. *Compos. Part B Eng.* **2009**, *40*, 613–618. [CrossRef]
7. Elsabbagh, A.; Attia, T.; Ramzy, A.; Steuernagel, L.; Ziegmann, G. Towards selection chart of flame retardants for natural fibre reinforced polypropylene composites. *Compos. Part B Eng.* **2018**, *141*, 1–8. [CrossRef]
8. Ding, S.; Liu, P.; Zhang, S.; Ding, Y.; Wang, F.; Gao, C.; Yang, M. Preparation and characterization of cyclodextrin microencapsulated ammonium polyphosphate and its application in flame retardant polypropylene. *J. Appl. Polym. Sci.* **2020**, *137*, 49001. [CrossRef]
9. Pan, M.; Mei, C.; Du, J.; Li, G. Synergistic effect of nano silicon dioxide and ammonium polyphosphate on flame retardancy of wood fiber–polyethylene composites. *Compos. Part A* **2014**, *66*, 128–134. [CrossRef]
10. Wu, K.; Zhang, Y.; Hu, W.; Lian, J.; Hu, Y. Influence of ammonium polyphosphate microencapsulation on flame retardancy, thermal degradation and crystal structure of polypropylene composite. *Compos. Sci. Technol.* **2013**, *81*, 17–23. [CrossRef]
11. Xu, B.; Wu, X.; Ma, W.; Qian, L.; Xin, F.; Qiu, Y. Synthesis and characterization of a novel organic-inorganic hybrid char-forming agent and its flame-retardant application in polypropylene composites. *J. Anal. Appl. Pyrolysis* **2018**, *134*, 231–242. [CrossRef]
12. Ma, D.; Li, J. Synthesis of a bio-based triazine derivative and its effects on flame retardancy of polypropylene composites. *J. Appl. Polym. Sci.* **2020**, *137*, 47267. [CrossRef]
13. Zheng, Z.; Liu, Y.; Zhang, L.; Wang, H. Synergistic effect of expandable graphite and intumescent flame retardants on the flame retardancy and thermal stability of polypropylene. *J. Mater. Sci.* **2016**, *51*, 5857–5871. [CrossRef]
14. Li, W.X.; Zhang, H.J.; Hu, X.P.; Yang, W.X.; Cheng, Z.; Xie, C.Q. Highly efficient replacement of traditional intumescent flame retardants in polypropylene by manganese ions doped melamine phytate nanosheets. *J. Hazard. Mater.* **2020**, *398*, 123001. [CrossRef]

15. Li, N.; Xia, Y.; Mao, Z.; Wang, L.; Guan, Y.; Zheng, A. Influence of antimony oxide on flammability of polypropylene/intumescent flame retardant system. *Polym. Degrad. Stab.* **2012**, *97*, 1737–1744. [CrossRef]
16. Almirón, J.; Roudet, F.; Duquesne, S. Influence of volcanic ash, rice husk ash, and solid residue of catalytic pyrolysis on the flame-retardant properties of polypropylene composites. *J. Fire Sci.* **2019**, *37*, 434–451. [CrossRef]
17. Doğan, M.; Yılmaz, A.; Bayramlı, E. Synergistic effect of boron containing substances on flame retardancy and thermal stability of intumescent polypropylene composites. *Polym. Degrad. Stab.* **2010**, *95*, 2584–2588. [CrossRef]
18. Shen, L.; Chen, Y.; Li, P. Synergistic catalysis effects of lanthanum oxide in polypropylene/magnesium hydroxide flame retarded system. *Compos. Part A* **2012**, *43*, 1177–1186. [CrossRef]
19. Zhao, W.; Cheng, Y.; Li, Z.; Li, X.; Zhang, Z. Improvement in fire-retardant properties of polypropylene filled with intumescent flame retardants, using flower-like nickel cobaltate as synergist. *J. Mater. Sci.* **2021**, *56*, 2702–2716. [CrossRef]
20. Zhu, C.; He, M.; Liu, Y.; Cui, J.; Tai, Q.; Song, L.; Hu, Y. Synthesis and application of a mono-component intumescent flame retardant for polypropylene. *Polym. Degrad. Stab.* **2018**, *151*, 144–151. [CrossRef]
21. Abdelkhalik, A.; Makhlouf, G.; Hassan, M.A. Manufacturing, thermal stability, and flammability properties of polypropylene containing new single molecule intumescent flame retardant. *Polym. Adv. Technol.* **2019**, *30*, 1403–1414. [CrossRef]
22. Qi, H.; Liu, S.; Chen, X.; Shen, C.; Gao, S. The flame retardancy and thermal performances of polypropylene with a novel intumescent flame retardant. *J. Appl. Polym. Sci.* **2020**, *137*, 49047. [CrossRef]
23. Lam, Y.L.; Kan, C.W.; Yuen, C.W.M. Objective measurement of hand properties of plasma pre-treated cotton fabrics subjected to flame-retardant finishing catalyzed by zinc oxide. *Fibers Polym.* **2014**, *15*, 1880–1886. [CrossRef]
24. Anjumol, K.S.; Sreenivasan, S.N.; Tom, T.; Mathew, S.S.; Maria, H.J.; Spatenka, P.; Thomas, S. Development of natural fiber-reinforced flame-retardant polymer composites. In *Bio-Based Flame-Retardant Technology for Polymeric Materials*; Elsevier: Amsterdam, The Netherlands, 2022; pp. 369–389.
25. Guo, W.; Kalali, E.N.; Wang, X.; Xing, W.; Hu, Y. Processing bulk natural bamboo into a strong and flame-retardant composite material. *Ind. Crops Prod.* **2019**, *138*, 111478. [CrossRef]
26. Chu, F.; Yu, X.; Hou, Y.; Mu, X.; Song, L.; Hu, W. A facile strategy to simultaneously improve the mechanical and fire safety properties of ramie fabric-reinforced unsaturated polyester resin composites. *Compos. Part A Appl. Sci. Manuf.* **2018**, *115*, 264–273. [CrossRef]
27. Zhang, L.; Li, Z.; Pan, Y.T.; Perez Yanez, A.; Hu, S.; Zhang, X.Q.; Wang, R.; Wang, D.Y. Polydopamine induced natural fiber surface functionalization: A way towards flame retardancy of flax/poly (lactic acid) biocomposites. *Compos. Part B Eng.* **2018**, *154*, 56–63. [CrossRef]
28. Pornwannachai, W.; Ebdon, J.R.; Kandola, B.K. Fire-resistant natural fibre-reinforced composites from flame retarded textiles. *Polym. Degrad. Stab.* **2018**, *154*, 115–123. [CrossRef]
29. Le Bras, M.; Duquesne, S.; Fois, M.; Griselb, M.; Poutch, F. Intumescent polypropylene/flax blends: A preliminary study. *Polym. Degrad. Stab.* **2005**, *88*, 80–84. [CrossRef]
30. Arzondo, L.M.; Vazquez, A.; Carella, J.M.; Pastor, J.M. A low-cost, low-fiber-breakage, injection molding process for long sisal fiber reinforced polypropylene. *Polym. Eng. Sci.* **2004**, *44*, 1766–1772. [CrossRef]
31. Joseph, P.V.; Joseph, K.; Thomas, S. Effect of processing variables on the mechanical properties of sisal-fiber-reinforced polypropylene composites. *Compos. Sci. Technol.* **1999**, *11*, 1625–1640. [CrossRef]
32. Oksman, K.; Mathew, A.P.; Långström, R.; Nyström, B.; Joseph, K. The influence of fibre microstructure on fibre breakage and mechanical properties of natural fibre reinforced polypropylene. *Compos. Sci. Technol.* **2009**, *69*, 1847–1853. [CrossRef]
33. Faruk, O.; Bledzki, A.K.; Fink, H.P.; Sain, M. Progress report on natural fiber reinforced composites. *Macromol. Mater. Eng.* **2014**, *299*, 9–26. [CrossRef]
34. Kerni, L.; Singh, S.; Patnaik, A.; Kumar, N. A review on natural fiber reinforced composites. *Mater. Today Proc.* **2020**, *28*, 1616–1621. [CrossRef]
35. Lu, Z.H.; Feng, W.L.; Kang, X.L.; Wang, J.L.; Xu, H.; Wang, Y.P.; Liu, B.Y.; Fang, X.M.; Ding, T. Synthesis of siloxane-containing benzoxazine and its synergistic effect on flame retardancy of polyoxymethylene. *Polym. Adv. Technol.* **2019**, *30*, 2686–2694. [CrossRef]
36. Li, H. Environmental Protection Flame-Retardant Fiber Reinforced Polypropylene and Its Combustion Behavior in Fire Environment. Master's Thesis, Henan University, Kaifeng, China, 2018.
37. Yang, G.; Guan, S.; Mehdi, S.; Fan, Y.; Liu, B.; Li, B. Co-CoOx supported onto TiO$_2$ coated with carbon as a catalyst for efficient and stable hydrogen generation from ammonia borane. *Green Energy Environ.* **2021**, *6*, 236–243. [CrossRef]
38. Xu, W.; Zhang, B.; Wang, X.; Wang, G.; Ding, D. The flame retardancy and smoke suppression effect of a hybrid containing CuMoO4 modified reduced graphene oxide/layered double hydroxide on epoxy resin. *J. Hazard. Mater.* **2018**, *343*, 364–375. [CrossRef] [PubMed]
39. Zhou, K.; Gao, R.; Qian, X. Self-assembly of exfoliated molybdenum disulfide (MoS$_2$) nanosheets and layered double hydroxide (LDH): Towards reducing fire hazards of epoxy. *J. Hazard. Mater.* **2017**, *338*, 343–355. [CrossRef]
40. Krauss, B.; Nemes-Incze, P.; Skakalova, V.; Biro, L.P.; Klitzing, K.V.; Smet, J.H. Raman scattering at pure graphene zigzag edges. *Nano Lett.* **2010**, *10*, 4544–4548. [CrossRef]
41. Kandola, B.K.; Mistik, S.I.; Pornwannachai, W.; Anand, S.C. Natural fibre-reinforced thermoplastic composites from woven-nonwoven textile preforms: Mechanical and fire performance study. *Compos. Part B* **2018**, *153*, 456–464. [CrossRef]

42. Cavalcanti, D.K.K.; Banea, M.D.; Neto, J.S.S.; Lima, R.A.A.; da Silva, L.F.M.; Carbas, R.J.C. Mechanical characterization of intralaminar natural fibre-reinforced hybrid composites. *Compos. Part B* **2019**, *175*, 107149. [CrossRef]
43. Prasad, A.V.R.; Rao, K.M. Mechanical properties of natural fibre reinforced polyester composites: Jowar, sisal and bamboo. *Mater. Des.* **2011**, *32*, 4658–4663. [CrossRef]
44. Ganeshan, P.; Kumaran, S.S.; Raja, K.; Venkateswarlu, D. An investigation of mechanical properties of madar fiber reinforced polyester composites for various fiber length and fiber content. *Mater. Res. Express* **2018**, *6*, 015303. [CrossRef]
45. Devi, L.U.; Bhagawan, S.S.; Thomas, S. Mechanical properties of pineapple leaf fiber-reinforced polyester composites. *J. Appl. Polym. Sci.* **1997**, *64*, 1739–1748. [CrossRef]

Disclaimer/Publisher's Note: The statements, opinions and data contained in all publications are solely those of the individual author(s) and contributor(s) and not of MDPI and/or the editor(s). MDPI and/or the editor(s) disclaim responsibility for any injury to people or property resulting from any ideas, methods, instructions or products referred to in the content.

Article

Using Heating and Cooling Presses in Combination to Optimize the Consolidation Process of Polycarbonate-Based Unidirectional Thermoplastic Composite Tapes

Janos Birtha [1,2,*], Eva Kobler [2], Christian Marschik [2], Klaus Straka [1] and Georg Steinbichler [1]

[1] Institute of Polymer Injection Moulding and Process Automation, Johannes Kepler University Linz, Altenbergerstraße 69, 4040 Linz, Austria
[2] Competence Center CHASE GmbH, Altenbergerstraße 69, 4040 Linz, Austria
* Correspondence: janos.birtha@chasecenter.at

Abstract: The main aim of this work was to optimize the consolidation of unidirectional fiber-reinforced thermoplastic composite tapes made of polycarbonate and carbon fibers using a heating press and a cooling press in combination. Two comprehensive studies were carried out to investigate the impact of process settings and conditions on the quality of the consolidated parts. The initial screening study provided valuable insights that informed the design of the second study, in which the experimental design was expanded and various modifications, including the implementation of a frame tool, were introduced. The second study demonstrated that the modifications in combination with a high heating press temperature and elevated heating and cooling pressures successfully achieved the desired goals: the desired thickness (2 mm), improved bonding strength (23% increase), and reduced void content (down to 4.64%) in the consolidated parts.

Keywords: thermoplastic composites; consolidation; optimization; unidirectional tapes

1. Introduction

Thermoplastic composites are prime candidates for the design of lightweight components. Compared to their thermoset counterparts, their advantages include that they can be remelted multiple times, are recyclable, and they offer high damage resistance and excellent vibration dampening [1,2]. One way of manufacturing or locally reinforcing structural plastic components is to use unidirectional (UD) fiber-reinforced thermoplastic tapes. They provide part designers with a high degree of flexibility, since they can be freely oriented and layered in an automated manner. Alongside the automated tape placement (ATP) process, compression molding using a hot plate to consolidate UD tapes is gaining in popularity due to the shorter cycle times required.

The consolidation process involves melting of the thermoplastic matrix of the UD tapes under compression, which causes the viscosity of the polymer to decrease. This allows the voids due to asperities of the tapes to be filled in. Once in full contact, the molecular chains of the matrix material can diffuse across the interface, which—after cooling—achieves bonding between the tapes [3]. The quality of the consolidated tape stack strongly depends on the processing conditions, such as temperature, pressure, and time. Producing a void-free, well-bonded part therefore requires finding optimal process settings.

The rich literature describes a variety of experimental approaches to optimizing the consolidation of polyether ether ketone (PEEK)-based thermoplastic composites using different processes. Khan et al. [4,5] used the ATP process to monitor the temperature profile below the first and seventh layer using thermocouples during consolidation, and to optimize various process settings based on the void content, the density, and the interlaminar shear strength using lap-shear tests. Similarly, Sonmez et al. [6] used PEEK with the ATP process to find optimal process settings by observing the peak residual stress, thermal degradation, and the degree of bonding. With three different processes, namely the vacuum-bag-only in oven (VBO) approach, laser-assisted automated fiber placement (AFP), and using a hot press, Saenz-Castillo et al. [7] measured the void content, the in-plane shear strength, and interlaminar shear stress (ILSS) to assess the optimal settings for producing PEEK-based composites.

The majority of studies have focused on PEEK matrices, while other materials have received less attention. Saffar et al. [8] used the VBO technique to investigate the bonding quality of polyetherketoneketone (PEKK)-based composites. Hoang et al. [9] performed a large-scale study employing the same type of material, but post-consolidated their parts using four different methods, including in situ consolidation, annealing, VBO, and using a hot press. Optimization of the consolidation process based on void content and mechanical properties in an autoclave with carbon/polyphenylene sulfide (CF/PPS) was investigated by Patou et al. [10]. Further materials, such as polyaryletherketone (PAEK) [11], polyamide-6 [12], and polypropylene-based [13] thermoplastic composites have also been used to optimize various processing techniques. To the best of our knowledge, research based on the consolidation of polycarbonate-based (PC) composites is scarce. Notably, Borowski et al. [14] used carbon-fiber-reinforced polycarbonate (PC/CF) to optimize their additive manufacturing-based in situ consolidation process by measuring the flexural strength and the porosity of the finished part. Asséko et al. focused on the temperature development during the joining of glass-fiber-reinforced polycarbonate (PC/GF) and confirmed that—even above the glass-transition temperature—perfect cohesion was not achieved [15].

Compared to the ATP and automated fiber-placement (AFP) processes, the consolidation of thermoplastic composites using a hot press has gained less attention. Schnell et al. [13] determined the robustness of the hot press process by the bonding quality of polypropylene (PP)/glass prepregs. They also used thermocouples to measure the temperature of the hot plates and at the interface between the prepregs to assess the stability of the process. In a study by Almeida et al. [16], PEEK laminates were used to assess the degradation of the matrix and the type of voids appearing in the semi-finished product. As previously mentioned, Saenz-Castillo et al. [7] used various processing techniques, including a hot press. In their study, they varied the temperature and the pressure of a hot press to assess how the void content and the ILSS are influenced by changing the process settings.

Most of these studies used the void content or interlaminar strength to assess the quality of the final product. Other mechanical tests used for optimization include the double cantilever beam test [17], the four-point bending test [18], and the mandrel peel test [19]. In the case of semi-crystalline matrix materials, the degree of crystallinity is also an important quality parameter [20–22]. By means of process modeling and experimental investigations, Sonmez et al. [3,6,23] measured the residual stress, among other quality parameters, to optimize the ATP process. The investigation of fiber waviness development during the consolidation process has recently started to attract attention, as it leads to a decline in compressive strength and surface quality [24–26]. Finally, the change in laminate thickness and the resulting squeeze flow has also been considered in some studies [27,28].

The main objective of this work was to optimize the consolidation process of PC/CF UD tapes using a combination of a hot and a cold press in three consecutive studies. The choice of material was based on the selection of an amorphous matrix-based thermoplastic composite, suitable for potential use in interior automotive applications. The overall goal was to obtain well-bonded, void-free consolidated parts made from 12 individual layers with a total thickness of 2 mm. From an optimization perspective, this involved (i) maximizing bond strength, (ii) minimizing void content, and (iii) achieving a 2 mm thickness to match the wall thickness of the injection molding machine's tool in the downstream process for an arbitrary amorphous matrix-based thermoplastic composite. First, consolidation experiments were performed to assess the thermal behavior of the core layer of the stack after changing the set temperature in both the heating and the cooling presses. For this purpose, Type K thermocouples were positioned in the center of the stack, and the information gathered was used to determine the cycle time. Second, a screening test was carried out to evaluate industrially relevant process windows. To this end, we investigated the influence of various process parameters—the cycle time, temperature, and pressure of the heating and cooling presses—on the quality of the consolidated part in terms of bonding strength, thickness, density, and warpage. In addition, we examined the change in part area after consolidation (as a measure of squeeze flow). Third, informed by the results of the screening test, we implemented a frame tool. Applying the findings of the previous analyses, we used a central composite design to identify optimal process parameters that produce the highest quality in terms of bonding, void content, thickness, and warpage.

2. Experimental

2.1. Material

We used Maezio CF GP 1003T UD tapes made from PC/CF with a fiber volume content of roughly 44% (according to the material supplier's data sheet) in our consolidation experiments provided by Covestro AG (Leverkusen, Germany). Table 1 shows an overview of the properties of the matrix and fiber materials. The tape stacks were positioned and spot-welded in a tape-laying cell based on the pick-and-place principle [29]. In each case, 12 layers of UD tapes were arranged in a stacking sequence of $[0°/90°/0°/90°/0°/90°]_s$. The nominal dimensions of the tape stack were $230 \times 150 \times 2.1$ mm. However, a previous study focusing on measuring the thickness of UD tapes at 648 different positions revealed an average tape thickness of 0.186 ± 0.0073 mm. This analysis confirms that stacking 12 layers of these UD tapes should yield a total thickness greater than 2.1 mm.

Table 1. Properties of the matrix and fiber materials in the UD tape.

Material	Property	Value	Unit
Matrix (Polycarbonate)	Melt mass flow rate	37	g/10 min (300 °C/1.2 kg)
	Density	1190	kg/m^3
	Glass-transition temperature	145	°C
	Tensile modulus	2400	MPa
	Yield stress	65	MPa, at 50 mm/s
Fiber (Carbon Fiber)	Density	1800	kg/m^3
	Denier	14,400	den
	Tensile modulus	250	GPa

2.2. Experimental Set-Up

The tape stacks were consolidated in a FILL SM-03 consolidation unit, as shown in Figure 1. The unit consists of two hydraulic presses, namely a heating press and a cooling press, along with a transport system for the material being processed. The tape stack is initially placed between two 5 mm steel tool plates (Figure 1a), which are first moved (Figure 1b) to the heating press (Figure 1c). In this stage, the lay-up is subjected to elevated temperatures above the glass-transition (T_g) (in case of amorphous polymers) or melting (T_m) temperature (in case of semi-crystalline polymers) of the matrix, while pressure is applied for a predetermined period. The steel plates (Figure 1d) carrying the molten material are then transported to the cooling press (Figure 1e) within approximately 5 s, where it is cooled to below the T_g or T_m of the matrix. Once the fully automatic process is completed, the consolidated plate can be retrieved by lifting the top steel plate using sucker pins and a mechanical locking mechanism.

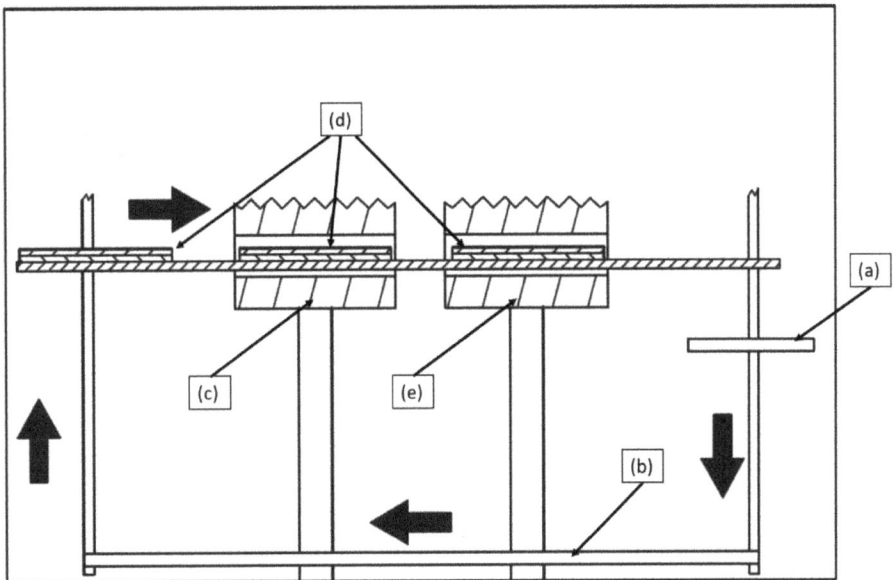

Figure 1. Consolidation unit used in the experiments: (**a**) transport system, (**b**) shuttle system, (**c**) heating press, (**d**) steel plate tools, (**e**) cooling press.

The heating press can reach a maximum temperature of 450 °C and apply a maximum force of 25 kN. In contrast, the cooling press has a maximum operating temperature of 140 °C and can exert a maximum force of 290 kN. These specifications define the upper temperatures and forces that the presses can achieve and maintain during operation. These limits also define the operation of the consolidation unit: in the heating press, minimal pressure is applied to prevent excessive squeeze flow, while in the cooling press a much higher force is applied to finalize the consolidation step.

Furthermore, pressure sensors installed at the hydraulic accumulators indirectly measure the force exerted on the parts. They are used to determine the time required to reach the maximum pressure during the consolidation process.

2.3. Experimental Design

In the first set of experiments, the preliminary tests, the temperature in the core of the laminate at 200 °C, 250 °C, and 300 °C heating press temperature and 60 °C cooling press temperature was assessed using thermocouples. These were positioned between the 6th and 7th layers of the tape stack to measure the minimum amount of time it takes for the core to reach the set temperature. In both presses, the pressure was kept low (1 bar and 10 bar for the heating and cooling presses, respectively) to ensure that the sensors remain intact.

Figure S1 illustrates the temperature evolution within the core of the samples at various heating press temperatures for the settings investigated. Heat is conducted from the plates via the outer layers to the core layer, which takes time. Consequently, we sought to investigate the time-dependent temperature behavior of the core layer. The heating process is completed when the core layer has reached its required temperature. As anticipated, higher set temperatures of the plate surfaces require more time for the core temperature to reach the set temperature, with approximately 104 s for 200 °C, 138 s for 250 °C, and 168 s for 300 °C. These values constitute the lower limits of the experimental design, representing the minimum periods of time required to reach the desired core temperature.

Furthermore, an additional test was conducted as part of the optimization trial, where two specimens were produced using the lowest (250 °C heating press temperature, 60 °C cooling temperature, 1 bar heating pressure, 10 bar cooling pressure) and highest (325 °C heating press temperature, 100 °C cooling press temperature, 6.5 bar heating pressure, 85 bar cooling pressure) process settings possible. Similarly, a thermocouple was positioned in the core layer of both stacks to monitor the temperature changes during consolidation. Pressure sensor data were additionally collected by an HBM data-recording device (QuantumX CS22B-W, HBM, Darmstadt, Germany) and analyzed in these experiments.

In the second set of experiments, the screening test, a definitive screening test design, was employed with 13 different settings. An overview of the experimental design is given in Table S1. The parameters were selected with the aim to (i) avoid thermal degradation of the matrix while (ii) covering a broad range of physical conditions. The consolidation pressures exerted on the parts were based on the set force and their nominal area. The holding time corresponds to an additional period that extends beyond the time necessary for the core of the part to reach the set temperature. This holding time parameter was established based on findings from preliminary tests (Figure S1). Twelve plates were produced within one consolidation parameter setting. No flow restrictions were applied to the tape stack under consolidation.

To analyze the quality of the consolidated parts, destructive and non-destructive methods were employed. Due to spatial constraints of some measurement systems, we used smaller plates (20 × 10 mm^2, obtained by water cutting) to analyze thickness, density, and apparent shear strength (ASS). The remaining parameters (e.g., warpage) were measured using the whole plate.

Position, size, and numbering of the cut samples are shown in Figure 2. The metrics used to analyze the quality of consolidation were: (i) projected area, (ii) thickness, (iii) warpage, (iv) density, and (v) ASS.

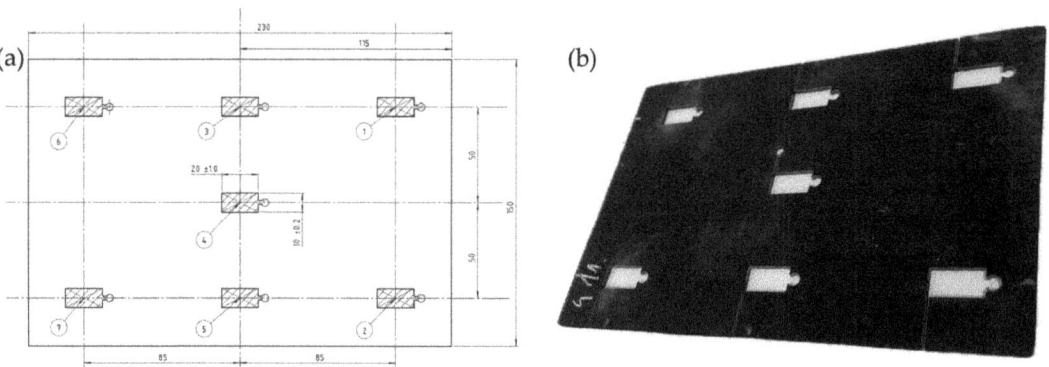

Figure 2. (**a**) Location and numbering of the small samples used in the screening and optimization trials; (**b**) photograph illustrating a consolidated plate with small samples cut out from it.

In the third set of experiments, the optimization test, several adjustments to the experimental design and setup were made based on the findings of the previous studies, such as expansion of the design space and introduction of a frame tool. A split-plot central composite design consisting of 26 different settings with four factors and multiple levels was employed, with the setting at the center point performed three times. Three plates were produced for each setting. In this new design, the holding time as an influencing factor was replaced by a constant cycle time of 3 min for both presses. Split-plot designs allow factors to be set that cannot be changed within an experimentally reasonable time frame, which imposes limitations on the randomization of the experiment's execution order. In particular, the temperature of the heating press—changing of which requires a significant amount of time—was considered as such a factor. The experimental design of the optimization test is shown in Table S2. Selection of the settings was guided by the insights gained from the screening test (see Section 3.1) and aimed to establish a configuration in which both the temperatures and pressures were maximized in relation to the recommended maximum processing temperatures and forces of the consolidation unit.

A frame tool with a thickness of 1.9 mm was installed between the two steel plates, as can be seen in Figure 3. A small opening was intentionally incorporated on the right side of the frame tool. Screening trials revealed that, when subjected to high pressure, the matrix material exhibited excessive flow in multiple directions, which made part removal difficult. Forcing the flow of the matrix material in one predefined direction made part handling easier. To assess the quality of the manufactured plates, (i) thickness, (ii) warpage, (iii) void content, and (iv) ASS were investigated. In analogy to the screening test, thickness, void content, and ASS measurements were performed using smaller samples. However, only areas 1, 4, and 7 (see Figure 2) were taken under consideration due to the high number of plates produced during the optimization trials. A summary of both experiments can be seen in Table 2.

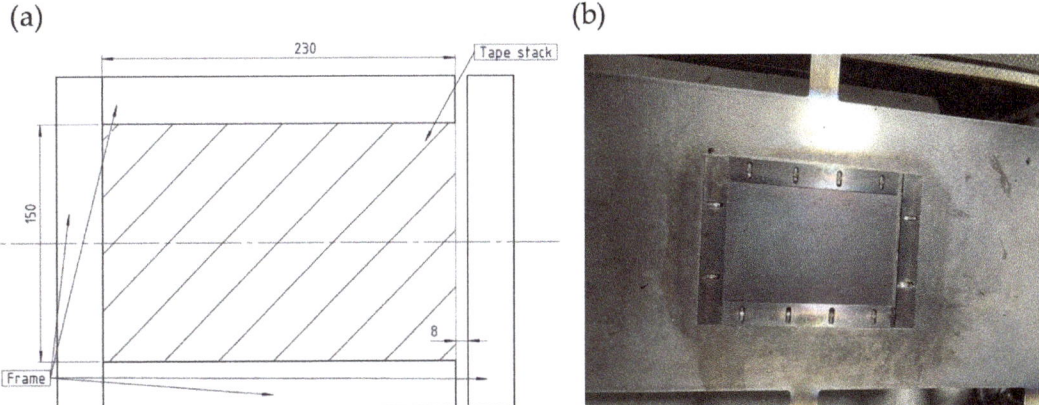

Figure 3. (a) Frame tool used to manufacture the samples for the optimization trials. The hatched area indicates the position of the tape stack under consolidation; (b) visual representation of the frame tool.

To analyze the influence of process settings on quality parameters, main influence graphs, cube plots, and surface plots were produced. To confirm the statistical significance of the results, we applied the analysis of variance (ANOVA) statistical method (with 95% statistical confidence) using the Design Expert software package (version: 22.0.4 64-bit). The resulting ANOVA table provides a summary regarding the statistical significance of factors through p-values. These p-values are calculated based on the sum of squares, which represents the squared differences between the overall average and the observed variation, degrees of freedom (df), denoting the number of estimated parameters used for computing the sum of squares, and the F-value, which serves as a test for comparing the calculated mean square to the residual mean square. A factor is considered influential with 95% statistical confidence when the corresponding p-value is below 0.05. The design model used in the screening trial was a reduced quadratic model that considered the main effects and the interaction between heating press temperature and pressure. In contrast, the optimization trial employed a quadratic model that incorporated all two-factor interactions. The summary of the ANOVA tables that served as a basis for interpretation of influencing factors can be found in the Supplementary.

Table 2. Summary of the screening and optimization trials.

Trial	Condition	Experimental Design	Number of Settings	Sample Type	Number of Test Samples Per Setting	Metrics
Screening	No frame tool	Definitive screening design	13	Whole plate	12	Projected area Warpage
				Small samples	14	Thickness Density ASS
Optimization	Frame tool	Split-plot central composite design	26	Whole plate	3	Warpage
				Small samples	9	Thickness Void content ASS

2.4. Measurement Methods

2.4.1. Projected Area

To measure the projected area of plates manufactured during the screening test, a high-resolution camera with 64 megapixels was used to record images. The camera was mounted on a stand, positioned above the consolidated plates placed on the ground. However, due to significant warpage observed in these analyses, the samples were flattened using a 500 × 500 × 6 mm^3 anti-reflective white glass. This glass helped to minimize distortions and ensured accurate measurement of the projected area. The schematic drawing of the setup is given in Figure 4.

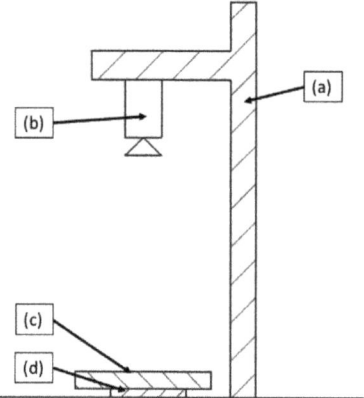

Figure 4. Schematic drawing of the experimental setup of the area measurement: (**a**) stand, (**b**) optical camera, (**c**) glass plate, and (**d**) specimen.

To determine the projected area of the parts, two prerequisites had to be fulfilled. First, the pictures captured had to exhibit sufficient contrast to allow thresholding of the part from the background. This was achieved by placing the black part against a white background. Second, the scale of pixels–mm had to be determined. To achieve this, three "calibration" pictures of the same 200 × 150 mm^2 part were taken. By carefully selecting the longitudinal length and converting the pixel distance to millimeters, a scale was obtained. To allow fast thresholding and measurement of the projected areas, the ImageJ software (version: 1.53e) was employed, which provides automated thresholding and enables quick computation of the areas.

2.4.2. Thickness

Thickness measurements of the consolidated plates were conducted using a micrometer with a precision of ±0.01 mm. Due to the limited reach of the measurement device, the small (i.e., cut) samples specified in Figure 2 were analyzed in both the screening and optimization trials.

2.4.3. Warpage

To assess the level of residual stress accumulated in a plate during the consolidation process, warpage tests were carried out. The test involved placing a part on a flat surface, securing it at one corner, and measuring the degree of lifting of the part using a caliper, as illustrated in Figure 5.

Figure 5. Warpage measurement of a consolidated plate.

2.4.4. Density and Void Content

The density of the small plates extracted from the consolidated plates (see Figure 2) was determined according to ASTM D792-13 [30] (November 2013). In addition, for the optimization tests, the void content was determined in accordance with ASTM D2734-16 [31] (December 2016). To obtain the fiber content, the matrix was removed by burning in a Gero HTK 8 high-temperature furnace according to ASTM D2584 [32]. Based on results from preliminary tests, the furnace was heated at a rate of 10 °C per 10 min until a temperature of 900 °C was reached. The parts were then held in the furnace at an isothermal temperature of 900 °C for one hour and subsequently cooled. Nitrogen was used throughout the process at a flow rate of 250 L/h. To obtain the void content, the following equation was used:

$$V_v = 100 - \left(\frac{M_f}{M_i} \times 100 \times \frac{\rho_c}{\rho_r}\right) - \left(\frac{M_i - M_f}{M_i} \times 100 \times \frac{\rho_c}{\rho_m}\right), \quad (1)$$

where M_f is the weight of the fiber that is retained after the matrix burn-off test, M_i is the initial weight of the composite, ρ_c is the density of the composite, ρ_r is the density of the reinforcement, and ρ_m is the density of the matrix material [9]. The densities of matrix and fiber were 1.19 g/cm^3 and 1.8 g/cm^3, respectively.

2.4.5. Apparent Shear Strength

The ASS tests were performed in accordance with the ISO 14130 [33] (1997) standard to evaluate the bonding strength of small samples. An MTS 852 Test Damper System with a 10 kN loading cell was used with a 1 mm/s displacement rate. The shear strength of the specimens was determined by considering the first maximum force achieved during the testing procedure.

3. Results and Discussion

3.1. Screening Test

3.1.1. Projected Area and Thickness

Figure 6 illustrates the consolidation settings (see Table S1) with the greatest influence on the projected area of the plates, while Figure 7 shows the effects of consolidation settings that led to significant variations in the thickness of parts. ANOVA reveals that both temperature and pressure of the heating press had a statistically significant influence on the projected area, with calculated p-values below 0.05 (see Table S3). Regarding the thickness, the heating press temperature showed a significant effect with a p-value of 0.0014. However, no other process parameters exhibited a discernible influence on either area or thickness.

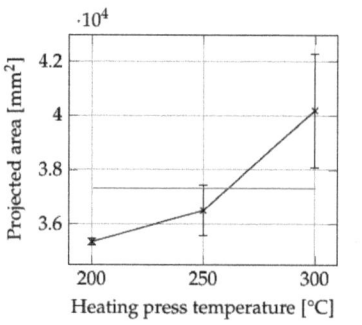
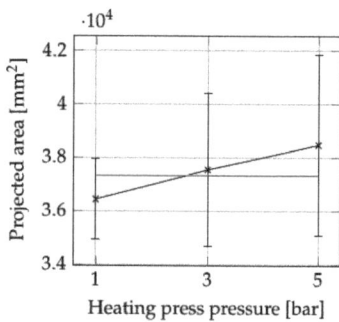

Figure 6. The consolidation settings with the greatest influence on the projected area of the plates as summarized in Table S1. The red line indicates the grand average of all values.

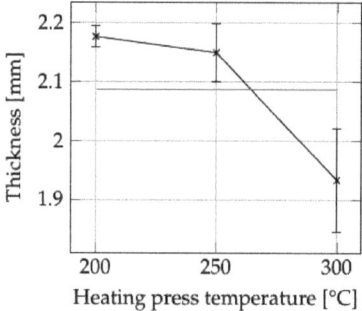

Figure 7. The influence of heating press temperature on the thickness of the plates. The red line indicates the grand average of all values.

At a heating press temperature of 200 °C, the plate showed no observable deviation from its initial size based on the projected area. At higher heating press temperatures, particularly at 300 °C, the plate exhibited an increased projected area and reduced thickness. This behavior can be attributed to the material's ability to flow freely in all directions, resembling a squeeze flow mechanism. Figure 8 shows a split open sample produced at a heating press temperature of 300 °C and a heating press pressure of 5 bar. It is evident that the squeeze flow not only significantly reduced the thickness, but also distorted the fiber structure of the plate. On the sides, the fibers appear slightly bent, reflecting the flow behavior of the matrix material. Consequently, a decrease in mechanical performance is expected in these areas.

Figure 8. A split-open sample produced at 300 °C heating press temperature and 5 bar heating press pressure. The fibers at the sides deviate from their original orientation.

The combined influence of heating press temperature and pressure is illustrated in Figure 9 for (a) area and (b) thickness. The ANOVA analysis reveals statistically significant effects of these two factors on the projected area and thickness. The influence of heating press pressure increased with increasing temperature. However, for the thickness, the p-value of 0.19 indicates no statistical significance. This lack of significance may be attributable to the considerable variations in tape thickness, as previously indicated. Nevertheless, it is evident that at higher heating press temperatures, the pressure applied by the heating press has an impact on the thickness achieved. Other two-factor interactions either lack statistical significance or are confounded with other factors, thus limiting the scope of analysis in this study.

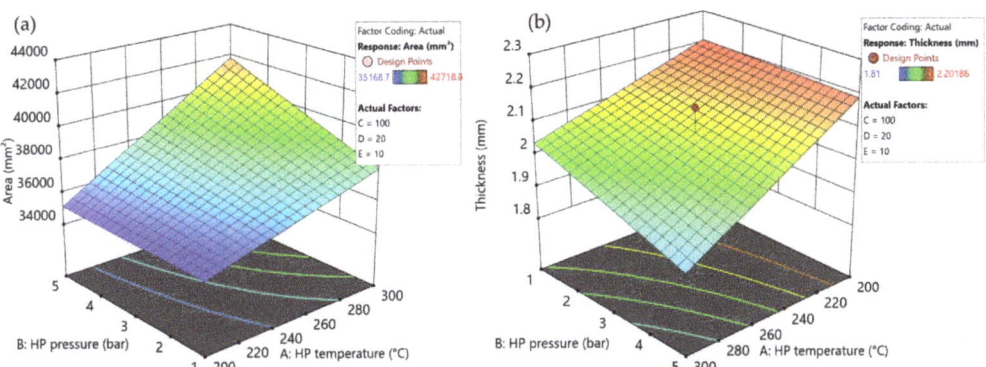

Figure 9. Surface plots showing the combined influence of heating press (HP) temperature and pressure on (**a**) the projected area and (**b**) the thickness in the screening test as summarized in Table S1. Set values for the remaining parameters: cooling press temperature: 100 °C (factor C); cooling press pressure: 20 bar (factor D); holding time: 10 s (factor E).

3.1.2. Warpage

Figure 10 illustrates the consolidation settings (see Table S1) with the greatest influence on the warpage of parts. Again, ANOVA showed that the influence of heating press temperature and pressure were significant, with p-values of 0.0005 and 0.033, respectively (see Table S5). Although the effect of the heating press pressure may not be immediately apparent in the main influence graphs, a closer examination of the results shown in Figure 11 reveals its significance at 300 °C.

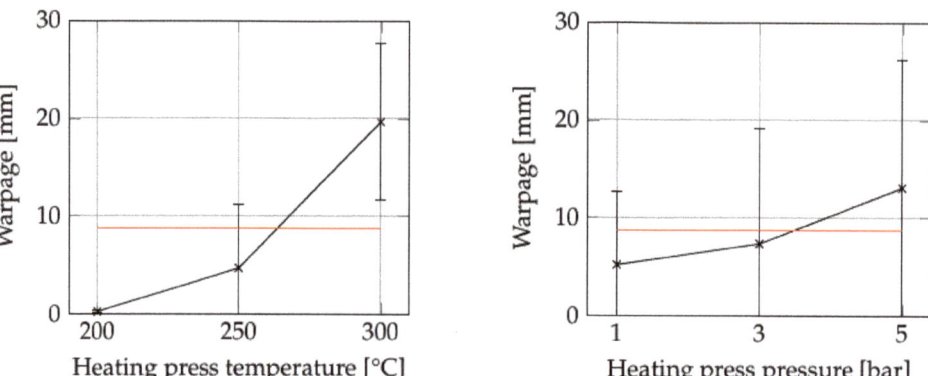

Figure 10. The influence of consolidation settings as summarized in Table S1 on the warpage of the plates. The red lines indicate the grand average of all values.

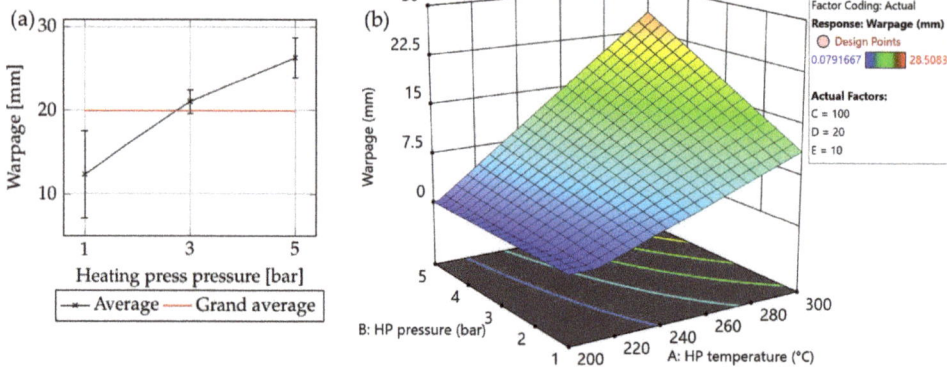

Figure 11. Detailed examination of the warpage results in the screening test: (**a**) the influence of heating press pressure at 300 °C heating press temperature; (**b**) surface plot showing the combined influences of heating press (HP) temperature and pressure on the warpage in the screening trial as summarized in Table S1. Set values for the other parameters: cooling press temperature: 100 °C (factor C); cooling press pressure: 20 bar (factor D); and holding time: 10 s (factor E).

As the part is squeezed during consolidation, the internal structure of the part also changes, as demonstrated in Figure 8. Increasing the heating press parameters leads to greater squeeze flow, resulting in a distorted laminate structure and subsequent warpage. Further, the dynamic nature of the process is noteworthy. During the screening experiments, the specimens were heated to the desired temperature within 2–3 min (depending on the set temperature) and subsequently cooled within the same time frame. We hypothesize that the rapid thermal cycling during the process may have induced higher levels of residual stress in the plate, thereby increasing warpage in the final part.

3.1.3. Density

The influence of the consolidation settings (see Table S1) on the density of the plate is illustrated in Figure 12. ANOVA shows that the heating press temperature had the greatest impact, followed by the heating press pressure and the cooling press pressure (see Table S6).

There is a slight correlation between the first two, but it is not statistically significant (see Figure 12d). With increasing heating press temperature and pressure, the matrix material exhibited enhanced flowability, which allowed the gaps and voids between the layers of the composite tape stacks to be filled more effectively. The high standard deviation of densities indicates that the press was unable to apply uniform pressure to the part. Additional investigation revealed that this issue was particularly prominent at cooling press pressures of 10 and 20 bar.

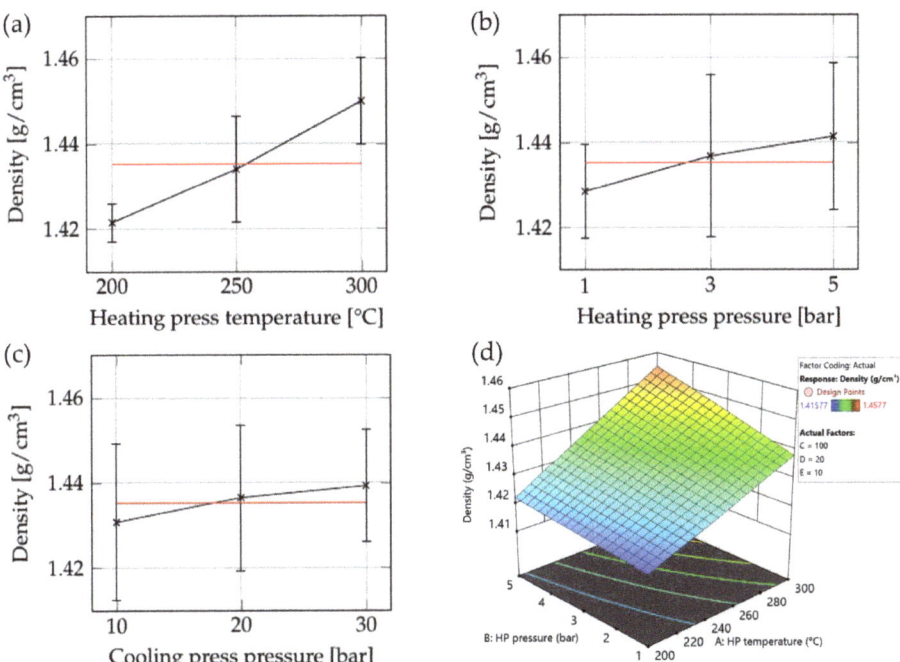

Figure 12. The influence of (**a**) heating press temperature, (**b**) heating pressure, and (**c**) cooling press temperature on the density of the plates (the red lines indicate the grand average of all samples). (**d**) Surface plot showing the combined influences of the heating press (HP) temperature and pressure on the density in the screening trial. Set values for the other parameters: cooling press temperature: 100 °C (factor C); cooling press pressure: 20 bar (factor D); and holding time: 10 s (factor E).

3.1.4. ASS

The heating press temperature had a significant impact on the apparent shear strength (see Figure 13 and Table S7). In addition, Figure 13 shows that the combined effect of heating press temperature and pressure—especially at 300 °C—influenced the ASS positively.

This behavior can be explained by examining the degree-of-bonding model [3]. As the temperature increases, the viscosity of the matrix material decreases, promoting higher macromolecular mobility. Consequently, stronger bonding between the layers of the laminate is facilitated as the interlayers diffuse. We assume that, although the heating press temperature was above the glass-transition temperature of the matrix material, a significant amount of time was required to achieve complete bonding, especially at 200 °C heating press temperature.

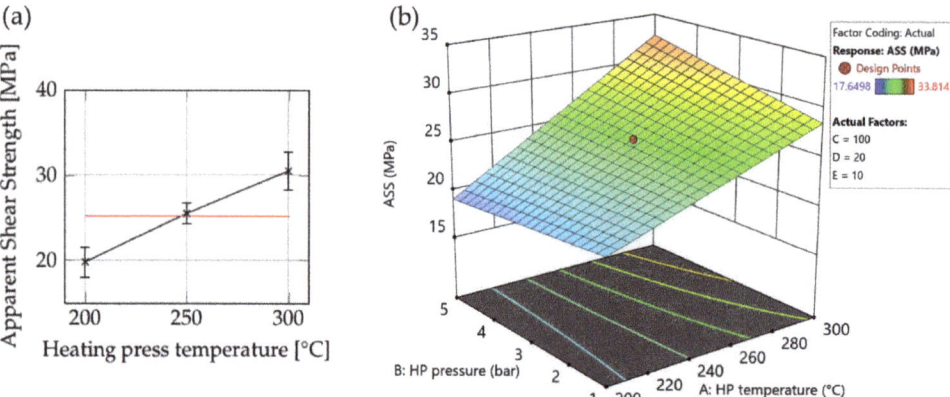

Figure 13. The influence of consolidation settings on the ASS of the plates: (**a**) the influence of heating press temperature; (**b**) surface plot of ASS with the heating press temperature and pressure in the screening trial. Set values for the other parameters: cooling press temperature: 100 °C (factor C); cooling press pressure: 20 bar (factor D); and holding time: 10 s (factor E).

The highest ASS value measured (33 MPa) fell well below the expected value. The low strength achieved could be attributed to the excessive squeezing of the material, which causes distortion in the inner structure of the composite plate, compromising the strength of the material. Additionally, high void content and/or defects could contribute to the low apparent shear strength. However, further investigation specifically focused on these factors was beyond the scope of this trial.

3.1.5. Screening Trial Discussion

The screening trial provided a first understanding of the key factors that influence the quality parameters of the composite plate, namely heating press temperature and pressure. The cooling press pressure was significant only in relation to density, while cooling press temperature and holding time did not exhibit any significant effect on the quality parameters measured.

Increasing the heating press parameters, specifically the temperature, enhanced the effect of squeeze flow. It promotes better bonding between the layers of the laminate, resulting in improved bonding and compaction, which can potentially lead to a lower void content. However, excessive squeezing is to be avoided, as it results in a significant increase in the projected area, a decrease in thickness of the samples, and an increase in warpage. This could cause problems in later manufacturing stages of the production cell, namely, in the mold of the injection molding machine. Therefore, precise control of the squeeze flow during consolidation is essential. The frame tool shown in Figure 3 partially contributes to achieving this control.

Significant density differences across different areas indicate pressure inhomogeneity during processing of the part. It remains unclear whether the variations in density can be attributed primarily to differences in carbon-fiber content resulting from the squeezing effect or whether void content played a significant role.

The definitive screening test design proved to be effective, providing a general understanding of the influences of process settings on the composite plate. However, due to the presence of aliased two-factor interactions, its usefulness for detailed analysis was found to be limited. Specifically, the design constraints hindered investigation of the combined effects of cooling press temperature and pressure.

In the optimization trial (see Table S2), a split-plot central composite design was implemented. First, the holding time was found to have no significant effect on the quality of the samples and was therefore removed from the design. A fixed cycle time of 180 s was instead used for both presses. Second, the process window of the heating press temperature was redefined to a range between 250 °C and 325 °C. Due to the time-consuming nature of heating and cooling the heating press, this factor was blocked in the experimental design. Although this led to the omission of information concerning the heating press temperature, the significant impact it demonstrated on the quality of the composite plates in the screening trial justified this decision. Third, the range for the heating press pressure was set to between 0.6 bar and 6.5 bar—the minimum and maximum forces achievable by the heating press, respectively. Fourth, the process window for the cooling press temperature was narrowed down to 40–100 °C. Since no significant effect of cooling press temperature was observed in the screening trial, with this adjustment we aimed to investigate whether lowering the temperature range would elicit any additional response in the quality parameters. Finally, the cooling pressure was also redefined to between 10 bar and 85 bar. The selection of 10 bar as the minimum pressure setting was based on its correlation with uneven pressure distribution in the screening trials. The objective was to examine whether this phenomenon would persist in the optimization trials conducted with a frame tool. The 85 bar corresponded to the maximum force the cooling press can exert on the part. These additional limits were chosen to enhance the responsiveness and capture the influence of the cooling press pressure.

3.2. Optimization Trials

3.2.1. Thickness

The influence of the consolidation settings on part thickness is plotted in Figure 14a,b, while the ANOVA results are shown in Table S8. Significant influences of heating press temperature, heating press pressure, and cooling press pressure were observed in all areas (see Table 2 and Figure 2) under investigation. In addition, a combined effect of heating press temperature and pressure was detected.

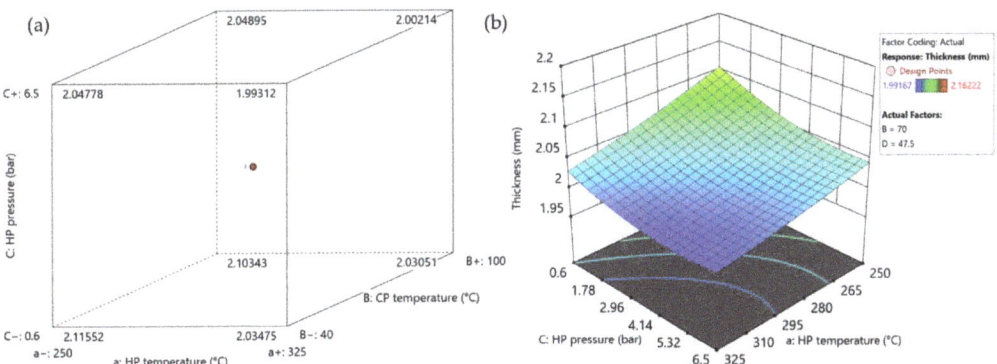

Figure 14. The influences of consolidation settings on part thickness in the optimization trial as summarized in Table S8: (**a**) cube plot representing the influence of heating press (HP) temperature and pressure and cooling press (CP) pressure; (**b**) surface plot showing the interaction between heating press temperature and pressure. Set values for the other parameters: cooling press temperature: 70 °C (factor B); cooling press pressure: 47.5 bar (factor D).

One notable achievement of the frame tool is that it ensured that part thickness remained above 2 mm even at the highest settings. This indicates that production of parts with consistent thickness is possible. Using high heating press temperatures and high heating- and cooling press pressures allows the target thickness of 2 mm to be achieved successfully.

3.2.2. Warpage

Figure 15 shows the influence of the consolidation settings on warpage, while Table S9 summarizes the ANOVA results. Both indicate that the process settings had no significant influence on warpage.

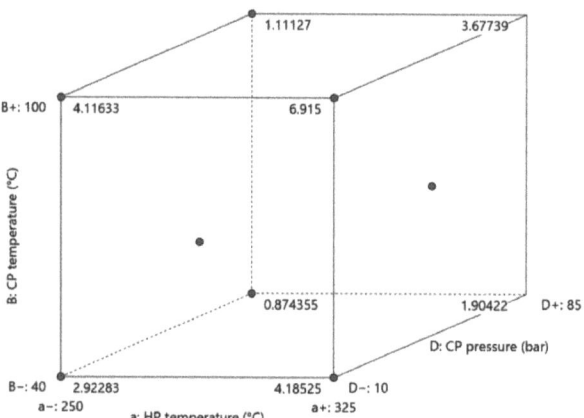

Figure 15. Influence of consolidation settings on warpage in the optimization trials as summarized in Table S9. For the cube plot, the heating press pressure was set to 0.6 bar.

We conclude that the limitation of squeeze flow led to minimal shape changes in the part, which resulted in a significant reduction in warpage. This is beneficial in terms of process stability, as constant part dimensions are also essential for downstream processing. However, the measurement method employed is unable to assess residual stress. We believe that the part retains a considerable amount of residual stress due to the dynamic nature of the process, but measuring it would require destructive methods.

3.2.3. Void Content

Figure 16 illustrates the impact of the consolidation settings on the void content, while Table S10 presents the ANOVA results. The effect of heating and cooling press temperature and cooling press pressure is significant. In addition, the combined effects of (i) cooling press temperature and heating pressure and (ii) heating and cooling pressure are significant. The average fiber, matrix, and void volume fraction can be found in Table S11.

Increasing the process settings led to a decrease in void content to some extent. Increasing the cooling press temperature also aids the removal of excessive void content. Employing a slower cooling rate gives the press more time to squeeze out effectively the voids, which results in a reduced void content in the consolidated parts due to lower viscosity and thus in higher molecular mobility. The minimum achieved void content was 4.64%. This value exceeds the upper limit in aerospace applications (1%) [34,35]. However, in some other applications, a maximum of 5% is allowed [36]. The inability to further reduce the void content may be attributed to the presence of inherent voids within the UD tapes themselves, which cannot be eliminated during the manufacturing process.

3.2.4. ASS

Table S12 summarizes the ANOVA results, while Figure 17 presents a box plot illustrating the influence of process settings on the ASS. The most influential parameter was the heating press temperature, followed by heating pressure and cooling pressure and temperature.

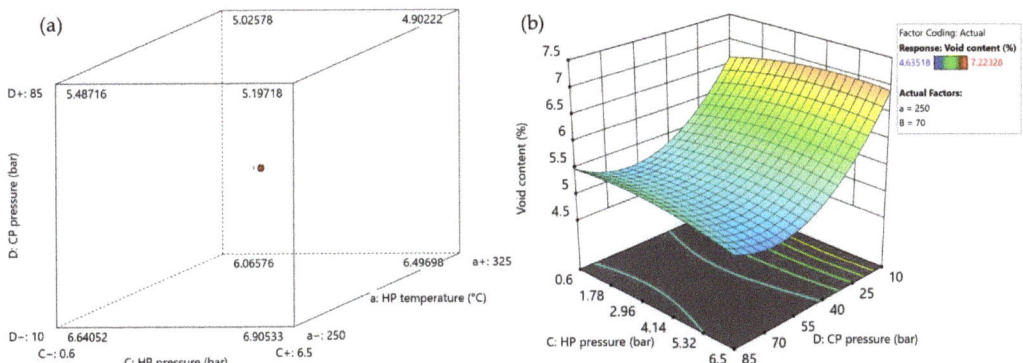

Figure 16. The influence of consolidation settings on the void content in the optimization trial (as summarized in Table S10: (**a**) cube plot representing the influence of heating press temperature and pressure and cooling pressure; (**b**) surface plot showing combined influence of heating pressure and cooling pressure. Set values for the other parameters: heating press temperature: 250 °C (factor a); cooling press temperature: 70°C (factor B).

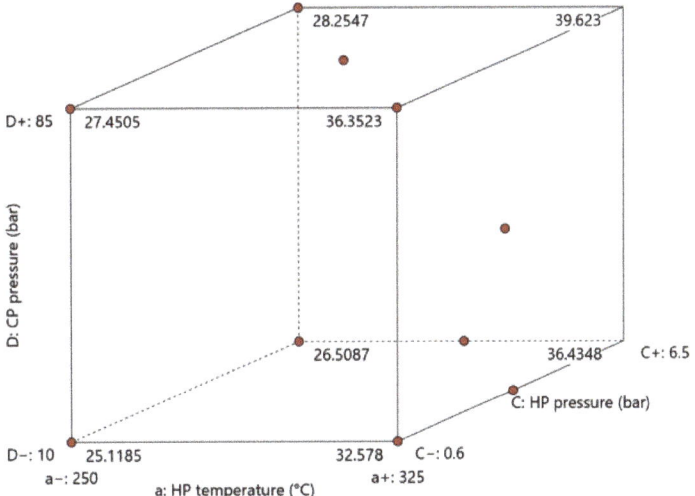

Figure 17. The influence of consolidation settings (as summarized in Table S2) on the apparent shear strength in the optimization trial. For the cube plot, a cooling press temperature of 100 °C was set.

Heating press temperature and pressure also had the greatest influence on the ASS, primarily due to the reduced viscosity of the matrix material. The low strength values obtained at low cooling pressures suggest that the tool does not establish full contact with the part under these conditions. Consequently, pressure and heat are inadequately transferred. This lack of contact is further supported by the void content measurements indicating inadequate quality. Implementation of the frame tool resulted in overall improvements in ASS values. Compared to the maximum bonding achieved in the screening test, an average increase of 23% in ASS was achieved at maximum settings under these conditions, which corresponds to an average value of 41.62 MPa.

3.2.5. Optimization Trials Discussion

In conclusion, the optimization trials resulted in an overall improvement in the quality of the parts produced. The implemented changes, including installation of the frame tool, contributed to maintaining a thickness above 2 mm, minimal changes in the projected area, and a significant enhancement in bonding quality. Controlling the squeeze flow allowed higher process settings to be used, which positively impacted the overall quality of the consolidated plates.

The extended experimental design provided a deeper understanding of the individual process settings and their combined effects. While the influence of the heating press was well understood from the screening test, aliasing of other two-factor interactions meant that understanding of the effects of the cooling press was incomplete.

In most cases, the cooling pressure had a noticeable impact on the quality of the parts, while the cooling press temperature had a minor influence in specific instances. Applying maximum pressure allowed the quality of the consolidated parts to be increased.

Higher void content and lower apparent shear strength values observed at lower settings can be attributed to three main factors. First, the construction of the frame tool consisted of four sections that formed the edges of the consolidated plate. The presence of an 8 mm gap on the right side of the tool allowed the matrix material to flow freely in that direction, resulting in increased thickness, higher void content, and weaker bonding in this area. On the opposite side of the part, we detected a small gap between the components of the frame tool that allowed some material flow, but not to the same extent as on the right side. Second, as hypothesized in the screening test, there may be pressure variations throughout the part. At low cooling pressures, the steel tool plate apparently does not make sufficient contact with the specimen or with the press during consolidation. This is supported by the fact that, at maximum cooling pressure, quality variations between different areas were minimal. Finally, note that the influence of the cooling press on part quality was generally lower than expected. For instance, in the center of the part, the impact of the heating pressure was more significant than that of the cooling pressure, as indicated by the p-values in the ANOVA results. This finding was unforeseen, as one would expect a higher impact from increasing the cooling pressure from 10 to 85 bar than from increasing the heating pressure from 0.6 to 6.5 bar.

To comprehensively assess the influence of the cooling press, tests were conducted to monitor the temperature changes within the core layer of the tape stack (specifically between the 6th and 7th layer) by means of a thermocouple. Additionally, the pressures applied by both the heating and cooling presses were measured and recorded. Figure 18 presents the temperature and pressure profiles during the pressing process, specifically focusing on the stage where the part underwent compression in the cooling press. At the lowest settings (Figure 18a), namely at 250 °C heating press temperature, 0.6 bar heating pressure, 40 °C cooling press temperature and 10 bar cooling pressure, the pressure had built up fully after 239 s of data recording. However, the part reached its glass-transition temperature after 242.4 s in the cooling press. This indicates that during the compression phase in the cooling press, the molten part experienced the defined pressure for only about 3 s. However, at the highest settings (325 °C heating press temperature, 6.5 bar, 100 °C cooling press temperature and 85 bar cooling pressure), the period under full pressure while at a temperature above glass-transition extended to approximately 41 s in the cooling press (Figure 18b). This phenomenon is the primary reason why the cooling press had a limited effect on the void content and ASS results, especially at low heating press temperatures. As indicated by the degree-of-bonding model, adhesion between the layers is a time-dependent process. At the lowest settings, when the part reached its glass-transition temperature in only 3 s, molecular movement between the layers ceased. Our findings demonstrate that this period is too short to achieve adequate bonding between the layers and to eliminate excessive void content.

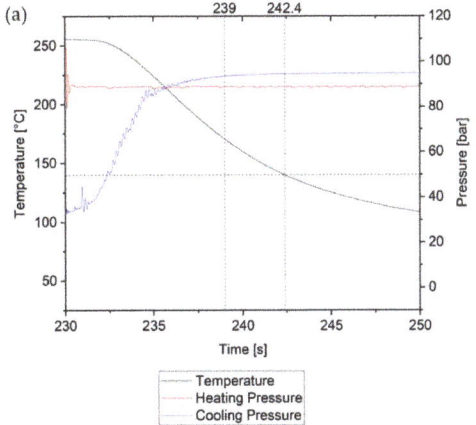 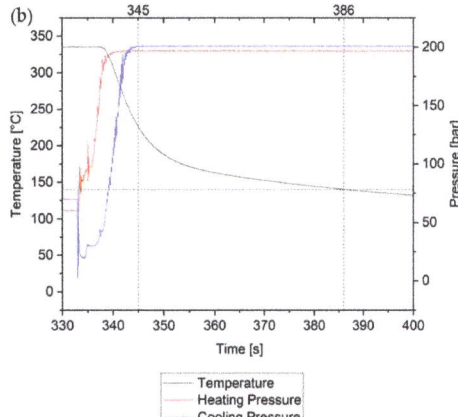

Figure 18. Temperature measurement at the core and the pressure development during consolidation at (**a**) low settings and at (**b**) high settings. The first black vertical line indicates the time at which the cooling press reached its maximum set force, while the second line indicates the time point at which the matrix material reached its T_g.

The numerical optimization tool of the Design Expert software, which combines a desirability function with a hill-climbing technique, can provide suggestions for optimal process settings. Based on our analysis of the results and using this tool, we consider the following process settings to be optimal:

- Heating press temperature: 325 °C;
- Heating pressure: 6.5 bar;
- Cooling press temperature: 100 °C;
- Cooling pressure: 85 bar.

Implementing these settings (the highest in the experimental design) is expected to achieve the desired thickness of 2 mm while minimizing void content and maximizing the apparent shear strength.

4. Conclusions

This work focused on optimizing the consolidation of PC/CF tape stacks using a combination of a heating press and a cooling press, with the aim to obtain well-bonded, void-free plates with a target thickness of 2 mm. Two studies were conducted to find optimal process settings and investigate the behavior of the process.

The screening test revealed that increasing the temperature and pressure of the heating press improved part quality, while the cooling press had minimal impact. Excessive squeeze flow resulted in a higher projected area and lower plate thickness than expected.

In the optimization trials, we expanded the experimental design and implemented a frame tool. These modifications had a positive impact on the consolidation process. We thus achieved a plate thickness of 2 mm and a 23% increase in bonding strength compared to the initial trials. The minimum measured void content was 4%. We observed that the cooling press influences the quality of the parts, and that this effect is more pronounced at high settings, where the material remains under pressure and above its T_g for a longer period. Based on our findings, we determined that setting the temperature and pressure at both presses as high as possible in this setup is crucial to obtaining parts with optimal properties.

Additional research is required to optimize the process further. We hypothesize that incorporating a fully enclosed frame tool, extending the cycle time, and using a cooling unit that can maintain temperatures above T_g of the matrix material would enhance the quality of the parts. Additionally, investigating the process settings' impact on the degree of crystallinity using a semi-crystalline matrix material would provide valuable insights.

Supplementary Materials: The following supporting information can be downloaded at: https://www.mdpi.com/article/10.3390/polym15234500/s1, Figure S1: Temperature development at the sample core at set temperatures of (a) 200°C, (b) 250°C and (c) 300°C. The first vertical black line indicates the start of the process, while the second shows the time at which the core has reached the set temperature; Table S1. Experimental design used in the screening test, Table S2. Experimental design used in the optimization test; Table S3. ANOVA for the influence of consolidation settings on the projected area in the screening trial; Table S4. ANOVA for the influence of consolidation settings on the thickness in the screening trial; Table S5. ANOVA for the influence of consolidation settings on the warpage in the screening trial; Table S6. ANOVA for the influence of consolidation settings on the density in the screening trial; Table S7. ANOVA for the influence of consolidation settings on the ASS in the screening trial; Table S8. ANOVA for the influence of consolidation settings on the thickness in the optimization trial; Table S9. ANOVA for the influence of consolidation settings on the warpage in the optimization trial; Table S10. ANOVA for the influence of consolidation settings on the void content in the optimization trial; Table S11. Average fiber, matrix, and void fraction in the optimization trials; Table S12. ANOVA for the influence of consolidation settings on the apparent shear strength in the optimization trial.

Author Contributions: Conceptualization, J.B. and C.M.; methodology, J.B. and E.K.; software, J.B.; validation J.B.; formal analysis, J.B., E.K. and K.S.; investigation, J.B., E.K. and K.S.; resources, K.S. and G.S.; data curation, J.B. and E.K.; writing—original draft preparation, J.B.; writing—review and editing, C.M.; visualization, J.B.; supervision, C.M., K.S. and G.S.; project administration, K.S. and C.M.; funding acquisition, G.S. and C.M. All authors have read and agreed to the published version of the manuscript.

Funding: This work was performed within the Competence Center CHASE GmbH, funded by the Austrian Research and Promotion Agency (grant number 868615). The authors acknowledge financial support by the COMET Centre CHASE (project No. 868615), which is funded within the framework of COMET—Competence Centers for Excellent Technologies—by BMVIT, BMDW, and the Federal Provinces of Upper Austria and Vienna. The COMET program is run by the Austrian Re-search Promotion Agency (FFG). Supported by Johannes Kepler Open Access Publishing Fund and the federal state of Upper Austria.

Institutional Review Board Statement: Not applicable.

Informed Consent Statement: Not applicable.

Data Availability Statement: The data presented in this study are available on request from the corresponding author.

Conflicts of Interest: Authors Janos Birtha, Eva Kobler and Christian Marschik were employed by the company Competence Center CHASE GmbH. The remaining authors declare that the research was conducted in the absence of any commercial or financial relationships that could be construed as a potential conflict of interest.

Abbreviations

The following abbreviations are used in this manuscript:

AFP	automated fiber placement
ANOVA	analysis of variance
ASS	apparent shear strength
ATP	automated tape laying
CF/PPS	carbon/polyphenylene sulfide
CP	cooling press
DoE	design of experiments
HP	heating press
PC	polycarbonate
PC/GF	polycarbonate/glass fiber
PC/CF	polycarbonate/carbon fiber

PAEK	polyaryl ether ketone
PEEK	polyetheretherketone
PEKK	polyetherketoneketone
PP	polypropylene
VBO	vacuum-bag-only approach
UD	unidirectional

References

1. Njuguna, J. *Lightweight Composite Structures in Transport: Design, Manufacturing, Analysis and Performance*; Elsevier: Amsterdam, The Netherlands, 2016; ISBN 978-1-78242-343-0.
2. Biron, M. *Thermosets and Composites: Material Selection, Applications, Manufacturing, and Cost Analysis*, 2nd ed.; Elsevier: Amsterdam, The Netherlands, 2013.
3. Sonmez, F.O.; Hahn, H.T. Analysis of the On-Line Consolidation Process in Thermoplastic Composite Tape Placement. *J. Thermoplast. Compos. Mater.* **1997**, *10*, 543–572. [CrossRef]
4. Khan, M.A.; Mitschang, P.; Schledjewski, R. Identification of Some Optimal Parameters to Achieve Higher Laminate Quality through Tape Placement Process. *Adv. Polym. Technol.* **2010**, *29*, 98–111. [CrossRef]
5. Khan, M.A.; Mitschang, P.; Schledjewski, R. Tracing the Void Content Development and Identification of Its Effecting Parameters during in Situ Consolidation of Thermoplastic Tape Material. *Polym. Polym. Compos.* **2010**, *18*, 1–15. [CrossRef]
6. Sonmez, F.O.; Akbulut, M. Process Optimization of Tape Placement for Thermoplastic Composites. *Compos. Part A Appl. Sci. Manuf.* **2007**, *38*, 2013–2023. [CrossRef]
7. Saenz-Castillo, D.; Martín, M.I.; Calvo, S.; Rodriguez-Lence, F.; Güemes, A. Effect of Processing Parameters and Void Content on Mechanical Properties and NDI of Thermoplastic Composites. *Compos. Part A Appl. Sci. Manuf.* **2019**, *121*, 308–320. [CrossRef]
8. Saffar, F.; Sonnenfeld, C.; Beauchêne, P.; Park, C.H. In-Situ Monitoring of the Out-Of-Autoclave Consolidation of Carbon/Poly-Ether-Ketone-Ketone Prepreg Laminate. *Front. Mater.* **2020**, *7*, 195. [CrossRef]
9. Hoang, V.-T.; Kwon, B.-S.; Sung, J.-W.; Choe, H.-S.; Oh, S.-W.; Lee, S.-M.; Kweon, J.-H.; Nam, Y.-W. Postprocessing Method-Induced Mechanical Properties of Carbon Fiber-Reinforced Thermoplastic Composites. *J. Thermoplast. Compos. Mater.* **2023**, *36*, 432–447. [CrossRef]
10. Patou, J.; Bonnaire, R.; De Luycker, E.; Bernhart, G. Influence of Consolidation Process on Voids and Mechanical Properties of Powdered and Commingled Carbon/PPS Laminates. *Compos. Part A Appl. Sci. Manuf.* **2019**, *117*, 260–275. [CrossRef]
11. Schiel, I.; Raps, L.; Chadwick, A.R.; Schmidt, I.; Simone, M.; Nowotny, S. An Investigation of In-Situ AFP Process Parameters Using CF/LM-PAEK. *Adv. Manuf. Polym. Compos. Sci.* **2020**, *6*, 191–197. [CrossRef]
12. Kropka, M.; Reichstein, J.; Neumeyer, T.; Altstaedt, V. Effect of the Pre-Consolidation Process on Quality and Mechanical Properties of Mono and Multi- Material Laminates Based on Thermoplastic UD Tapes. In Proceedings of the 21st International Conference on Composite Materials, Xi'an, China, 20–25 August 2017.
13. Schell, J.S.U.; Guilleminot, J.; Binetruy, C.; Krawczak, P. Computational and Experimental Analysis of Fusion Bonding in Thermoplastic Composites: Influence of Process Parameters. *J. Mater. Process. Technol.* **2009**, *209*, 5211–5219. [CrossRef]
14. Borowski, A.; Vogel, C.; Behnisch, T.; Geske, V.; Gude, M.; Modler, N. Additive Manufacturing-Based In Situ Consolidation of Continuous Carbon Fibre-Reinforced Polycarbonate. *Materials* **2021**, *14*, 2450. [CrossRef] [PubMed]
15. Akué Asséko, A.C.; Cosson, B.; Lafranche, É.; Schmidt, F.; Le Maoult, Y. Effect of the Developed Temperature Field on the Molecular Interdiffusion at the Interface in Infrared Welding of Polycarbonate Composites. *Compos. Part B Eng.* **2016**, *97*, 53–61. [CrossRef]
16. Almeida, O.D.; Bessard, E.; Bernhart, G. Influence of Processing Parameters and Semi-Finished Product on Consolidation of Carbon/Peek Laminates. In Proceedings of the 15th European Conference on Composite Materials ECCM15, Venise, Italy, 24–28 June 2012.
17. Qureshi, Z.; Swait, T.; Scaife, R.; El-Dessouky, H.M. In Situ Consolidation of Thermoplastic Prepreg Tape Using Automated Tape Placement Technology: Potential and Possibilities. *Compos. Part B Eng.* **2014**, *66*, 255–267. [CrossRef]
18. Henneberg, A.; Transier, G.; Sinapius, M. Consolidated Fibre Placement (CFP)—Adhesive Joining of Consolidated Fibre Tapes. *Compos. Struct.* **2016**, *140*, 337–343. [CrossRef]
19. Grouve, W.J.B.; Warnet, L.L.; Rietman, B.; Visser, H.A.; Akkerman, R. Optimization of the Tape Placement Process Parameters for Carbon–PPS Composites. *Compos. Part A Appl. Sci. Manuf.* **2013**, *50*, 44–53. [CrossRef]
20. Zhang, C.; Duan, Y.; Xiao, H.; Wang, B.; Ming, Y.; Zhu, Y.; Zhang, F. Effect of Porosity and Crystallinity on Mechanical Properties of Laser In-Situ Consolidation Thermoplastic Composites. *Polymer* **2022**, *242*, 124573. [CrossRef]
21. Lee, W.I.; Talbott, M.F.; Springer, G.S.; Berglund, L.A. Effects of Cooling Rate on the Crystallinity and Mechanical Properties of Thermoplastic Composites. *J. Reinf. Plast. Compos.* **1987**, *6*, 2–12. [CrossRef]
22. Manson, J.-A.E.; Schneider, T.L.; Seferis, J.C. Press-Forming of Continuous-Fiber-Reinforced Thermoplastic Composites. *Polym. Compos.* **1990**, *11*, 114–120. [CrossRef]
23. Sonmez, F.O.; Hahn, H.T.; Akbulut, M. Analysis of Process-Induced Residual Stresses in Tape Placement. *J. Thermoplast. Compos. Mater.* **2002**, *15*, 525–544. [CrossRef]

24. Krämer, E. The Formation of Fiber Waviness during Thermoplastic Composite Laminate Consolidation: Identification and Description of the Underlying Mechanisms. Ph.D. Thesis, University of Twente, Enschede, The Netherlands, 2021.
25. Sitohang, R.D.R.; Grouve, W.J.B.; Warnet, L.L.; Akkerman, R. Effect of In-Plane Fiber Waviness Defects on the Compressive Properties of Quasi-Isotropic Thermoplastic Composites. *Compos. Struct.* **2021**, *272*, 114166. [CrossRef]
26. Wilhelmsson, D.; Asp, L.E. A High Resolution Method for Characterisation of Fibre Misalignment Angles in Composites. *Compos. Sci. Technol.* **2018**, *165*, 214–221. [CrossRef]
27. Valverde, M.A.; Belnoue, J.P.-H.; Kupfer, R.; Kawashita, L.F.; Gude, M.; Hallett, S.R. Compaction Behaviour of Continuous Fibre-Reinforced Thermoplastic Composites under Rapid Processing Conditions. *Compos. Part A Appl. Sci. Manuf.* **2021**, *149*, 106549. [CrossRef]
28. Shuler, S.F.; Advani, S.G. Transverse Squeeze Flow of Concentrated Aligned Fibers in Viscous Fluids. *J. Non-Newton. Fluid Mech.* **1996**, *65*, 47–74. [CrossRef]
29. Birtha, J.; Marschik, C.; Kobler, E.; Straka, K.; Steinbichler, G.; Schlecht, S.; Zwicklhuber, P. Optimizing the Process of Spot Welding of Polycarbonate-Matrix-Based Unidirectional (UD) Thermoplastic Composite Tapes. *Polymers* **2023**, *15*, 2182. [CrossRef]
30. *ASTM D792-13*; Standard Test Methods for Density and Specific Gravity (Relative Density) of Plastics by Displacement. ASTM: West Conshohocken, PN, USA, 2013.
31. *ASTM D2734-16*; Standard Test Methods for Void Content of Reinforced Plastics. ASTM: West Conshohocken, PN, USA, 2016.
32. *ASTM D2584-18*; Standard Test Method for Ignition Loss of Cured Reinforced Resins. ASTM: West Conshohocken, PN, USA, 2018.
33. *ISO 14130:1997*; Fibre-reinforced plastic composites: Determination of apparent interlaminar shear strength by short-beam method. ISO: Geneva, Switzerland, 1997.
34. Liu, L.; Zhang, B.-M.; Wang, D.-F.; Wu, Z.-J. Effects of Cure Cycles on Void Content and Mechanical Properties of Composite Laminates. *Compos. Struct.* **2006**, *73*, 303–309. [CrossRef]
35. Sloan, J. Out-of-Autoclave Processing: <1% Void Content? Available online: https://www.compositesworld.com/articles/out-of-autoclave-processing-1-void-content (accessed on 20 September 2023).
36. Mehdikhani, M.; Gorbatikh, L.; Verpoest, I.; Lomov, S.V. Voids in Fiber-Reinforced Polymer Composites: A Review on Their Formation, Characteristics, and Effects on Mechanical Performance. *J. Compos. Mater.* **2019**, *53*, 1579–1669. [CrossRef]

Disclaimer/Publisher's Note: The statements, opinions and data contained in all publications are solely those of the individual author(s) and contributor(s) and not of MDPI and/or the editor(s). MDPI and/or the editor(s) disclaim responsibility for any injury to people or property resulting from any ideas, methods, instructions or products referred to in the content.

Article

Influence of Different Hot Runner-Systems in the Injection Molding Process on the Structural and Mechanical Properties of Regenerated Cellulose Fiber Reinforced Polypropylene

Jan-Christoph Zarges [1,*], André Schlink [1], Fabian Lins [1], Jörg Essinger [2], Stefan Sommer [2] and Hans-Peter Heim [1]

1. Institute of Material Engineering, Polymer Engineering, University of Kassel, 34125 Kassel, Germany
2. Günther Heisskanaltechnik GmbH, 35066 Frankenberg (Eder), Germany
* Correspondence: zarges@uni-kassel.de; Tel.: +49-561-804-2544

Citation: Zarges, J.-C.; Schlink, A.; Lins, F.; Essinger, J.; Sommer, S.; Heim, H.-P. Influence of Different Hot Runner-Systems in the Injection Molding Process on the Structural and Mechanical Properties of Regenerated Cellulose Fiber Reinforced Polypropylene. *Polymers* **2023**, *15*, 1924. https://doi.org/10.3390/polym15081924

Academic Editors: Abderrahmane Ayadi, Patricia Krawczak and Chung Hae Park

Received: 6 March 2023
Revised: 3 April 2023
Accepted: 12 April 2023
Published: 18 April 2023

Copyright: © 2023 by the authors. Licensee MDPI, Basel, Switzerland. This article is an open access article distributed under the terms and conditions of the Creative Commons Attribution (CC BY) license (https:// creativecommons.org/licenses/by/ 4.0/).

Abstract: The increasing demand for renewable raw materials and lightweight composites leads to an increasing request for natural fiber composites (NFC) in series production. In order to be able to use NFC competitively, they must also be processable with hot runner systems in injection molding series production. For this reason, the influences of two hot runner systems on the structural and mechanical properties of Polypropylene with 20 wt.% regenerated cellulose fibers (RCF) were investigated. Therefore, the material was processed into test specimens using two different hot runner systems (open and valve gate) and six different process settings. The tensile tests carried out showed very good strength for both hot runner systems, which were max. 20% below the reference specimen processed with a cold runner and, however, significantly influenced by the different parameter settings. Fiber length measurements with the dynamic image analysis showed approx. 20% lower median values of GF and 5% lower of RCF through the processing with both hot runner systems compared to the reference, although the influence of the parameter settings was small. The X-ray microtomography performed on the open hot runner samples showed the influences of the parameter settings on the fiber orientation. In summary, it was shown that RCF composites can be processed with different hot runner systems in a wide process window. Nevertheless, the specimens of the setting with the lowest applied thermal load showed the best mechanical properties for both hot runner systems. It was furthermore shown that the resulting mechanical properties of the composites are not only due to one structural property (fiber length, orientation, or thermally induced changes in fiber properties) but are based on a combination of several material- and process-related properties.

Keywords: viscose fiber; natural fiber composites; hot runner system; injection molding; fiber length; fiber orientation; X-ray microtomography analysis; mechanical properties; dynamic image analysis

1. Introduction

Natural fiber composites (NFC) with polypropylene (PP) as matrix material and short natural fibers for reinforcement are increasingly used in series production, e.g., in the field of automotive parts and other engineering applications [1–3]. The reason for this is the increasing demand for renewable raw materials and the possibility to combine good mechanical properties with the lightweight potential of low-density fibers [4,5]. A large number of studies regarding the processing and characterization of NFC used compression molding for specimen production [6–9], while injection molding tends to be more important for industrial applications, especially for the production of serial parts. In most natural fiber composites, polyolefins are used as matrix material, and natural plant fibers are used for reinforcement [7,8,10,11]. The reason for using polyolefins is their comparatively low processing temperature of approx. 200 °C, which allows processing with almost no thermal degradation of the sensitive natural fibers (NF). Degradation, the rate of which depends strongly on the exposure temperature, starts at approx. 170 °C, depending on the cellulose content of the NF [12,13], and results in a reduction in mechanical properties [14–18]. The pure cellulose fibers are thermally more stable and show significant material degradation in thermogravimetric analyses only from 240 °C due to a measured loss of mass [19–22].

A major disadvantage of NFC is the deviating geometrical and mechanical properties of the natural fibers used, which makes it considerably difficult to predict the fracture and failure behavior of NFC [2,23]. For this reason, regenerated cellulose fibers (RCF) have been used as reinforcement in addition to plant fibers (hemp, flax, jute, sisal, etc.) in previous studies. RCFs, also known as viscose or rayon, produced by the viscose process have very constant geometric and mechanical properties in addition to their lower density compared to conventional glass fibers and the significant lightweight potential associated with them [24–26]. Another reason for using RCFs as reinforcement is their higher elongation at break (approx. 13%) compared to that of glass fibers (approx. 2%), resulting in a more ductile, less spontaneous, and more predictable failure behavior of the composites [9,27]. This toughness and the good mechanical properties of RCF composites in general can be attributed to the fiber length distribution and fiber-matrix adhesion [28–30]. Due to the lower bending stiffness of RCF compared to GF, the RCF is less shortened by the shear stresses in the compounding and injection molding process and thus still exceed the critical fiber length in some cases even in the test specimens or components, while the lengths of GF in the components are usually significantly below the critical fiber length. This leads to a good reinforcement effect of the RCF and at the same time to a good toughness due to friction-intensive fiber pullouts at failure [9,30–32].

The above-mentioned properties of RCF lead to a significant increase in mechanical properties, especially notched impact strength, and fracture toughness [12,33,34]. More specifically, compared with glass fiber reinforced composites (GFC), the values of notched impact strength are about four times higher at the same fiber weight content, and the values of fracture toughness are about three times higher [31,35], while the density is 20% lower.

With regard to the structure, in particular, the fiber orientation, which is induced by the injection molding process, the RCF composites, similar to all other short glass fiber reinforced thermoplastics, are significantly influenced by the injection molding parameters. As a result, the short fiber reinforced thermoplastics exhibited locally different microstructures, such as crystallinity [36] and fiber orientations [37–39], due to the processing influences, which can have a significant influence on mechanical and fatigue properties [36,39–44]. Several publications have characterized the influence of fiber orientation in injection molded composites, showing that the more fibers oriented in the loading direction, the better the mechanical properties [44–47]. With regard to the process influences during injection molding, a significant influence of the melt and mold temperature as well as the volume flow rate could be shown. It was demonstrated that higher melt temperatures and volume flow rates have a positive effect on the strength, stiffness, and fatigue properties of the resulting GFC due to the higher number of fibers oriented in the direction of loading [47,48].

In order to make the natural or viscose fiber-reinforced plastics competitive as well as to increase efficiency and reduce material waste, as mentioned above, it must be possible to process the RCF composites in the injection molding process even with hot runner systems. This, however, is complicated by the thermal sensitivity of RCF and the longer dwell time of the melt in the hot runner at a higher temperature, which negatively affects the mechanical properties of RCF, such as tensile strength and elongation at break [10,14,18,19,27,34]. At the same time, the small cross-sections in a hot runner system can lead to higher shear and thus to a significant reduction in RCF length [8,17,28,31]. Both result in a reduction of the mechanical properties of the RCF composites. In addition, the high fiber length combined with the flexural properties of the RCF could lead to a plugging of the narrow cross sections in a hot runner system, which would make a trouble-free series production considerably more difficult.

As explained in detail here, RCF composites have not yet been processed with hot runner systems in series-production injection molding processes in previous investigations. The reason for this is that it was not clear until now to what extent processing is possible without exposing the RCF composites to excessive thermal and mechanical stress. The objective of this paper is to demonstrate the feasibility of this processing and to present a process-structure-property correlation. For this purpose, two different hot runner systems and a conventional cold runner system (as reference) are used, combined with different injection molding process parameters. In addition to mechanical properties, structural properties such as fiber lengths and orientations are characterized and presented in correlation to the process parameters. To illustrate the material differences of an RCF composite, reference composites of a glass fiber-reinforced PP are characterized in parallel.

2. Materials, Processing, and Characterization

2.1. Used Materials

The polypropylene PP 575P used as the matrix material, was provided by the company Sabic (Riyadh, Saudi Arabia). The PP used has a melt flow rate (MFR) of 10.5 g/10 min at 230 °C and 2.16 kg and a density of 0.905 g/cm³. According to the manufacturer's datasheet, the processing temperature of the material is 200–225 °C, while the molecular weight distribution is given as broad.

Chopped, regenerated cellulose fibers provided by the company Cordenka GmbH & Co. KG (Obernburg, Germany) with an average filament diameter of approx. 12 µm and an average initial length of approx. 2.3 mm was used for reinforcement. The fibers are produced by the viscose process, which is still considered the main large-scale RCF production process, although both the lyocell and carbamate processes are far less environmentally harmful than the viscose process because of the chemicals used [49,50]. Compared to the Lyocell fibers (Tencel), the rayon fibers from Cordenka have a significantly higher Young's modulus and tensile strength (see Table 1). The pure cellulose fibers have only very short fiber lengths, which, compared to the Lyocell and rayon fibers, achieve only a very small strengthening effect in the composite [51–53].

Table 1. Lengths and mechanical properties of different cellulosic fibers [51–53].

	Initial Length [mm]	Diameter [µm]	Young's Modulus [GPa]	Tensile Strength [MPa]	Elongation [%]
Cordenka Rayon (Viscose Fiber)	2.3	12	22	778	13
Tencel FCP (Lyocell Fiber)	0.4–6	f	15	556	11
Arbocel BC1000 (pure Cellulose)	0.7	20	[-] *	[-] *	[-] *

* not measurable due to short fiber length.

The very good mechanical properties of RCF by Cordenka are also reflected in the mechanical properties of the composites, which were already shown in a large number of publications and independent of the matrix material. In addition to the good quasi-static properties of the RCF composites, this always led to a significant increase in the notched impact strength, both in comparison with glass fibers and other natural fibers [34,35,54–57]. The used RCF were coated with a PPL-sizing (aqueous polyvinyl alcohol solution) of approx. 10 wt.% by the manufacturer, which was applied to increase the pourability for use in a gravimetric feeding systems. The RCF exhibits a density of 1.5 g/cm³ and an elongation at a break of 13%, a Young's modulus of approx. 22 GPa, and a tensile strength of approx. 800 MPa [58–60].

Regenerated cellulose fibers (RCF) are referred to as cellulose type II due to their molecular structure caused by the manufacturing process, which usually has larger non-crystalline regions [19,50,61,62]. A major difference from cellulose type I is its behavior in absorbing and releasing moisture. The tensile strength of cellulose type II (RCF fibers) increases with the release of water, e.g., residual moisture at higher temperatures, while the elongation at break decreases [62–64]. For that reason, the RCF was dried prior to compounding.

In addition to that, glass fibers CS 7952 provided by the company Lanxess AG (Cologne, Germany) with a diameter of 14 µm, an initial length of 4.5 mm, a density of 2.6 g/cm³ and a sizing suitable for polypropylene were investigated for reference purposes.

2.2. Compounding

To compare the properties of the glass and cellulose fiber reinforced composites, compounds with 20 wt.% of each fiber were compounded using the twin screw extruder ZSE 18 HPE (Leistritz Extrusionstechnik GmbH, Nuremberg, Germany) that has a screw diameter of 18 mm and a process length of 40 D. The fiber content of 20 wt.% was chosen because 20 or 30 wt.% glass fibers are currently mainly used for reinforcement in technical components. Due to the significantly lower density of RCF (1.5 g/cm³) compared to glass fibers (2.6 g/cm³), this results in a significantly higher fiber volume content of RCF for the same fiber weight content. Since it was not clear at the beginning of the test series whether the high fiber volume content at 30 wt.% RCF could lead to a problem during processing with the hot runner systems, 20 wt.% RCF was used.

Prior to the compounding process, the cellulosic fibers were dried in an air convection oven until their moisture content was less than 1%. All materials were fed into the extruder via a gravimetric feeding system (Brabender Technologie, Duisburg, Germany). After compounding, the strand was cooled down on a discharge conveyer using compressed air before being pelletized to a length of approx. 3 mm by a Scheer SGS 25-E strand pelletizer (Maag Germany GmbH, Grossostheim, Germany).

The screw configuration was optimized regarding less shear stress in previous investigations. For the named reason the configuration only consists of conveying elements after the fiber feeding zone, which reduces the shear stress but still realizes a homogeneous distribution of the fibers. The screw speed was set to 200 rpm while the processing temperatures were also set to a lower and more gentle level (below 200 °C, see Table 2) to reduce the thermal load on the cellulosic fibers [31,35,51].

Table 2. Process temperatures of twin screw compounding.

	Feeding Zone	Zone 1	Zone 2	Zone 3	Zone 4	Zone 5	Zone 6	Zone 7
Temperature [°C]	200	200	180	180	160	140	140	160

2.3. Injection Molding

Prior to the injection molding process, the compounds containing viscose fibers were dried using an air dryer TORO-systems TR–Dry–Jet EASY 15 (Gfk Thomas Jakob und Robert Krämer GbR, Igensdorf, Germany) for 4 h at 80 °C.

The test specimens (see Figure 1) were produced on two different injection molding machines. In connection with the valve gate hot runner system, a hybrid injection molding machine from Engel VC200/80 Electric (ENGEL AUSTRIA GmbH, Schwertberg, Austria) was used. This machine is equipped with an electric injection unit and a hydraulic clamping unit. The clamping force is 500 kN, the screw diameter is 30 mm with a resulting max. metering volume of 85 cm³ and the used open machine nozzle has a diameter of 6 mm.

For comparison, test specimens were also produced with an open hot runner system, for which a hydraulic injection molding machine Arburg A270S (Arburg GmbH + Co KG, Loßburg, Germany) was used. This machine has a clamping force of 250 kN, and a screw diameter of 22 mm with a resulting max. metering volume of 30 cm³ and a machine nozzle with a diameter of 6 mm. The mold temperature on both machines was set to 40 °C and the temperatures of the plasticizing unit were set according to the following Table 3.

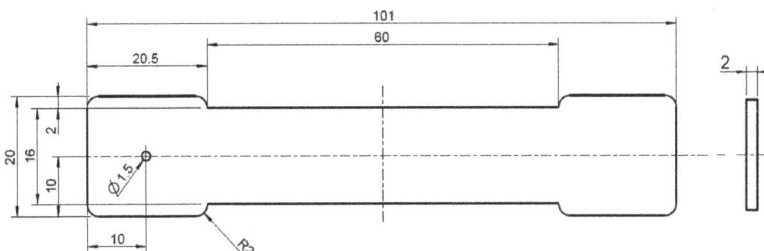

Figure 1. Schematic representation of the produced specimen with point gate of the hot runner systems.

Table 3. Process temperatures of injection molding.

	Feeding Zone	Zone 1	Zone 2	Zone 3	Zone 4
Temperature [°C]	40	200	200	200	200

With both hot runner systems, the melt is injected via the point gate and spreads out from there in a circular shape in the cavity (see Figure 1).

After adjusting the process (Setting 01), the composites with cellulose fibers (PP 20RCF) were then subjected to successive increases in temperature and dwell time (see Table 4) in order to show the influence of the higher thermal load, which results from a combination of the temperature and the dwell time, on the structural and mechanical properties.

Table 4. Process parameters of injection molding.

Material	Setting	Hot Runner Temperature [°C]	Cooling Time [s]	Injection Time [s]	Holding Pressure [bar]	Holding Time [s]	Dwell Time [s]
PP 20GF	01	200	8	0.50	300	3	16
PP 20RCF	01	200	8	0.65	300	3	16
PP 20RCF	02	200	8	0.30	300	3	16
PP 20RCF	03	200	20	0.30	300	3	28
PP 20RCF	04	220	20	0.30	300	3	28
PP 20RCF	05	240	20	0.30	300	3	28
PP 20RCF	06	240	40	0.30	300	3	48

In cooperation with Günther Heisskanaltechnik GmbH (Frankenberg, Germany) the following two hot runner systems (see Figure 2) were used to characterize the feasibility of processing cellulose fiber reinforced composites with different hot runner systems and their influence on the structural and mechanical properties of the cellulose and glass fiber reinforced plastics:

(a) Open hot runner system with tip (5SHF50) with a gate diameter of 1.5 mm
(b) Valve gate hot runner system (nozzle 6NHF50 LA-1.4; needle 3NHP175-1.4 (clamping force of the needle: 800 N)

Both nozzle typesthe fiber length distribution and orientation (see are designed with the Blueflow® heaters by Günther Heisskanaltechnik. In conjunction with the two-part shaft (steel vs. titanium alloy with low thermal conductivity) of the nozzles, this results in a very homogeneous temperature profile in the nozzle and reduces the heat transfer from the nozzle to the mold. The BlueFlow® technology involves heating elements that are manufactured on the basis of thick-film technology. Here, the dielectric layers and the heating conductor are applied under clean room conditions using the screen printing process. The manufacturer declares the following advantages compared to similar systems:

(a) Precise and homogeneous power distribution over the entire length of the nozzle
(b) Avoidance of temperature peaks in the melt-carrying material tube
(c) High power concentration in the front nozzle area.
(d) Rapid thermal reaction, thereby lower energy consumption

Due to the rapid thermal reaction of these heating elements, appropriate control technology must also be used. Therefore, a DPT control device from Günther Heisskanaltechnik GmbH was used for the tests. Experience shows that the homogeneous temperature profile of the applied hot runner and control technology is particularly suitable for the processing of thermally and shear-sensitive bioplastics or compounds with natural fibers in order to avoid thermal damage to the melt or the natural fibers.

In order to be able to quantify the influence of processing with the hot runner systems on the mechanical properties in a comparative manner, reference test specimens were produced with a cold runner system on a hydraulic injection molding machine Allrounder 320C Golden Edition (Arburg GmbH + Co KG, Loßburg, Germany) with a screw diameter of 25 mm and a clamping force of 500 kN. The cycle time was approximately 43 s, including a cooling time of 20 s.

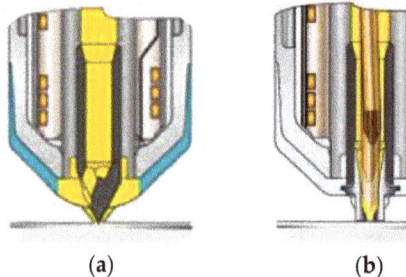

Figure 2. Schematic representation of the open hot runner system with tip (**a**) and the valve gate hot runner system (**b**).

2.4. Characterization

All composites were characterized while in a dry state and in a standardized climate (23 °C, 50% relative humidity).

2.4.1. Tensile Test

Tensile tests were carried out at a speed of 5 mm/min according to EN ISO 527 using a UPM 1446 testing machine (Zwick Roell, Ulm, Germany) with a 10 kN load cell. During the tests, Young's modulus, tensile strength, and elongation at break were evaluated. Five specimens were tested for each material.

2.4.2. Color Measurement

Any thermal loads to the cellulose fibers resulting from processing lead to a darkening of the composites due to the degradation that occurs.

To objectively quantify this discoloration and its influence by individual process parameters, color measurements were carried out on the test specimens using the Ultra Scan Pro spectrophotometer (Hunterlab, Reston, VA, USA). This system uses the L*a*b*-color-model for color measurement, in which the brightness of a color is indicated by the L-value. The higher the L-value, the brighter the color. More precisely, an L value of 100 means that the color is white, while an L-value of 0 means a black color.

2.4.3. Fiber Length Measurement

The resulting fiber length in the tensile specimen was measured using the dynamic image analysis QicPic R06 (Sympatec GmbH, Clausthal-Zellerfeld, Germany) with a Mixcel liquid dispersion unit. The fibers of representative parts of the specimens were separated from the matrix using xylene with a temperature of 80 °C for a duration of min. six hours. Afterward, the fibers were dispersed in isopropanol and filled into the liquid dispersion unit of the measuring system, which provided a constant flow of the dispersed fibers. This isopropanol flow with the fibers passed a cuvette with a thickness of 2 mm and a window at which a high-speed camera captured images of each fiber. Subsequently, the software Windox calculated the length and diameter of the fibers. The objective M7 with a resolution of 4.2 µm was employed, which realizes a minimal fiber size of 12.6 µm and a maximum fiber size of 8.66 mm. The distribution of the calculated fiber lengths is number-based (q0), which is well-suited for representing broad ranges of size distributions.

2.4.4. X-ray Microtomography Analysis of the Composite Structure

For a structural characterization by X-ray microtomography (µCT), RCF-reinforced specimens produced with Settings 01, 02, 03, 05, 06 and the open hot runner system were selected. The intention of these analyses is to determine the correlation between fiber orientations within the parts and their mechanical properties.

High-resolution results have been obtained using an X-ray microtomograph Zeiss Xradia Versa 520 (Carl Zeiss, Oberkochen, Germany). These allow individual fibers to be examined separately and evaluated quantitatively. The measurements were performed at a voltage of 72 kV and a current of 83 µA using the 0.4x objective and no filter. The number of images acquired with an exposure time of 3.5 s for each image was 1601. Binning setting 1 resulted in a voxel size of 4.13 µm. These settings were chosen to obtain an image section in the center of the specimens of half the cross-section with an adequate voxel size. The subsequent reconstruction was performed with the Zeiss XMReconstructor software. A 3D data visualization and analysis software system Avizo 9.4 (Thermo Fisher Scientific, Waltham, MA, USA) with the XFiber extension for the quantitative analysis of fiber properties was utilized to generate the required data. After the preparation of the volume data, primarily the software modules "Cylinder Correlation" and "Trace Correlation Lines" were employed to detect the individual fibers regarding their fiber orientation with the settings from Table 5. The minimum continuation quality parameter was chosen in order to ensure that the resulting fiber lengths of the fiber tracing model have the same median as the fiber lengths of the QicPic fiber length measurements (see Section 2.4.3).

Table 5. XFiber extension settings for the individual evaluated samples.

XFiber Parameter		Setting				
		01	02	03	05	06
Cylinder length	[µm]	38	38	38	38	38
Angular sampling	[-]	5	5	5	5	5
Mask cylinder radius	[µm]	8.3	8.3	8.3	8.3	8.3
Outer cylinder radius	[µm]	6.3	6.3	6.3	6.3	6.3
Minimum seed correlation	[-]	203	203	203	196	198
Minimum continuation quality	[-]	140	123	107	98	89
Direction coefficient	[-]	0.4	0.4	0.4	0.4	0.4
Minimum distance	[µm]	3	3	3	3	3
Minimum length	[µm]	38	38	38	38	38

The investigated volume within the samples is shown in Figure 3. Within this region, the orientation angle theta (Θ) was evaluated. It describes the angle between the x-axis and the yz-plane. In the case of $\Theta = 0°$, a fiber is positioned exactly in the direction of flow respectively in the direction of the load during tensile tests, in the case of $\Theta = 90°$ perpendicular to this.

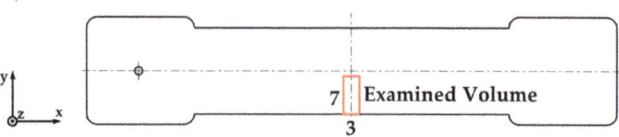

Figure 3. Position and size (in mm) of the examined sample volume by means of X-ray-microtomography (red).

3. Results

3.1. Mechanical Properties

The comparison of the mechanical properties of glass and cellulose fiber reinforced composites in Figure 4 initially shows the expected significantly higher stiffness and lower elongation at the break of the GF composites.

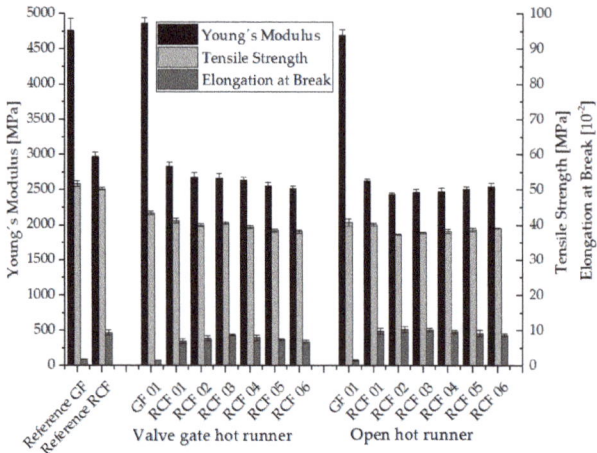

Figure 4. Mechanical properties of the RCF and GF reinforced composites.

Furthermore, with regard to the glass fiber reinforced composites, it can be seen that the values of stiffness (Young's modulus) and elongation at break obtained during processing with the two hot runner systems are very close to those of the reference composites processed with a cold runner system. Regarding the tensile strength, a drop in the values of the GF composites of both hot runner systems of approx. 20% compared to the reference specimen is noticeable. The extent to which this drop is caused by damage to the structures, in particular a reduction in fiber length, by the hot runner systems is described with the characterized fiber length distributions in Section 3.3.

For the cellulose fiber reinforced composites, a reduction in Young's modulus of 5% for the valve gate hot runner system and 10% for the open hot runner with tip is shown in comparison between the reference specimen and the specimen of RCF 01 (Setting 01). The tensile strength drops by 20% with both hot runner systems compared to the reference specimen. There are also further differences between the composites RCF 01 to RCF 06, which were processed with the valve gate hot runner system. For example, Young's modulus and tensile strength decrease with increasing composite number. This can be explained by the combination of higher hot runner temperature and higher dwell time of the melt in the hot runner due to the higher cooling time in the injection molding process (see Table 4), which leads to higher thermal load and damage of the fibers. The decreasing elongation at break also indicates higher thermal damage to the fibers. The lowest tensile strength and stiffness values of specimen of RCF 06 are in total approx. 30% below the cellulose fiber-reinforced reference specimen.

The mechanical properties resulting from the different process settings can be attributed to the thermal properties of the RCF in addition to the fiber length distribution and orientation (see Sections 3.3 and 3.4). Higher temperature exposure leads to degradation phenomena, which occur in RCF from about 220 °C due to the breaking of chemical bonds [19,65]. Overall, pure cellulose (e.g., RCF) exhibits higher heat resistance compared to conventional natural fibers [12,13,66]. With the onset of the degradation processes, the tensile strength of the RCF and especially the elongation at break decreases. In addition, the brittleness of the fibers and the associated higher number of fiber breaks in the process leads to a reduction in fiber lengths [67,68].

The tensile strengths and stiffnesses of the RCF-reinforced composites processed by means of an open hot runner system with tip are rather different. After a reduction in tensile strength (3 MPa) and stiffness (200 MPa) from batch RCF 01 to RCF 02, the values increase again up to batch RCF 06, although this cannot be described as a statistically significant change. At the same time, the elongation at the break of the composites decreases. In detail, this means that a higher melt temperature and the associated low viscosity in combination with a higher volume flow rate (at a shorter injection time) leads to higher tensile strengths and stiffnesses of the test specimens, since the fibers tend to be oriented in the flow direction and thus in the loading direction [41,47,69]. This will be validated with the results of the μCT analysis and the resulting fiber orientation in Section 3.4. The correlation between the shear rate and the viscosity of PP reinforced with glass and cellulose fibers has been shown in previous work. Here, a decreasing viscosity with a higher shear rate could be shown, whereby in particular the higher fiber volume fraction due to the RCF´s low density led to higher viscosities of the RCF composites overall [51,70].

The increasing dark coloration of the test specimens with increasing thermal loading (see Figure 5) is described by the results of the color measurement in Section 3.2, while the influence of the fiber length distribution and orientation is described in Sections 3.3 and 3.4.

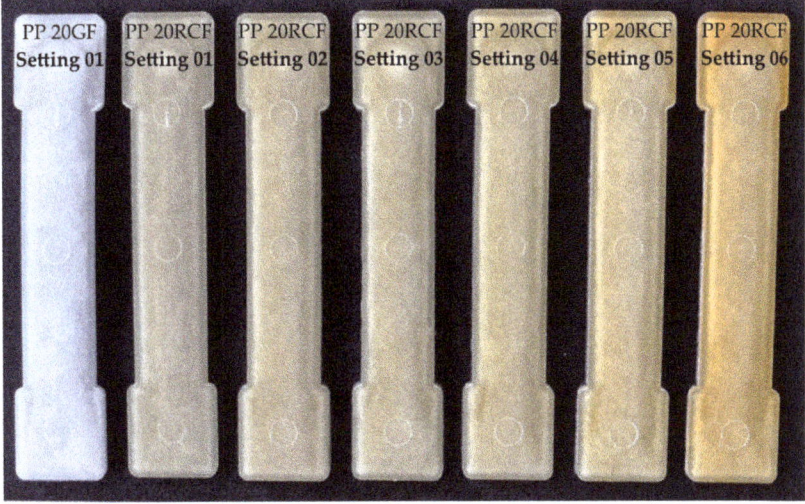

Figure 5. Representative specimens of PP 20GF (Setting 01) and PP 20RCF (Setting 01 to Setting 06).

3.2. Discoloration

Figure 5 shows representative specimens processed with the settings of Table 4, while Figure 6 shows the results of the color measurement described in Section 2.4.2.

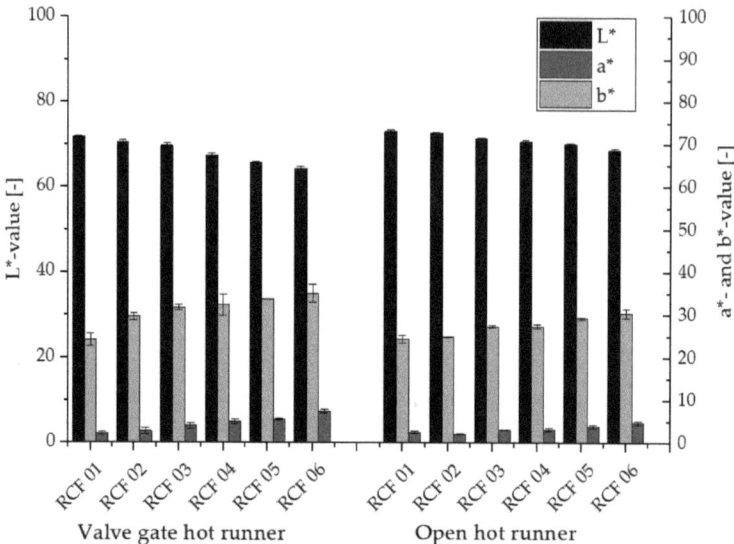

Figure 6. L*a*b*-values of the RCF und GF reinforced composites.

It can be seen that in the tests with both hot runner systems, the composites show a lower L-value and thus a stronger dark coloration with the increasing number and thus with increasing thermal load. Cellulose is a polysaccharide and thus a sugar in which the process of caramelization takes place at higher temperatures (approx. 140 °C for saccharides). This process is accompanied by a darker coloration of the cellulose and thus also of the RCF, as well as the development of a roasted aroma (caramel odor). Since, depending on the thermal stability of the RCF, degradation processes do not yet occur immediately, the mechanical properties initially decrease only to a minor extent. With further increasing temperatures or longer dwell times, degradation of the RCF then leads to a reduction in tensile strength and elongation at break of the fibers. [18,27,35]. In addition to the decreasing L-value, an increase in the a-value and b-value can be seen with increasing thermal stress, which is indicative of an overall more intense coloration of the RCF composites.

3.3. Fiber Length Distribution

The results of the fiber length distributions determined with the help of the dynamic image analysis system QicPic, described in Section 2.4.3, are presented and explained in the following. Figures 7 and 8 show the fiber length distributions of the glass fiber reinforced PP and the results of the six parameter settings for the RCF reinforced PP processed with the hot runner system (see Table 4) as well as the two reference composites processed with the cold runner system.

The fiber lengths of the materials processed with the open hot runner system, shown in Figure 7, initially show the two distribution curves typical for GF and RCF. The glass fibers show a Gaussian distribution with a median of about 411 µm. A comparison of the two GF distributions shows that the fiber lengths of the material processed with the open hot runner are slightly below the values of the reference batch (median: 357 µm). Thus, there is shortening of the GF in the hot runner of approx. 15%, which is, however, sufficient to bring about the 20% reduction in tensile strength described in Section 3.1.

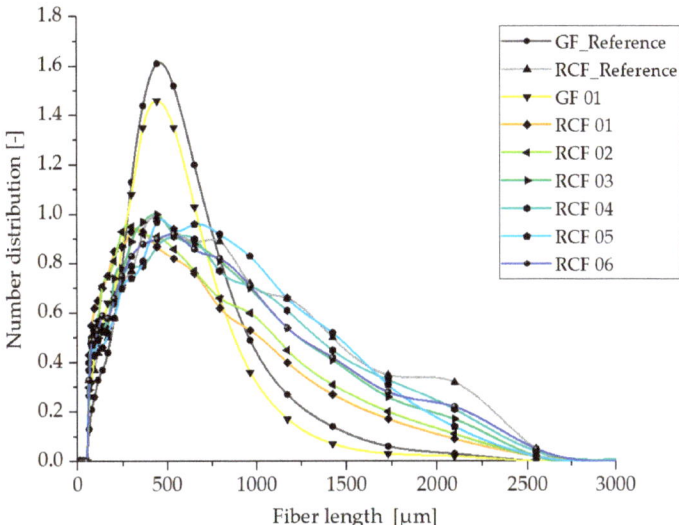

Figure 7. Fiber length distribution in specimen produced with the open hot runner system.

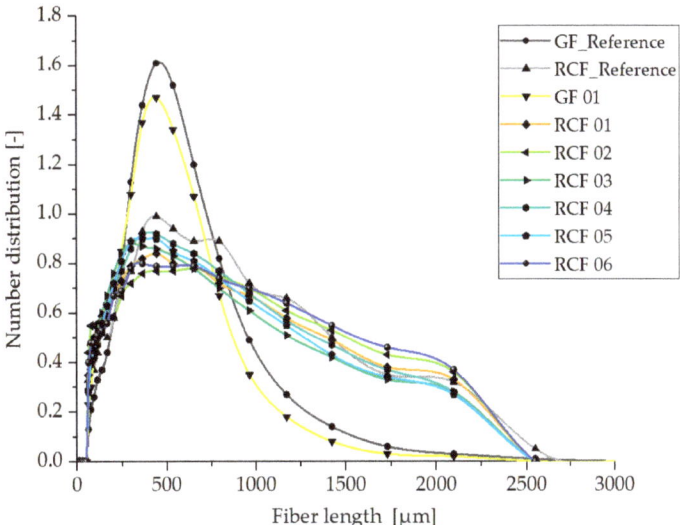

Figure 8. Fiber length distribution in specimen produced with the valve gate hot runner system.

In contrast to the GF, the distribution of the RCF lengths does not turn out to be a Gaussian distribution. Due to the already described lower stiffness and higher elongation at break, the RCF are not shortened as much and are therefore present in greater numbers with higher lengths. This allows the good mechanical properties of these composites described above. Compared to the six hot runner composites, the reference composite has a similar fiber length distribution and only slightly more long fibers (10% higher median). Contrary to the assumption, this could be due to a similar shear in the cold runner geometry of the reference composite, which is different from that of the hot runner. The differences between the RCF reinforced composites of Settings 03 to 06 are very small and insignificant and can be attributed to minor differences in viscosity due to the process parameters from

Table 4. In comparison, the RCF lengths of Settings 01 and 02 show lower values of fiber lengths, which can be attributed to the melt viscosity due to the low temperatures and dwell times and the resulting high viscosity (see Table 4).

Table 6 shows the values of the 10th, 50th and 90th percentile of the fiber length distribution of both hot runner systems for comparison.

Table 6. Percentile values (10th, 50th and 90th) from the fiber length distributions of the characterized composites.

Composite	System	x10 [μm]	x50 [μm]	x90 [μm]
Reference GF	cold runner	108.6	436.1	1311.3
Reference RCF	cold runner	143.1	411.7	832.9
PP 20GF 01	open hot runner	110.4	357.1	742.0
PP 20RCF 01	open hot runner	90.9	303.4	979.9
PP 20RCF 02	open hot runner	98.9	324.6	1033.8
PP 20RCF 03	open hot runner	103.3	386.9	1155.4
PP 20RCF 04	open hot runner	94.8	390.9	1233.9
PP 20RCF 05	open hot runner	102.0	422.7	1192.9
PP 20RCF 06	open hot runner	98.7	391.9	1232.2
PP 20GF 01	valve gate hot runner	111.4	361.1	747.1
PP 20RCF 01	valve gate hot runner	96.5	404.8	1444.3
PP 20RCF 02	valve gate hot runner	106.9	384.9	1311.5
PP 20RCF 03	valve gate hot runner	106.4	399.9	1324.1
PP 20RCF 04	valve gate hot runner	102.4	371.8	1310.9
PP 20RCF 05	valve gate hot runner	90.9	398.5	1444.5
PP 20RCF 06	valve gate hot runner	97.6	388.6	1392.8

Figure 8 shows the fiber length distributions of the materials processed with the valve gate hot runner system. The typical distribution curves for GF and RCF can also be clearly seen here. When comparing the two normal distributions of the glass fibers, the valve gate hot runner also shows only a slight shortening of the GF. Similar to the open hot runner system, there is a shortening of the GF of approx. 13%, which nevertheless leads to the 20% reduction in tensile strength described in Section 3.1.

The distribution of RCF lengths slightly deviates from that of the open hot runner. Contrary to the assumption the geometry and cross-sections of the valve gate hot runner does not result in a greater shortening of the RCF. Compared to the fiber length distribution of the open hot runner and the reference specimen processed with cold runner this is reflected in a larger number of longer fibers and a smaller number of shorter fibers. This result reflects very well the better mechanical properties (see Figure 4), in which the composites from the processing with valve gate hot runner show the better mechanical properties, which can now be attributed, at least in part, to the longer fiber lengths.

The comparison of the six process settings shows only very small differences in the fiber length distributions, so that the influence of the valve gate hot runner is greater than the influence of the process parameters used (see Table 4).

Based on the shown results, the reduction in tensile strength of RCF composite specimens from Setting 01 to 06 of both hot runner systems compared to the reference cannot only be attributed to the reduced fiber lengths but rather a combination with a reduction in tensile strength of the RCF due to thermal loading of the fibers. This will be discussed in more detail in the following sections.

3.4. Fiber Orientation

Figure 9 shows the fiber orientations of the specimens produced with different Settings and with open hot runner system, which were detected with the trace algorithm (see Section 2.4.4). It should be noted that the individual fibers are represented by lines in three-dimensional space and that white areas indicate that there are only few fibers. On the one hand, this can be explained by the lower fiber content in the edge areas of the sample. On the other hand, this area is located at the outer edge of the μCT measurement volume, where the image sharpness is lower, so that fewer fibers can be detected. Here, the fibers that are present in the flow direction (x-direction) and thus have an angle of $\Theta = 0°$ are colored blue. The fibers oriented perpendicular to the flow direction and thus aligned closer to the yz-plane ($\Theta = 90°$) are shown in red. Each image represents the measuring area marked in Figure 3.

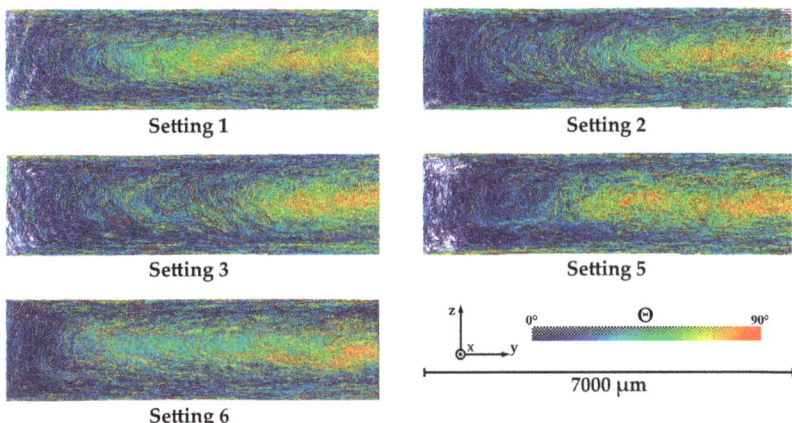

Figure 9. Illustrations of fiber orientations of RCF reinforced specimens produced with open hot runner system obtained by fiber tracing evaluation of the μCT data.

In this representation, the core of the sample is located on the right, a mold wall is positioned on the left, as well as at the upper and lower edges. The blue areas on the mold walls (edge regions) mark a high degree of orientation along the flow direction, while the fibers in the core region are rather unoriented or oriented perpendicular to the x-axis (red). The strong orientation of the fibers in the edge regions is due to classical layered models, which provide an explanation for the parallel orientation (0°) at high shear rates [71–73]. The fibers aligned in this way cool and freeze particularly quickly at the mold wall, so that this alignment is conserved. The core layers are influenced by the swell flow and lead to this (approximately) perpendicular alignment. The size of the respective areas can be attributed to the different process parameters within the settings used.

The frequency distributions of the angle Θ derived from fiber tracing are shown in Figure 10. Here, the test specimens of Settings 01, 02, and 03 show the highest values in the range of $\Theta = 5–15°$ and thus have the most fibers with an orientation in the flow direction. This can be attributed to the short injection time and the resulting high shear rate, which results in a wider edge area.

Contrary to the expectations, the increase in melt temperature and the associated decrease in viscosity in samples 5 and 6 do not lead to the increase in fiber orientation in the flow direction described in Section 3.1.

Based on the results of the process-induced fiber orientation of the open hot runner system, which was characterized by means of μCT, again no clear correlation to the mechanical properties of the composites shown in Figure 4 can be concluded. This will be discussed in detail in the following section.

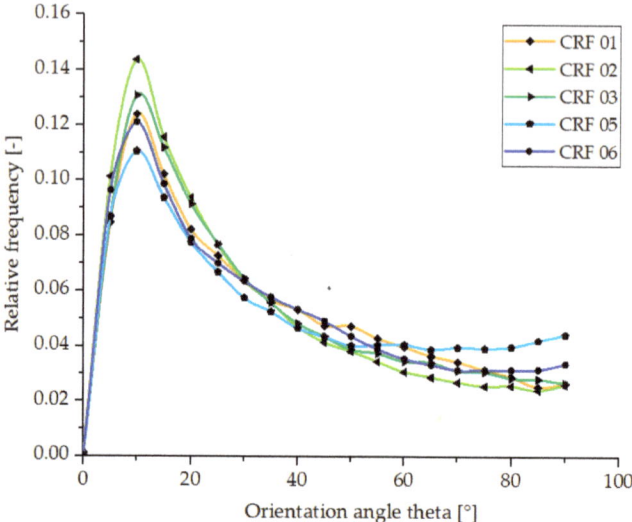

Figure 10. Quantitative plot of fiber orientations of RCF reinforced specimens produced with open hot runner system obtained by fiber tracing evaluation of the µCT data.

3.5. Discussion

As described in the previous paragraphs, the mechanical properties of the characterized composites could not be attributed only to one structural feature, such as the thermal load, fiber length distribution, or fiber orientation. This will be discussed in the following.

Figure 5 shows that the RCF-reinforced composites darken with increasing thermal load due to higher process temperatures and dwell times. From this, it can be concluded that with increasing thermal load from Setting 01 to Setting 06, the fiber tensile strength and consequently the tensile strength of the composites is also reduced.

Regarding the fiber lengths, it was shown that these are influenced by the material shear stress resulting from the melt temperature and shear rate. Thus, also here the fiber lengths are highest at high melt temperatures (Setting 04, 05, and 06) and thus low viscosity for both hot runner systems. For this reason, especially with the open hot runner, the average fiber lengths of Settings 05 and 06 are higher than the values of Settings 01 and 02.

Thus, the fiber length distribution, which achieves larger values at higher temperatures, represents an opposing effect with the fiber strength, which is reduced more at high temperatures. This fact provides the explanation for the fact that the RCF-reinforced specimen of Setting 01 processed with the open hot runner system achieves the highest mechanical properties due to the lowest thermal stress on the fibers despite the lowest fiber lengths present.

As already described in Section 3.1, tensile strengths are generally higher at higher melt temperatures and high injection volume flow rates due to better fiber orientation. This effect with e.g., higher injection velocities at Settings 02-06 and higher melt temperatures at Settings 04, 05, and 06 thus also partially counteracts the effects of fiber length distribution and fiber strength.

The qualitative curves in Figure 11 from the values of fiber strength [74] and average fiber length for the valve gate hot runner and additionally the fiber orientation (Θ-values between 5° and 15°) for the open hot runner system of the various settings show that a curve qualitatively composed of these variables describes very well the course of the tensile strengths of the composites from Figure 4.

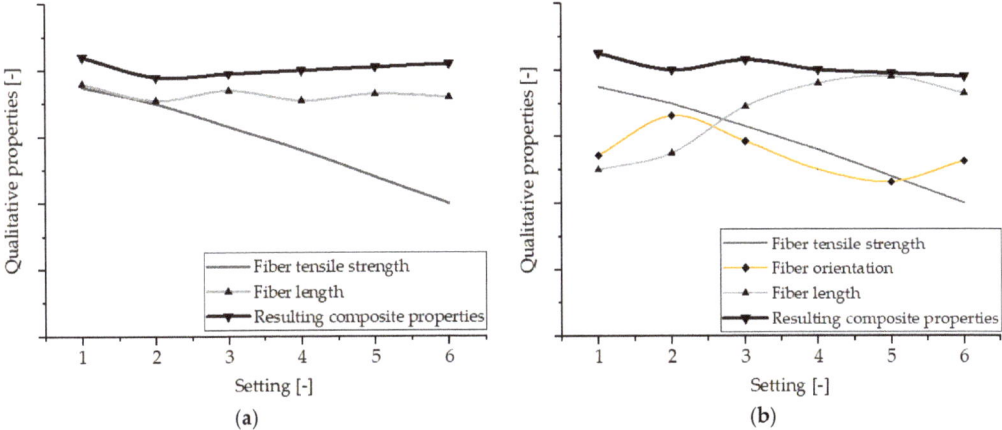

Figure 11. Qualitative curves of structural properties (fiber tensile strength, fiber length, and fiber orientation) and resulting composite properties of the characterized setting for the valve gate (**a**) and the open hot runner system (**b**).

4. Conclusions

In summary, the investigations carried out here showed that cellulose fiber-reinforced composites from PP can be processed very well with various hot runner systems in the injection molding process.

Regarding the influence of hot runner processing, it was shown that the tensile strength of the GF-composites and the RCF-composites is about 20% lower than the reference samples even at the lowest thermal load (Setting 01).

As expected, the tensile strengths of the RCF composites decrease with high thermal load (Setting 06) up to 30% for the valve gate hot runner and 22% for the open hot runner. The elongation at break of the RCF composites at Setting 06 is about 28% (valve gate) and 10% (open hot runner) lower than for the reference specimen.

These differences are partly due to the resulting fiber length distributions, whose median in both hot runner systems is 20% lower than that of the reference batches for the GF composites and only about 5% lower for the RCF composites. With regard to the fiber orientation, it was found that the different parameter settings have an influence on the orientation of the RCF in the core and edge areas. Especially at the higher shear rates, the fibers in the edge region are more oriented in the flow direction and thus also in the loading direction.

Summarized it was shown that the resulting mechanical properties of the composites are not only attributable to one structural property but are rather based on a combination of several process-related properties. The influence of the process parameters was also presented, showing the temperature-time influence on the properties of the RCF composites. From this, it can generally be deduced that the thermal load as a combination of temperature and cycle time should be kept as low as possible.

However, in the context of this temperature-time influence, it also shows that the process window for achieving good mechanical properties is quite large. This increases the fault tolerance in a later series production and thus also the acceptance of the partially bio-based composites. Thus, the results represent a further step towards the possible series production of components made from alternative material systems by means of injection molding.

Author Contributions: Conceptualization, J.-C.Z.; methodology, J.-C.Z., A.S. and F.L.; validation, S.S., J.E. and J.-C.Z.; fiber length analysis, F.L.; fiber orientation investigation, A.S.; mechanical characterization, J.-C.Z.; resources, S.S. and H.-P.H.; data curation, J.-C.Z.; writing—original draft preparation, J.-C.Z. and A.S.; writing—review and editing, S.S. and J.E.; visualization, A.S., F.L. and J.-C.Z.; supervision, S.S. and H.-P.H.; All authors have read and agreed to the published version of the manuscript.

Funding: This research received no external funding.

Institutional Review Board Statement: Not applicable.

Informed Consent Statement: Not applicable.

Data Availability Statement: Not applicable.

Conflicts of Interest: The Authors Jörg Essinger and Stefan Sommer were employed by the company Günther Heisskanaltechnik GmbH. The remaining authors declare that the research was conducted in the absence of any commercial or financial relationships that could be construed as a potential conflict of interest.

References

1. Tirupathi; Kumar, J.S.; Hiremath, S.S. Investigation of Mechanical Characterisation and Thermal Performance of Hybrid Natural Fiber Composites for Automotive Applications. *Fibers Polym.* **2022**, *23*, 3505–3515. [CrossRef]
2. Jagadeesh, P.; Puttegowda, M.; Boonyasopon, P.; Rangappa, S.M.; Khan, A.; Siengchin, S. Recent developments and challenges in natural fiber composites: A review. *Polym. Compos.* **2022**, *43*, 2545–2561. [CrossRef]
3. Dehury, J.; Mohanty, J.R.; Nayak, S.; Dehury, S. Development of natural fiber reinforced polymer composites with enhanced mechanical and thermal properties for automotive industry application. *Polym. Compos.* **2022**, *43*, 4756–4765. [CrossRef]
4. Seculi, F.; Espinach, F.X.; Julián, F.; Delgado-Aguilar, M.; Mutjé, P.; Tarrés, Q. Comparative Evaluation of the Stiffness of Abaca-Fiber-Reinforced Bio-Polyethylene and High Density Polyethylene Composites. *Polymers* **2023**, *15*, 1096. [CrossRef] [PubMed]
5. Sayeed, M.M.A.; Sayem, A.S.M.; Haider, J.; Akter, S.; Habib, M.M.; Rahman, H.; Shahinur, S. Assessing Mechanical Properties of Jute, Kenaf, and Pineapple Leaf Fiber-Reinforced Polypropylene Composites: Experiment and Modelling. *Polymers* **2023**, *15*, 830. [CrossRef] [PubMed]
6. Silva, R.V.; Spinelli, D.; Bose Filho, W.W.; Claro Neto, S.; Chierice, G.O.; Tarpani, J.R. Fracture toughness of natural fibers/castor oil polyurethane composites. *Compos. Sci. Technol.* **2006**, *66*, 1328–1335. [CrossRef]
7. Ranganathan, N.; Oksman, K.; Nayak, S.K.; Sain, M. Regenerated cellulose fibers as impact modifier in long jute fiber reinforced polypropylene composites: Effect on mechanical properties, morphology, and fiber breakage. *J. Appl. Polym. Sci.* **2015**, *132*, 41301. [CrossRef]
8. Ranganathan, N.; Oksman, K.; Nayak, S.K.; Sain, M. Structure property relation of hybrid biocomposites based on jute, viscose and polypropylene: The effect of the fibre content and the length on the fracture toughness and the fatigue properties. *Compos. Part A Appl. Sci. Manuf.* **2016**, *83*, 169–175. [CrossRef]
9. Zarges, J.-C.; Feldmann, M.; Heim, H.-P. Influence of the compounding process on bio-based polyamides with cellulosic fibers. In Proceedings of the 30th International Conference of the Polymer Processing Society, Cleveland, OH, USA, 6–12 June 2014.
10. Furtado, S.C.R.; Araújo, A.L.; Silva, A.; Alves, C.; Ribeiro, A.M.R. Natural fibre-reinforced composite parts for automotive applications. *IJAUTOC* **2014**, *1*, 18. [CrossRef]
11. Clemons, C.M.; Caulfield, D.F.; Giacomin, A.J. Dynamic Fracture Toughness of Cellulose-Fiber-Reinforced Polypropylene: Preliminary Investigation of Microstructural Effects. *J. Elastomers Plast.* **1999**, *31*, 367–378. [CrossRef]
12. Weigel, P.; Ganster, J.; Fink, H.P.; Gassan, J.; Uihlein, K. Polypropylene-cellulose compounds—High strength cellulose fibres strengthen injection moulded parts. *Kunstst. Plast Eur.* **2002**, *92*, 95.
13. Teuber, L.; Militz, H.; Krause, A. Dynamic particle analysis for the evaluation of particle degradation during compounding of wood plastic composites. *Compos. Part A Appl. Sci. Manuf.* **2016**, *84*, 464–471. [CrossRef]
14. Feldmann, M. *Biobasierte Polyamide mit Cellulosefasern: Verfahren-Struktur-Eigenschaften*; Kassel University Press: Kassel, Germany, 2013.
15. Holbery, J.; Houston, D. Natural-fiber-reinforced polymer composites in automotive applications. *JOM* **2006**, *58*, 80–86. [CrossRef]
16. Rowell, R.M. *Handbook of Wood Chemistry and Wood Composites*, 2nd ed.; CRC Press: Boca Raton, FL, USA, 2013.
17. Bledzki, A.K.; Sperber, V.E.; Faruk, O. *Natural and Wood Fibre Reinforcement in Polymers*; Smithers Rapra Technology: Shrewsbury, UK, 2002; Volume 13.
18. Reußmann, T. Entwicklung Eines Verfahrens zur Herstellung von Langfasergranulat mit Naturfaserverstärkung. 2003. Available online: https://www.researchgate.net/publication/28670519_Entwicklung_eines_Verfahrens_zur_Herstellung_von_Langfasergranulat_mit_Naturfaserverstarkung (accessed on 7 March 2023).

19. Liu, Q.; Lv, C.; Yang, Y.; He, F.; Ling, L. Study on the pyrolysis of wood-derived rayon fiber by thermogravimetry–mass spectrometry. *J. Mol. Struct.* **2005**, *733*, 193–202. [CrossRef]
20. Gassan, J.; Bledzki, A.K. Thermal degradation of flax and jute fibers. *J. Appl. Polym. Sci.* **2001**, *82*, 1417–1422. [CrossRef]
21. Feldmann, M. The effects of the injection moulding temperature on the mechanical properties and morphology of polypropylene man-made cellulose fibre composites. *Compos. Part A Appl. Sci. Manuf.* **2016**, *87*, 146–152. [CrossRef]
22. Carrillo, F.; Colom, X.; Suñol, J.J.; Saurina, J. Structural FTIR analysis and thermal characterisation of lyocell and viscose-type fibres. *Eur. Polym. J.* **2004**, *40*, 2229–2234. [CrossRef]
23. Sälzer, P.; Feldmann, M.; Heim, H.-P. Wood-Polypropylene Composites: Influence of Processing on the Particle Shape and Size in Correlation with the Mechanical Properties Using Dynamic Image Analysis. *Int. Polym. Process.* **2018**, *33*, 677–687. [CrossRef]
24. Fernandez, R.; Luis, J.; Thomason, J. Characterisation of the mechanical performance of natural fibers for lightweight automotive applications. In Proceedings of the 15th European Conference on Composite Materials, Venice, Italy, 24–28 June 2012; p. 1231.
25. Pandey, J.K.; Ahn, S.H.; Lee, C.S.; Mohanty, A.K.; Misra, M. Recent Advances in the Application of Natural Fiber Based Composites. *Macromol. Mater. Eng.* **2010**, *295*, 975–989. [CrossRef]
26. Njuguna, J.; Wambua, P.; Pielichowski, K.; Kayvantash, K. Natural Fibre-Reinforced Polymer Composites and Nanocomposites for Automotive Applications. In *Cellulose Fibers: Bio- and Nano-Polymer Composites*; Springer: Berlin/Heidelberg, Germany, 2011; pp. 661–700.
27. Feldmann, M.; Bledzki, A.K. Bio-based polyamides reinforced with cellulosic fibres—Processing and properties. *Compos. Sci. Technol.* **2014**, *100*, 113–120. [CrossRef]
28. Zarges, J.-C.; Kaufhold, C.; Feldmann, M.; Heim, H.-P. Single fiber pull-out test of regenerated cellulose fibers in polypropylene: An energetic evaluation. *Compos. Part A Appl. Sci. Manuf.* **2018**, *105*, 19–27. [CrossRef]
29. Graupner, N.; Rößler, J.; Ziegmann, G.; Müssig, J. Fibre/matrix adhesion of cellulose fibres in PLA, PP and MAPP: A critical review of pull-out test, microbond test and single fibre fragmentation test results. *Compos. Part A Appl. Sci. Manuf.* **2014**, *63*, 133–148. [CrossRef]
30. Zarges, J.-C.; Heim, H.-P. Influence of cyclic loads on the fiber-matrix-interaction of cellulose and glass fibers in polypropylene. *Compos. Part A Appl. Sci. Manuf.* **2021**, *149*, 106491. [CrossRef]
31. Zarges, J.-C.; Minkley, D.; Feldmann, M.; Heim, H.-P. Fracture toughness of injection molded, man-made cellulose fiber reinforced polypropylene. *Compos. Part A Appl. Sci. Manuf.* **2017**, *98*, 147–158. [CrossRef]
32. Thomason, J.L.; Vlug, M.A. Influence of fibre length and concentration on the properties of glass fibre-reinforced polypropylene: 4. Impact properties. *Compos. Part A Appl. Sci. Manuf.* **1997**, *28*, 277–288. [CrossRef]
33. Bledzki, A.K.; Jaszkiewicz, A. Mechanical performance of biocomposites based on PLA and PHBV reinforced with natural fibres—A comparative study to PP. *Compos. Sci. Technol.* **2010**, *70*, 1687–1696. [CrossRef]
34. Ganster, J.; Fink, H.-P.; Uihlein, K.; Zimmerer, B. Cellulose man-made fibre reinforced polypropylene—Correlations between fibre and composite properties. *Cellulose* **2008**, *15*, 561–569. [CrossRef]
35. Feldmann, M.; Heim, H.-P.; Zarges, J.-C. Influence of the process parameters on the mechanical properties of engineering biocomposites using a twin-screw extruder. *Compos. Part A Appl. Sci. Manuf.* **2016**, *83*, 113–119. [CrossRef]
36. Raphael, I.; Saintier, N.; Robert, G.; Béga, J.; Laiarinandrasana, L. On the role of the spherulitic microstructure in fatigue damage of pure polymer and glass-fiber reinforced semi-crystalline polyamide 6.6. *Int. J. Fatigue* **2019**, *126*, 44–54. [CrossRef]
37. Hessman, P.A.; Riedel, T.; Welschinger, F.; Hornberger, K.; Böhlke, T. Microstructural analysis of short glass fiber reinforced thermoplastics based on x-ray micro-computed tomography. *Compos. Sci. Technol.* **2019**, *183*, 107752. [CrossRef]
38. Lafranche, E.; Krawczak, P.; Ciolczyk, J.-P.; Maugey, J. Injection moulding of long glass fiber reinforced polyamide 66: Processing conditions/microstructure/flexural properties relationship. *Adv. Polym. Technol.* **2005**, *24*, 114–131. [CrossRef]
39. Vincent, M.; Giroud, T.; Clarke, A.; Eberhardt, C. Description and modeling of fiber orientation in injection molding of fiber reinforced thermoplastics. *Polymer* **2005**, *46*, 6719–6725. [CrossRef]
40. Mortazavian, S.; Fatemi, A. Fatigue behavior and modeling of short fiber reinforced polymer composites including anisotropy and temperature effects. *Int. J. Fatigue* **2015**, *77*, 12–27. [CrossRef]
41. SadAbadi, H.; Ghasemi, M. Effects of Some Injection Molding Process Parameters on Fiber Orientation Tensor of Short Glass Fiber Polystyrene Composites (SGF/PS). *J. Reinf. Plast. Compos.* **2007**, *26*, 1729–1741. [CrossRef]
42. Santharam, P.; Marco, Y.; Le Saux, V.; Le Saux, M.; Robert, G.; Raoult, I.; Guévenoux, C.; Taveau, D.; Charrier, P. Fatigue criteria for short fiber-reinforced thermoplastic validated over various fiber orientations, load ratios and environmental conditions. *Int. J. Fatigue* **2020**, *135*, 105574. [CrossRef]
43. Abdo, D.; Gleadall, A.; Silberschmidt, V. Damage and damping of short-glass-fibre-reinforced PBT composites under dynamic conditions: Effect of matrix behaviour. *Compos. Struct.* **2019**, *226*, 111286. [CrossRef]
44. Fouchier, N.; Nadot-Martin, C.; Conrado, E.; Bernasconi, A.; Castagnet, S. Fatigue life assessment of a Short Fibre Reinforced Thermoplastic at high temperature using a Through Process Modelling in a viscoelastic framework. *Int. J. Fatigue* **2019**, *124*, 236–244. [CrossRef]
45. Dar, U.A.; Xu, Y.J.; Zakir, S.M.; Saeed, M.-U. The effect of injection molding process parameters on mechanical and fracture behavior of polycarbonate polymer. *J. Appl. Polym. Sci.* **2017**, *134*, 44474. [CrossRef]

46. Farotti, E.; Natalini, M. Injection molding. Influence of process parameters on mechanical properties of polypropylene polymer. A first study. *Procedia Struct. Integr.* **2018**, *8*, 256–264. [CrossRef]
47. Bernasconi, A.; Davoli, P.; Filippi, A. Effect of fibre orientation on the fatigue behaviour of a short glass fibre reinforced polyamide-6. *Int. J. Fatigue* **2007**, *29*, 199–208. [CrossRef]
48. Mrzljak, S.; Delp, A.; Schlink, A.; Zarges, J.-C.; Hülsbusch, D.; Heim, H.-P.; Walther, F. Constant Temperature Approach for the Assessment of Injection Molding Parameter Influence on the Fatigue Behavior of Short Glass Fiber Reinforced Polyamide 6. *Polymers* **2021**, *13*, 1569. [CrossRef]
49. Makarov, I.S.; Golova, L.K.; Smyslov, A.G.; Vinogradov, M.I.; Palchikova, E.E.; Legkov, S.A. Flax Noils as a Source of Cellulose for the Production of Lyocell Fibers. *Fibers* **2022**, *10*, 45. [CrossRef]
50. Klemm, D.; Heublein, B.; Fink, H.-P.; Bohn, A. Cellulose: Faszinierendes Biopolymer und nachhaltiger Rohstoff. *Angew. Chem.* **2005**, *117*, 3422–3458. [CrossRef]
51. Zarges, J.-C. *Charakterisierung des Bruchverhaltens von Polypropylen-Celluloseregeneratfaser-Verbunden*; Kassel University Press: Kassel, Germany, 2018.
52. Fink, H.-P.; Ganster, J. Novel Thermoplastic Composites from Commodity Polymers and Man-Made Cellulose Fibers. *Macromol. Symp.* **2006**, *244*, 107–118. [CrossRef]
53. Adusumali, R.-B.; Reifferscheid, M.; Weber, H.; Roeder, T.; Sixta, H.; Gindl, W. Mechanical Properties of Regenerated Cellulose Fibres for Composites. *Macromol. Symp.* **2006**, *244*, 119–125. [CrossRef]
54. Ganster, J.; Fink, H.-P.; Pinnow, M. High-tenacity man-made cellulose fibre reinforced thermoplastics—Injection moulding compounds with polypropylene and alternative matrices. *Compos. Part A Appl. Sci. Manuf.* **2006**, *37*, 1796–1804. [CrossRef]
55. Bax, B.; Müssig, J. Impact and tensile properties of PLA/Cordenka and PLA/flax composites. *Compos. Sci. Technol.* **2008**, *68*, 1601–1607. [CrossRef]
56. Jiang, G.; Huang, W.; Li, L.; Wang, X.; Pang, F.; Zhang, Y.; Wang, H. Structure and properties of regenerated cellulose fibers from different technology processes. *Carbohydr. Polym.* **2012**, *87*, 2012–2018. [CrossRef]
57. Erdmann, J.; Ganster, J. Einfluss des Faserdurchmessers auf die Struktur und Mechanik Cellulosefaser verstärkter PLA-Komposite. *Lenzing. Ber.* **2011**, *89*, 91–102.
58. CORDENKA GmbH & Co. KG. *High-Tenacity Rayon Filament: Profile of Premium Rayon Reinforcement for High Performance Tires*; CORDENKA GmbH & Co. KG.: Obernburg am Main, Germany, 2013.
59. CORDENKA GmbH & Co. KG. *Performance Rayon Reinforcement: For Brake Hoses and Industrial Hoses*; CORDENKA GmbH & Co. KG.: Obernburg am Main, Germany, 2013.
60. Ganster, J.; Lehmann, A.; Erdmann, J.; Fink, H.-P. *Biobasierte Technische Fasern und Composite*; Fraunhofer Society: Wolfsburg, Germany, 2012.
61. Klemm, D.; Schmauder, H.-P.; Heinze, T. Cellulose. In *Biopolymers: Biology, Chemistry, Biotechnology, Applications: Lignin, Humic Substances and Coal*; Steinbüchel, A., Matsumura, S., Hofrichter, M., Koyama, T., Vandamme, E.J., Baets, S.d., Fahnestock, S.R., Eds.; Wiley-VCH: Weinheim, Germany, 2005.
62. Hatakeyama, T.; Ikeda, Y.; Hatakeyama, H. Effect of bound water on structural change of regenerated cellulose. *Makromol. Chem.* **1987**, *188*, 1875–1884. [CrossRef]
63. Müssig, J. *Industrial Applications of Natural Fibres: Structure, Properties and Technical Applications*; John Wiley & Sons: New York, NY, USA, 2010.
64. Lee, S.B.; Kim, I.H.; Ryu, D.D.; Taguchi, H. Structural properties of cellulose and cellulase reaction mechanism. *Biotechnol. Bioeng.* **1983**, *25*, 33–51. [CrossRef]
65. Le Baillif, M.; Oksman, K. The Effect of Processing on Fiber Dispersion, Fiber Length, and Thermal Degradation of Bleached Sulfite Cellulose Fiber Polypropylene Composites. *J. Thermoplast. Compos. Mater.* **2009**, *22*, 115–133. [CrossRef]
66. Caulfield, D.F.; Jacobson, R.E.; Sears, K.D.; Underwood, J.H. Fiber reinforced engineering plastics. In Proceedings of the 2nd International Conference on Advanced Engineered Wood Composites, Bethel, ME, USA, 14–16 August 2001.
67. Ho, M.-P.; Wang, H.; Lee, J.-H.; Ho, C.-K.; Lau, K.-T.; Leng, J.; Hui, D. Critical factors on manufacturing processes of natural fibre composites. *Compos. Part B Eng.* **2012**, *43*, 3549–3562. [CrossRef]
68. Thomas, S.; John, M. Biofibres and biocomposites. *Carbohydr. Polym.* **2008**, *71*, 343–364.
69. Zhou, Y.; Mallick, P.K. Effects of melt temperature and hold pressure on the tensile and fatigue properties of an injection molded talc-filled polypropylene. *Polym. Eng. Sci.* **2005**, *45*, 755–763. [CrossRef]
70. Zarges, J.-C.; Sälzer, P.; Feldmann, M.; Heim, H.-P. Determining Viscosity Directly in the Injection Molding Process. *Kunstst. Int.* **2016**, *106*, 106–109.
71. Tseng, H.-C.; Chang, R.-Y.; Hsu, C.-H. Predictions of fiber concentration in injection molding simulation of fiber-reinforced composites. *J. Thermoplast. Compos. Mater.* **2018**, *31*, 1529–1544. [CrossRef]
72. Mondy, L.A.; Brenner, H.; Altobelli, S.A.; Abbott, J.R.; Graham, A.L. Shear-induced particle migration in suspensions of rods. *J. Rheol.* **1994**, *38*, 444–452. [CrossRef]

73. Folgar, F.; Tucker, C.L. Orientation behavior of fibers in concentrated suspensions. *J. Reinf. Plast. Compos.* **1984**, *3*, 98–119. [CrossRef]
74. Cosmi, F.; Bernasconi, A. Fatigue Behaviour of Short Fibre Reinforced Polyamide: Morphological and Numerical Analysis of Fibre Orientation Effects. *Forni Sopra.* **2009**, *17*, 6–10.

Disclaimer/Publisher's Note: The statements, opinions and data contained in all publications are solely those of the individual author(s) and contributor(s) and not of MDPI and/or the editor(s). MDPI and/or the editor(s) disclaim responsibility for any injury to people or property resulting from any ideas, methods, instructions or products referred to in the content.

Article

Modification of Polyamide 66 for a Media-Tight Hybrid Composite with Aluminum

Fabian Lins, Christian Kahl *, Jan-Christoph Zarges and Hans-Peter Heim

Institute of Material Engineering, Polymer Engineering, University of Kassel, 34125 Kassel, Germany
* Correspondence: c.kahl@uni-kassel.de

Abstract: Metal–plastic composites are becoming increasingly important in lightweight construction. As a combination, e.g., for transmission housings in automobiles, composites made of die-cast aluminum housings and Polyamide 66 are a promising material. The interface between metal and plastic and the properties of the plastic component play an important role with regard to media tightness against transmission oil. The mechanical properties of the plastic can be matched to aluminum by glass fibers and additives. In the case of fiber-reinforced plastics, the mechanical properties depend on the fiber length and their orientation. These structural properties were investigated using computer tomography and dynamic image analysis. In addition to the mechanical properties, the thermal expansion coefficient was also investigated since a strongly different coefficient of the joining partners leads to stresses in the interface. Polyamide 66 was processed with 30 wt% glass fibers to align the mechanical and thermal expansion properties to those of aluminum. In contrast to the reinforcement additives, an impact modifier to improve the toughness of the composite, and/or a calcium stearate to exert influence on the rheological behavior of the composite, were used. The combination of the glass fibers with calcium stearate in Polyamide 66 led to high stiffnesses (11,500 MPa) and strengths (200 MPa), which were closest to those of aluminum. The coefficient of thermal expansion was found to be 6.6×10^{-6}/K for the combination of Polyamide 66 with 30 wt% glass fiber and shows a low expansion exponent compared to neat Polamid 66. It was detected that the use of an impact modifier led to less orientated fibers along the injection direction, which resulted in lower modulus and strength in terms of mechanical properties.

Keywords: plastic–metal hybrids; fiber orientation; X-ray microtomography; dynamic image analysis; thermal expansion exponent

Citation: Lins, F.; Kahl, C.; Zarges, J.-C.; Heim, H.-P. Modification of Polyamide 66 for a Media-Tight Hybrid Composite with Aluminum. *Polymers* **2023**, *15*, 1800. https://doi.org/10.3390/polym15071800

Academic Editors: Abderrahmane Ayadi, Patricia Krawczak and Chung Hae Park

Received: 6 March 2023
Revised: 27 March 2023
Accepted: 3 April 2023
Published: 6 April 2023

Copyright: © 2023 by the authors. Licensee MDPI, Basel, Switzerland. This article is an open access article distributed under the terms and conditions of the Creative Commons Attribution (CC BY) license (https://creativecommons.org/licenses/by/4.0/).

1. Introduction

The combination of two materials makes it possible to combine the positive properties of both partners [1]. Metal–plastic hybrid composites are therefore a promising way to reduce the weight of, for example, automobiles and, thus, CO_2 emissions [2]. Due to their high mechanical properties, fiber-reinforced plastics and aluminum, with their low densities, play a major role in this material combination [3].

In-mold technology has become established for the production of metal–plastic composites. A metal insert is placed in an injection mold and the plastic is injected with an injection molding machine [4,5]. One challenge in combining the two materials is the interface between the metal and the plastic, as these have different mechanical and thermal characteristic values [3]. When metal–plastic joints are produced by injection molding, the quality of the finished part significantly depends on the adhesion of the composite. The metal part is often pretreated by chemical etching or sandblasting to generate a rough surface [6]. The rough surface creates strong bonding by mechanically anchoring the melt to the pretreated surface [7].

Especially for the automotive industry, a strong bond between metal and plastic is essential. The substitution of non-load-bearing areas in a die-cast housing with plastic must also provide tightness against media such as transmission oil on the interface between plastic and metal.

The influence of additives, such as an impact modifier or a lubricant, has already been demonstrated in combination with Polyamide 6 [8]. The viscosity of the polymer can be influenced by the impact modifier and the calcium stearate. The mechanical properties of the polyamide are increased and brought as close as possible to those of aluminum by the addition of glass fibers. This study investigated to what extent the impact modifier and calcium stearate influence the fiber length and the fiber orientation.

For a media-tight bond between Polyamide 66 and aluminum, the interface plays an important role. In order to bond the two joining partners as strongly as possible, the surface is roughened and/or an coupling agent is used. Ultimately, however, the failure of the bond is caused by stress peaks caused by brittle properties or different expansion coefficients. In this project, it will later be investigated whether a less brittle material can enable a long-lasting, tight connection between the joining partners, which can guarantee tightness even with small movements and vibrations due to higher ductility. For this reason, the fiber-reinforced material will be provided with additives such as a lubricant and/or an impact modifier and investigated with regard to the mechanical properties and the thermal expansion coefficient. This study can be seen as preliminary work for the further investigation of composite Polyamide 66 and aluminum.

The examination of glass fibers in a selected volume with regard to fiber length and fiber orientation is possible through 3-dimensional analysis by X-ray microtomography [9,10]. Due to the high density of the glass fiber of 2.6 g/cm^3, it can be easily separated from the polyamide with a density of 1.15 g/cm^3. In 2015, Nguyen Thi et al. investigated glass fibers in Polyamide 6 using X-ray microtomography and were able to visualize and evaluate the fibers in terms of length and their orientation at different locations within a test specimen [11].

In addition to the mechanical properties, the coefficient of thermal expansion plays a major role with regard to the bond strength. If the coefficient of thermal expansion of the two components is very different, the different behavior during cooling has a negative effect on the bond strength. A similar behavior of the two joining partners means a low interfacial stress. For this reason, fiber-reinforced polymers are used for applications in joining metal and plastic in order to obtain high dimensional stability under thermal influences, and during cooling after injection molding [12,13]. Heckert et al. were able to show that the coefficient of thermal expansion decreased as the weight content of fibers in a fiber-reinforced polymer increased. The coefficient of thermal expansion for Polyamide 6 is 85×10^{-6}/K and could be reduced to 50×10^{-6}/K by adding 60 wt% glass fiber [14].

2. Materials and Methods

2.1. Materials

The base polymer used for this study was Polyamide 66 (PA66) from BASF (Ludwigshafen, Germany). The grade Ultramid A27E has a density of 1.14 g/cm^3 and is a special grade for compounding. Before processing, the polyamide was dried in a Dry Jet Easy Dryer from TORO-Systems (Igensdorf, Germany) for at least 4 h at 80 °C to achieve a moisture content of max 0.2%. The PA66 was used as a non-colored grade.

To reinforce the polyamide, glass fibers were added as short fibers with a content of 30 wt% in the compounding process. The glass fibers of type CS 7928 were purchased from Lanxess Germany GmbH (Cologne, Germany). The fibers had an initial length of 4.5 mm and a fiber diameter of 11 μm. A sizing on the surface of this fiber type promised good adhesion to polyamides.

The use of an impact modifier leads to an increase in the toughness of a plastic and was, therefore, used here to modify the properties. An impact modifier from Kraton Polymers LLC (Houston, TX, USA) was added to the polyamide in two components. One component of the modifier was type FG1901, which consists of a linear triblock copolymer based on styrene and ethylene/butylene with a polystyrene content of 30%. This component was added with a weight content of 8%. The second component was type G1657, which is a linear triblock copolymer based on styrene and ethylene/butylene with a polystyrene content of 13%. This component was added with a weight content of 12%. Both components had a density of 0.9 g/cm^3. The weight content of the impact modifier was 20% in total. The components and their weight contents were selected according to a recommendation of the manufacturer [15].

The lubricant calcium stearate from the Faci Group (Genova, Italy) was also used to reduce the viscosity of the polyamide melt. The salt is also used as a lubricant in pharmaceutical products and as a lubricant in the paper and metal processing industries. It has a melting point of 150 °C and a density of 2.6 g/cm^3. The stearate was used with a weight content of 0.1%. After consultation with the manufacturer, it was advised that a higher concentration would have a negative effect on the adhesion properties between the plastic compound and the aluminum.

2.2. Compounding

The compounds were produced on a co-rotating twin-screw extruder from Leistritz (Nuremberg, Germany), type ZSE18 HPe. The barrel has a diameter of 18 mm and an L/D ratio of 40. The extruder is equipped with a screw configuration that has a high proportion of conveying elements. This configuration is intended to generate lower shear energy in the melt to prevent severe fiber shortening. The impact modifier and the lubricant were mixed into the PA66 granules and added to the compounding process by a gravimetric feeder from Brabender (Duisburg, Germany). After melting of the granules, the glass fibers were also added to the process by a gravimetric feeder. The melt temperature during compounding was 280 °C and the screw speed was 200 rpm. The extruded material strand was cooled with compressed air after exiting the die and then cut into granules with a length of 3 mm in a pelletizer of the type Scheer SGS 25-E (Grossostheim, Aschaffenburg, Germany). The material composition of the PA66, the additives, and the glass fibers can be seen in Table 1.

Table 1. Compositions of charges produced by compounding.

	PA66	GF	Impact Modifier		Calcium Stearate
			Kraton FG1901	Kraton G1657	CaSt
	wt%	wt%	wt%	wt%	wt%
PA66/30GF	70	30	-	-	-
PA66/30GF/IM	50	30	8	12	-
PA66/30GF/CaSt	69.9	30	-	-	0.1
PA66/30GF/IM/CaSt	49.9	30	8	12	0.1

In the following table and in the results section, the compounds are described with the abbreviations glass fiber (GF), impact modifier (IM), and calcium stearate (CaSt).

2.3. Injection Molding

The specimens, according to DIN EN ISO 527 Type 1A, were injection-molded for mechanical characterization and X-ray microtomography (μ-CT). An injection molding machine from Arburg (Loßburg, Germany) type 320C Golden Edition was used for this purpose. The machine has a clamping force of 500 kN. The temperature along the screw was set to 290 °C and the injection speed was set to 16 cm^3/s. The mold temperature was set to 80 °C. All compounds were dried at 80 °C for at least 4 h before processing.

2.4. Tensile Testing

Tensile tests to DIN EN ISO 527 were performed on a universal testing machine, type Z010 from Zwick Roell (Ulm, Germany). The tests were performed at a speed of 5 mm/min and the Young's modulus, tensile strength, and elongation at break were evaluated.

2.5. Dynamic Image Analysis (DIA)

For dynamic image analysis, the fibers were separated from the matrix material by ashing the samples at 600 °C for 6 h. The system QICPIC/R06 (Sympatec, Clausthal-Zellerfeld, Germany) with a MIXCEL wet dispersion unit was used to measure the fiber length distribution. The fibers were dispersed in isopropanol. The images were acquired at a rate of 175 Hz and a resolution of 4.2 MP. A cuvette with a width of 0.5 mm was used and the fibers were measured with an M5 objective. This objective records fibers with a length of 1.8 µm to 3700 µm. For each sample, 3 measurements of 60 s were performed. Approximately 10,000 fibers were measured for each measurement. The fibers were evaluated with the LEFI (length of fiber) module, which measures the shortest distance between the endpoints of the fiber.

2.6. X-ray Microtomography

A volume shown in Figure 1 was examined by X-ray microtomography to compare the fiber length in the volume with the fiber lengths from DIA and to examine the fiber orientation along the flow direction. For this purpose, a Zeiss Xradia Versa 520 (Oberkochen, Germany) was used. The measurements were performed at a voltage of 80 kV and a current of 87.2 µA. A volume of size 4 mm × 5 mm × 12 mm from the middle of a tensile specimen was recorded, as shown in Figure 1, with a 5279 µm field of view and a pixel size of 5.2 µm. Avizo 9.4 software was used to compose the 1600 taken images and a volume of 5 mm × 2 mm × 2 mm was used to display the optical fibers and to evaluate them in terms of orientation and length. The flow direction during mold filling corresponds to the z-axis. The deviation of the fiber from the z-axis is represented with an angle theta (θ). Thus, a fiber with an angle $\theta = 0°$ is oriented along the z-axis and an angle of $\theta = 90°$ shows that the fiber is oriented orthogonally to the z-axis. About 60,000 fibers were evaluated in the volume of the specimen. Fiber length in X-ray microtomography was evaluated to provide a reference for the results of the DIA.

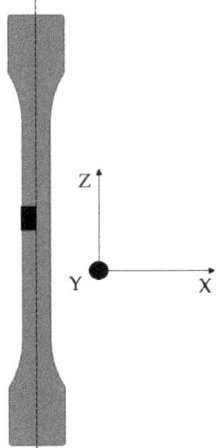

Figure 1. Evaluated volume of tensile specimen according to DIN EN ISO 527 for orientation and length of glass fiber.

2.7. Dynamic Mechanical Analysis

The coefficient of thermal expansion was determined by dynamic mechanical analysis (DMA) with variation in the temperature and a constant load. A Q 800 module from TA Instruments (Hüllhorst, Germany) was used for this purpose. The specimens for the measurements had dimensions of 30 mm × 10 mm × 4 mm and the clamp distance was 20 mm. The coefficient of expansion was measured in a temperature range from 25 to 100 °C and the heating rate was 3 °C/min. A constant load of 10 N was applied to the specimen. For the evaluation, the average coefficient of linear expansion ($\bar{\alpha}$) was determined using the following formula:

$$\bar{\alpha} = dL/(dT * L_0) \tag{1}$$

The mean coefficient of thermal expansion is represented by $\bar{\alpha}$ taken from the first and last measuring points (dL) according to Figure 2. This is divided by the temperature difference (dT) and the original length (L_0). Three specimens were tested for each compound.

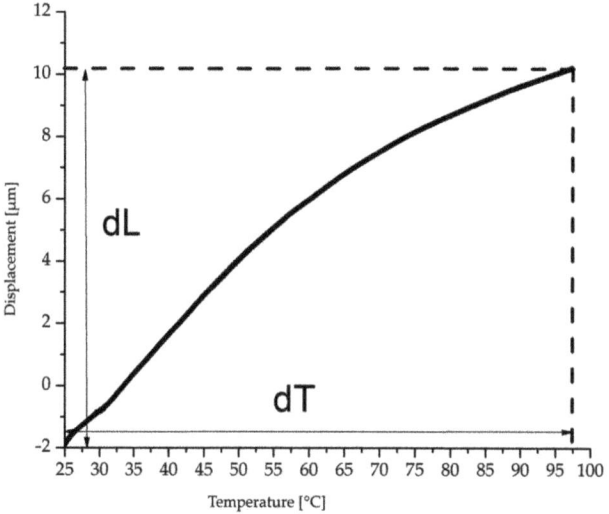

Figure 2. Displacement and temperature course of DMA measurements with constant load at 10 N.

3. Results and Discussion

3.1. Tensile Test

For the evaluation of the results, tensile test specimens were first manufactured according to DIN EN ISO 527 and tested in tensile tests. The use of various additives, such as an impact modifier or a lubricant, changes the viscosity of the melt and the flow properties. In this study, the mechanical properties as well as the glass fiber orientation and the length of the fibers in the test specimens were demonstrated on Polyamide 66 with the use of the individual additives and considering the interaction of both additives.

Figure 3 shows the results of the tensile tests. An increase in the elongation at break was clearly visible between the samples PA66/30GF and PA66/30GF/IM. By adding a weight fraction of 20 wt% of the impact modifier to the compound PA66/30GF, the elongation at break increased from 3.1% to 3.7%, while the tensile strength decreased from 185 MPa to 137 MPa. On the one hand, this may have been due to the change in the chemical structure caused by the impact modifier. The modifier largely consisted of styrene and formed a two-phase mixture due to its incompatibility with polyamide [16]. On the other hand, it is possible that the impact modifier changed the bond of the fiber to the matrix. If the fiber is in contact with the impact modifier, the adhesion to it is worse than to the PA66. As a result, the reinforcing effect is reduced due to lower force transmission.

The elongation at break of the PA66/30GF/IM/CaSt compound was also similar to that of the compound PA66/30GF/IM, accompanied by a reduction in tensile strength. The use of both additives was, therefore, dominated by the high degree of toughness of the impact modifier. The compounds contained 0.1% by weight of CaSt. Such a small amount was sufficient to have an influence on the flow properties of the polyamide melt. In terms of mechanical properties, a slight increase in tensile strength was seen for the compound

PA66/30GF/CaSt at 200 MPa compared to the PA66/30GF at 185 MPa. The Young's modulus was also increased from 10,650 MPa to 11,500 MPa by using calcium stearate. The flow properties of the compound with calcium stearate led to better orientation of the fiber to the flow direction, which was reflected in enhanced mechanical properties [17].

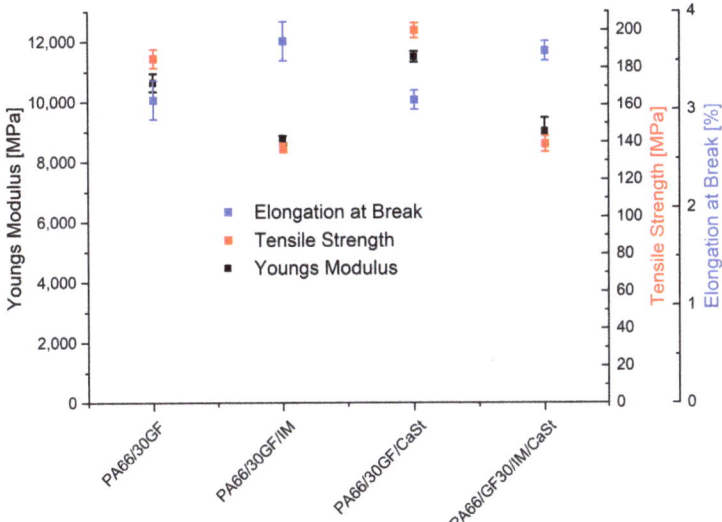

Figure 3. Results of tensile test with PA66/30GF-based specimen.

For a media-tight composite of aluminum and PA66, the approach used was to make the reinforced PA66 tougher by means of additives so that the media-tight composite could be maintained even in the case of vibrations and small movements in the interface. For the later tests of the composite, variation in the mechanical properties was, therefore, possible with the additives used here. The use of an impact modifier was able to increase the toughness of the plastic even with a fiber weight content of 30%.

3.2. Dynamic Image Analysis (DIA)

The fibers detached from the matrix by ashing were measured by DIA to compare the results with the fiber lengths measured from μ-CT analysis. The results of the DIA are shown in the diagrams of Figure 4. Overall, it can be seen that the fibers based on the results of DIA were longer than the fiber lengths from the X-ray microtomography. This was evident at the 75th and 90th percentiles. For the compound PA66/30GF, the 75th percentile was 260 μm when evaluated by X-ray microtomography and approximately 310 μm in the DIA. The results of the other measurements with IM or CaSt were similar. In the evaluation of the DIA results, as well as in the evaluation by the X-ray microtomography, no great change in fiber length was seen. The compound PA66/30GF/CaSt only showed slightly shorter fibers, especially at the high percentiles (75th and 90th), compared to the PA66/30GF/IM/CaSt and PA66/30GF/IM compounds. This can be explained by a decrease in the viscosity [11]. This resulted in more fiber breakage and, thus, in shorter fiber lengths.

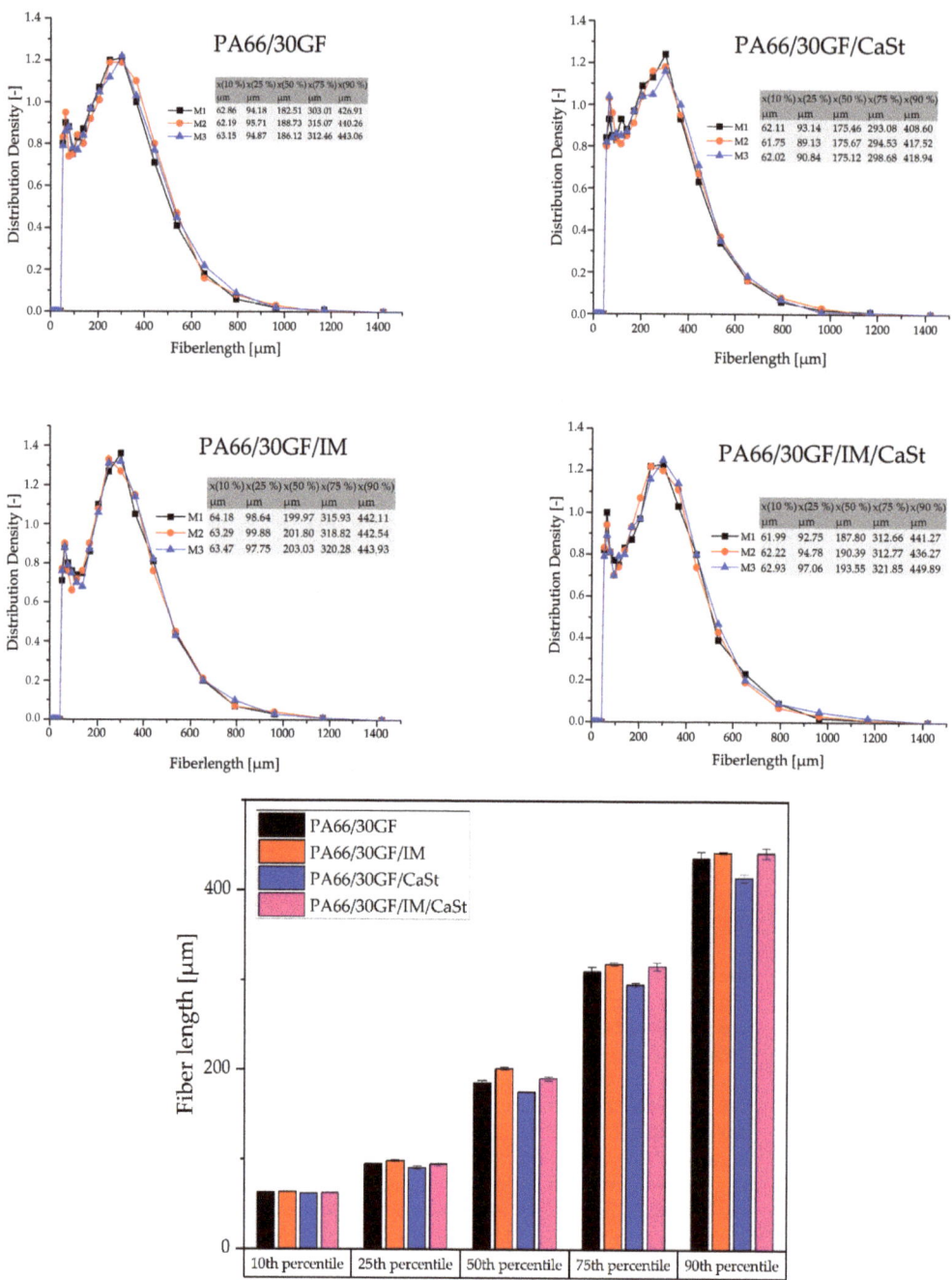

Figure 4. Results of the distribution density of measured fiber length from dynamic image analysis.

3.3. X-ray Microtomography

The investigated volume from a tensile specimen included the edge region as well as the center of the specimen in order to investigate all regions with respect to orientation and fiber length.

Figure 5 shows the fiber lengths determined by X-ray microtomography. The differences between the four compounds were very small. The results visualized as boxplots showed a slight fiber shortening in the compound PA66/30GF/CaSt. In particular, a difference was observed at the 75th and 90th percentiles. The differences, especially in the long fibers of the 75th and 90th percentiles, showed that the use of an impact modifier influenced the process-induced shortening of the fiber length. In this compound, the fibers were the shortest. On the other hand, the compounds with impact modifier PA66/30GF/IM and PA66/30GF/IM/CaSt had longer fibers, especially at the 75th and 90th percentiles, due to the tough properties of the impact modifier. Fiber shortening was less for these compounds.

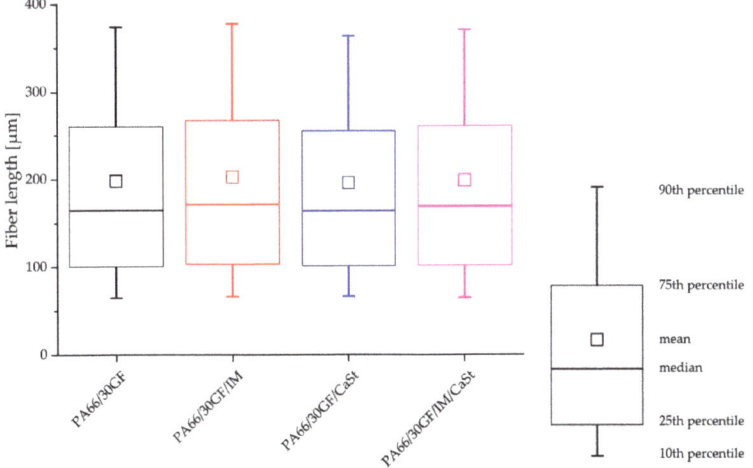

Figure 5. Fiber length in tensile specimen shown in boxplots from 90th percentile to 10th percentile.

The figure shows that a difference in fiber length along all compounds was observed in the higher values of the percentiles. At the 10% and 25% percentiles, no difference in fiber length was visible.

In addition to the fiber length, the fiber orientation was investigated since the fiber orientation also has a strong influence on the mechanical properties. In Figure 1, the z-axis specifies the flow direction. The orientation of a fiber is described by an axis of its cylindrical shape. The deviation of this axis from the flow direction is described by the angle theta (θ). The smaller the angle theta, the better the fiber is oriented in the flow direction [18].

The orientation of the fibers can be seen in Figure 6, in which the fibers are colored according to their orientation. Blue shows the fibers well-aligned to the z-axis and those red-colored are the fibers with an orientation orthogonal to the flow direction. The figure shows a volume that was separated from a tensile specimen and examined according to Figure 1. In all four images in Figure 6, a very good orientation of the fibers can be seen in the edge region. This can be explained by the swelling flow during injection molding and has already been reported in numerous publications. In the volumes from specimens without an impact modifier, good orientation of the blue colored fibers can be seen over the whole examined area. In the case of the specimens from PA66/30GF/IM and PA66/30GF/IM/CaSt, it can be seen that the fibers were less oriented to the flow direction in the center of the specimen than in the edge region. A fiber with an orientation perpendicular to the flow direction absorbed less stress than a fiber with an orientation in the flow direction since the flow direction also corresponds to the loading direction in the tensile test. The orientation to the flow direction decreased towards the center for the samples mentioned.

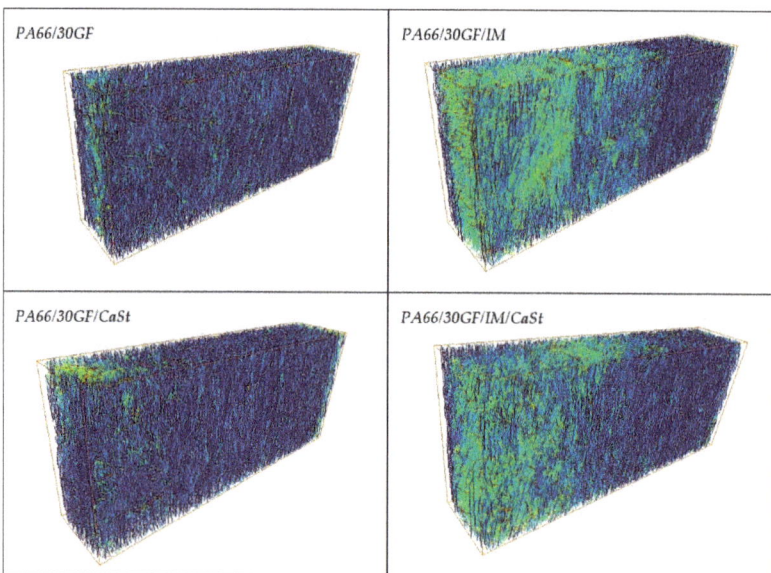

Figure 6. Fiber orientation of different PA66 compositions.

The fiber orientation is also illustrated quantitatively in bar graphs in Figure 7. The fibers oriented to the flow direction are shown on the left of the horizontal axis in these diagrams; the perpendicular oriented fibers are on the right. It can also be seen that a high number of fibers in the specimens without impact modifier were oriented along the flow direction. With this orientation, the fibers can absorb the forces under a tensile load in the z-direction (Figure 1) much better than fibers that are oriented perpendicular to the flow direction. The bar graphs also show that a smaller number of fibers were oriented along the flow direction in the specimens with impact modifier. This is also confirmed by the pictures in Figure 6.

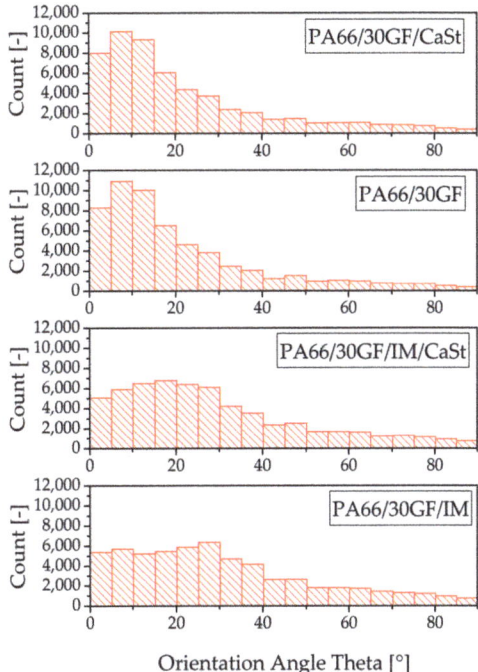

Figure 7. Fiber orientation in tensile specimen shown from 0° to 90°.

The fiber orientation of the compounds with and without impact modifier also provides an explanation for the results of the tensile test in Figure 3. Here, the compounds with impact modifier showed lower stiffness and strength than the compounds without impact modifier. This was due, on the one hand, to the lower number of fibers oriented to the loading direction, and, on the other hand, to the toughness of the impact modifier. The mechanical properties of differently oriented fibers had previously been investigated by Zarges et al. In their study, specimens with fibers oriented along the direction of loading were also observed to be able to absorb a higher force than specimens with a fiber orientation transverse to the direction of loading [19].

3.4. Dynamic Mechanical Analysis

The thermal expansion of the two joining partners is of great importance after processing by injection molding. Plastic has a significantly higher coefficient of thermal expansion than aluminum. This difference results in stresses in the interface between the two joining partners, which, in turn, had a negative effect on adhesion, and, as a result, on the tightness at the interface between the plastic and the metal. For this reason, the coefficient of thermal expansion of the plastic should be brought into line with that of the aluminum [12,14,20]. Figure 8 shows the coefficients of expansion of the modified PA66, and, in comparison, the coefficient of expansion of neat PA66. In the temperature range from 25 °C to 100 °C, aluminum has a very low thermal expansion coefficient compared with plastics [21]. The coefficient can be reduced by fiber reinforcement with GF [12,20]. It can be seen that the compound PA66/30GF achieved the lowest coefficient of thermal expansion with 6.6×10^{-6}/K. The specimens for the determination of the coefficient of thermal expansion were prepared in the same way as for the tensile tests for the injection-molded specimens according to DIN

EN ISO 527. The fiber orientation in these tests was, therefore, exactly as shown in Figure 7. The addition of additives to the compound PA66/30GF led to an increase in the coefficient of expansion. As expected, the neat PA66 had the highest coefficient of thermal expansion.

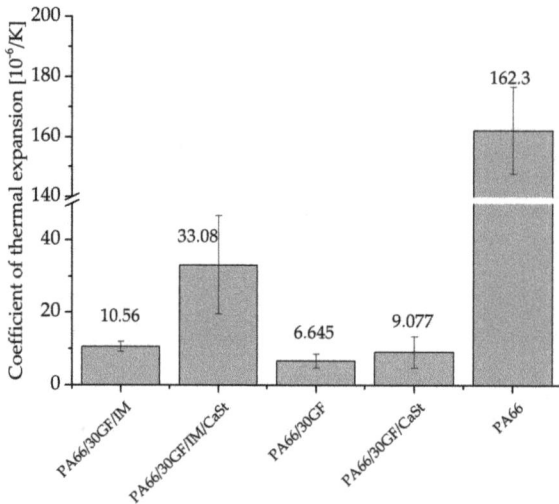

Figure 8. Thermal expansion exponent of compounds.

4. Conclusions

In this study, specimens were manufactured to investigate the tensile and thermal properties of PA66 reinforced with 30 wt% glass fibers. In addition to glass fibers, an impact modifier and/or a lubricant was added to vary the properties for investigation of a media-tight plastic-metal bond. To explain the mechanical results, a volume of a specimen was characterized by X-ray microtomography and the fibers were examined in terms of length and orientation. In addition, the coefficient of thermal expansion was investigated to compare the expansion of the modified PA66 with aluminum. The following conclusions can be drawn:

- The addition of an impact modifier to GF-reinforced PA66 resulted in a higher elongation at break and a reduction in the strength and stiffness of the compound. The addition of calcium stearate led to a small increase in stiffness and strength, which was closest to the properties of aluminum.
- The fiber length measured with X-ray microtomography was only very slightly affected by the addition of the additives. In the case of the short fibers with a length of approx. 100 µm, no influence of the additives could be detected. For longer fibers of 250 µm to 400 µm, the addition of an impact modifier had a minor effect on increase in fiber length. The results of DIA produced similar results to those of X-ray microtomography. The fiber length was slightly higher in the DIA results. In summary, no significant change in fiber length due to the addition of the additives used in this study was responsible for changes in the mechanical properties.
- The impact modifier had an influence on the fiber orientation. Towards the center of specimens with impact modifier, the fibers were oriented in parallel less than in specimens without impact modifier. At the edge region, the fibers were well oriented along the flow direction due to the swell flow. The fiber orientation to the flow direction showed a clear dependence on the mechanical properties. Fibers oriented to the direction of flow showed higher strength and stiffness in the composite than fibers whose orientation deviated from the direction of flow.

- The use of individual additives led to an increase in the coefficient of thermal expansion. As a result, the very low coefficient of thermal expansion of PA66/30GF (6.6×10^{-6}/K) was able to be brought into line with the very low coefficient of aluminum, which led to low stresses in the interface and contributed to a media-tight bond. For low stresses in the interface between aluminum and GF reinforced PA66, the addition of the additives used here was counterproductive as they increased the thermal expansion coefficient.

Author Contributions: Conceptualization, C.K.; methodology, F.L.; validation, C.K. and F.L.; formal analysis, F.L.; investigation, C.K. and F.L.; resources, H.-P.H.; data curation, F.L.; writing—original draft preparation, C.K.; writing—review and editing, J.-C.Z.; visualization, C.K. and F.L.; supervision, H.-P.H. All authors have read and agreed to the published version of the manuscript.

Funding: The results presented in this paper were developed as part of the "TightHybrid" research project funded by the German Federal Ministry of Education and Research (BMBF). The Grant No is 03XP0366C.

Institutional Review Board Statement: Not applicable.

Data Availability Statement: Not applicable.

Conflicts of Interest: The authors declare no conflict of interest.

References

1. Bräuer, M.; Edelmann, M.; Häußler, L.; Kühnert, I. Metall-Kunststoff-Verbunde: Untersuchungen zur Wirkungsweise einer Adhäsionsschicht aus Uretdionpulverlacksystemen. *Mater. Und Werkst.* **2012**, *6*, 534–543. [CrossRef]
2. Arai, S.; Sugawara, R.; Shimizu, M.; Inoue, J.; Horita, M.; Nagaoka, T.; Itabashi, M. Excellent bonding strength between steel and thermoplastic resin using roughened electrodeposited Ni/CNT composite layer without adhesives. *Mater. Lett.* **2020**, *263*, 127241. [CrossRef]
3. Bonpain, B.; Stommel, M. Influence of surface roughness on the shear strength of direct injection molded plastic-aluminum hybrid-parts. *Int. J. Adhes. Adhes.* **2018**, *82*, 290–298. [CrossRef]
4. Kleffel, T.; Drummer, D. Investigating the suitability of roughness parameters to assess the bond strength of polymer-metal hybrid structures with mechanical adhesion. *Compos. Part B Eng.* **2017**, *117*, 20–25. [CrossRef]
5. Grujicic, M.; Sellappan, V.; Omar, M.A.; Seyr, N.; Obieglo, A.; Erdmann, M.; Holzleitner, J. An overview of the polymer-to-metal direct-adhesion hybrid technologies for load-bearing automotive components. *J. Mater. Process. Technol.* **2008**, *1–3*, 363–373. [CrossRef]
6. Yang, H.; Ng, B.C.; Yu, H.C.; Liang, H.H.; Kwok, C.C.; Lai, F.W. Mechanical properties study on sandwich hybrid metal/(carbon, glass) fiber reinforcement plastic composite sheet. *Adv. Compos. Hybrid. Mater.* **2022**, *1*, 83–90. [CrossRef]
7. Bula, K.; Sterzyński, T.; Piasecka, M.; Różański, L. Deformation Mechanism in Mechanically Coupled Polymer-Metal Hybrid Joints. *Materials* **2020**, *11*, 2512. [CrossRef] [PubMed]
8. Malakhov, S.N.; Belousov, S.I.; Bakirov, A.V.; Chvalun, S.N. Electrospinning of Non-Woven Materials from the Melt of Polyamide-6 with Added Magnesium, Calcium, and Zinc Stearates. *Fibre Chem.* **2015**, *1*, 14–19. [CrossRef]
9. Kahl, C.; Feldmann, M.; Sälzer, P.; Heim, H.-P. Advanced short fiber composites with hybrid reinforcement and selective fiber-matrix-adhesion based on polypropylene—Characterization of mechanical properties and fiber orientation using high-resolution X-ray tomography. *Compos. Part A Appl. Sci. Manuf.* **2018**, *111*, 54–61. [CrossRef]
10. Mrzljak, S.; Delp, A.; Schlink, A.; Zarges, J.-C.; Hülsbusch, D.H.H.-P.; Walther, F. Constant Temperature Approach for the Assessment of Injection Molding Parameter Influence on the Fatigue Behavior of Short Glass Fiber Reinforced Polyamide 6. *Polymers* **2021**, *10*, 1569. [CrossRef] [PubMed]
11. Nguyen Thi, T.B.; Morioka, M.; Yokoyama, A.; Hamanaka, S.; Yamashita, K.; Nonomura, C. Measurement of fiber orientation distribution in injection-molded short-glass-fiber composites using X-ray computed tomography. *J. Mater. Process. Technol.* **2015**, *219*, 1–9. [CrossRef]
12. Yin, S.; Xie, Y.; Li, R.; Zhang, J.; Zhou, T. Polymer–Metal Hybrid Material with an Ultra-High Interface Strength Based on Mechanical Interlocking via Nanopores Produced by Electrochemistry. *Ind. Eng. Chem. Res.* **2020**, *27*, 12409–12420. [CrossRef]
13. Hirahara, T.; Hidai, H. Fiber Implantation for Interfacial Joining of Polymer to Metal. *ACS Appl. Polym. Mater.* **2020**, *8*, 3049–3053. [CrossRef]
14. Heckert, A.; Zaeh, M.F. Laser surface pre-treatment of aluminum for hybrid joints with glass fiber reinforced thermoplastics. *J. Laser Appl.* **2015**, *S2*, S29005. [CrossRef]
15. Kraton Performance Polymers. Blends with Polyamides 2013; Data-Sheet. Available online: www.kraton.de (accessed on 30 March 2023).

16. Naeim Abadi, A.; Garmabi, H.; Hemmati, F. Toughening of polyamide 12/nanoclay nanocomposites by incompatible styrene-butadiene-styrene rubber through tailoring interfacial adhesion and fracture mechanism. *Adv. Polym. Technol.* **2018**, *6*, 2303–2313. [CrossRef]
17. Jiang, Q.; Liu, H.-s.; Wang, X.-c.; Hu, Z.-x.; Zhou, H.-y.; Nie, H.-r. Three-dimensional numerical simulation on fiber orientation of short-glass-fiber-reinforced polypropylene composite thin-wall injection-molded parts simultaneously accounting for wall slip effect and pressure dependence of viscosity. *J. Appl. Polym. Sci.* **2022**, *45*, e53129. [CrossRef]
18. Kahl, C.; Schlink, A.; Heim, H.-P. Pultruded Hybrid Reinforced Compounds with Glass/Cellulose Fibers in a Polybutylene Terephthalate Matrix: Property Investigation. *Polymers* **2022**, *6*, 1149. [CrossRef] [PubMed]
19. Zarges, J.-C.; Minkley, D.; Feldmann, M.; Heim, H.-P. Fracture toughness of injection molded, man-made cellulose fiber reinforced polypropylene. *Compos. Part A Appl. Sci. Manuf.* **2017**, *98*, 147–158. [CrossRef]
20. Kleffel, T.; Drummer, D. Electrochemical treatment of metal inserts for subsequent assembly injection molding of tight electronic systems. *J. Polym. Eng.* **2018**, *7*, 675–684. [CrossRef]
21. Kim, J.; Kordijazi, A.; Rohatgi, P. Thermal Expansion of Pressure Infiltrated Aluminum/Hollow Cenosphere Particulate Composites. *JOM* **2023**, *1*, 209–217. [CrossRef]

Disclaimer/Publisher's Note: The statements, opinions and data contained in all publications are solely those of the individual author(s) and contributor(s) and not of MDPI and/or the editor(s). MDPI and/or the editor(s) disclaim responsibility for any injury to people or property resulting from any ideas, methods, instructions or products referred to in the content.

Review

Rapid Impregnating Resins for Fiber-Reinforced Composites Used in the Automobile Industry

Mei-Xian Li [1,2,†], Hui-Lin Mo [1,†], Sung-Kwon Lee [3], Yu Ren [1,2], Wei Zhang [1,2] and Sung-Woong Choi [3,*]

1. School of Textile and Clothing, Nantong University, Nantong 226019, China; lmx321@ntu.edu.cn (M.-X.L.)
2. National and Local Joint Engineering Research Center of Technical Fiber Composites for Safety and Protection, Nantong University, Nantong 226019, China
3. Department of Mechanical System Engineering, Gyeongsang National University, Tongyeong-si 53064, Gyeongsangnam-do, Republic of Korea
* Correspondence: younhulje@gnu.ac.kr
† These authors contributed equally to this work.

Abstract: As environmental regulations become stricter, weight- and cost-effective fiber-reinforced polymer composites are being considered as alternative materials in the automobile industry. Rapidly impregnating resin into the reinforcing fibers is critical during liquid composite molding, and the optimization of resin impregnation is related to the cycle time and quality of the products. In this review, various resins capable of rapid impregnation, including thermoset and thermoplastic resins, are discussed for manufacturing fiber-reinforced composites used in the automobile industry, along with their advantages and disadvantages. Finally, vital factors and perspectives for developing rapidly impregnated resin-based fiber-reinforced composites for automobile applications are discussed.

Keywords: rapid impregnation; thermoset resins; thermoplastic resins; fiber-reinforced composites

Citation: Li, M.-X.; Mo, H.-L.; Lee, S.-K.; Ren, Y.; Zhang, W.; Choi, S.-W. Rapid Impregnating Resins for Fiber-Reinforced Composites Used in the Automobile Industry. *Polymers* 2023, 15, 4192. https://doi.org/10.3390/polym15204192

Academic Editors: Patricia Krawczak, Chung Hae Park and Abderrahmane Ayadi

Received: 11 August 2023
Revised: 16 October 2023
Accepted: 18 October 2023
Published: 23 October 2023

Copyright: © 2023 by the authors. Licensee MDPI, Basel, Switzerland. This article is an open access article distributed under the terms and conditions of the Creative Commons Attribution (CC BY) license (https://creativecommons.org/licenses/by/4.0/).

1. Introduction

Recently, lightweight automobiles have been increasingly used to save energy and reduce pollution in light of strict regulations. It has been reported that reducing automobile weight by 10 wt.% could decrease fuel consumption by 6–8% and reduce CO_2 emissions [1]. The key to solving this problem is to replace metal components with high-performance, fiber-reinforced, polymer-based composites. As for environmental issues, extensive research has been conducted on biocompatible and environmentally-friendly resins, including the use of the poly (ethylene glycol) diacrylate (PEGDA) monomer [2], their applications in dental materials [3], and their utilization in 3D printed materials [4–6]. This research focuses on polymer composite molding processes, and understanding these molding processes is essential.

Polymer composites, including thermosets and thermoplastic composites, have been used in panels, modules, structures, and other parts of automobiles after being reinforced with continuous or discontinuous fibers that have undergone liquid composite molding (LCM) processes, such as resin transfer molding (RTM), vacuum infusion, and reaction injection molding [7]. Takahashi et al. [8] reported that the weight of carbon fiber reinforced polymer (CFRP) composites is one third that of steel panels, and the flexural strength of CFRPs is approximately three times higher. As for the polymer composite, fiber-reinforced thermoset composites (FRTSCs) generally exhibit good mechanical properties, thermal stability, and dimensional stability [9]. Thermoset resins have a relatively low viscosity compared with thermoplastic resins, which is an important factor in the RTM process. Traditionally, depending on the part size and geometry, the cycle time of a standard RTM is 30–60 min with a 10–20 bar injection pressure, whereas that of high-pressure resin transfer molding is less than 10 min with a 20–120 bar injection pressure.

Fiber-reinforced thermoplastic composites (FRTPCs) are also extensively used because of their high processability and recyclability. However, the high viscosity of thermoplastic resins requires high temperatures and pressures for the materials to be impregnated with fiber reinforcements. For example, thermosetting resins, such as epoxy, can be impregnated with fiber reinforcements and cured below 200 °C, whereas thermoplastic resins must be heated above the melting temperature, which is typically well above 200 °C, and impregnated with high pressures (10–50 bar). Moreover, the high pressure and viscosity of the resins may cause a misalignment of the fiber reinforcements.

In the automobile industry, cost and cycle time reductions are key issues. In the case of thermoset resins, there is a need for research into fast-curing resins like rapid curing epoxies and endo-dicyclopentadiene. On the other hand, for thermoplastic resins, research is required on the development of rapidly impregnating resins that can be applied at low temperatures and low pressures. Regarding fast-curing thermoset resins, recent research by Zhang et al. [10], Odom et al. [11], Gan et al. [12], and Reichanadter et al. [13] presented the application of fast-curing epoxy resins and its processes. Boros, Róbert, et al. [14], Ota et al. [15], and Willicombe, K., et al. [16] showed the rapid impregnation process and feasibility of a new approach using thermoplastic resin. In many instances, research tends to emphasize methodologies and approaches that are focused on specific manufacturing systems. Consequently, it can be challenging to find examples and studies that provide a comprehensive exploration of various types of resins.

The main objective of the present study is to provide a comprehensive overview of various types of fast-curing and rapidly impregnating resins using a broad array of resin cases. In the present study, a comprehensive overview of commonly used, rapidly impregnating, low-viscosity resins in the automobile industry was provided, including thermosets and thermoplastic resins. Additionally, various reactive processes, and the parameters that influence them, were presented, alongside an examination of the properties associated with Fast-Curing Resin Thermosetting Composites (FRTSCs) and Fast-Curing Resin Thermoplastic Composites (FRTPCs). Finally, the current status of related research and insights into future perspectives in this field were addressed.

2. Rapidly Impregnating Resins and Their Fiber-Reinforced Composites
2.1. Knowledge Gaps and Current Challenges

Numerous methods for rapidly impregnating thermoset and thermoplastic resins in fiber-reinforced composites have focused their attention on various aspects requiring further development. This section presents the main knowledge gaps that need to be filled in order to address the current challenges regarding the rapid impregnation of thermoset and thermoplastic resins and their fiber-reinforced composites, to understand the development method and the its applications.

Thermoset resins, such as epoxy, polyester, and vinyl ester, are used in a wide range of automobile parts, such as headlamp housings, battery covers, and frames for windows or sunroofs. Thermoset composites have excellent dimensional and chemical stabilities and high impact strengths, which are necessary for the interior and exterior parts of automobiles. In this section, thermoset resins and their fiber-reinforced composites are discussed.

If the manufacturing cycle time can be reduced by using thermoset resin, several advantages become particularly noteworthy. To address these issues, various state-of-the-art methods have been introduced to achieve the fast impregnation of thermoset resin, as shown in Figure 1. The current state-of-the-art method for the rapid impregnation of thermoset resins and their fiber-reinforced composites primarily involves material selection, including the choice of resin and curing agent, as well as the utilization of advanced mixing techniques. Utilizing optimal materials (such as the epoxy resin with phenolic novolac epoxy and bisphenol-A epoxy, and the curing agent aliphatic polyamine dicyandiamide (DICY)), a specific manufacturing process was applied to achieve the rapid impregnation of the resin into the reinforcing fibers for resin transfer molding (RTM), vacuum infusion, and the reaction injection molding process [7]. The approach to achieving a fast and cost-effective impregnation and curing process was closely related to factors such as curing time, gel time for the resin, and curing kinetics. Consequently, finding most suitable resin and curing agent ultimately resulted in a substantial reduction in manufacturing costs (10~20% reduction in fiber cost, mandrel cost, tooling cost, system set up cost/process time under 240 min.) [17].

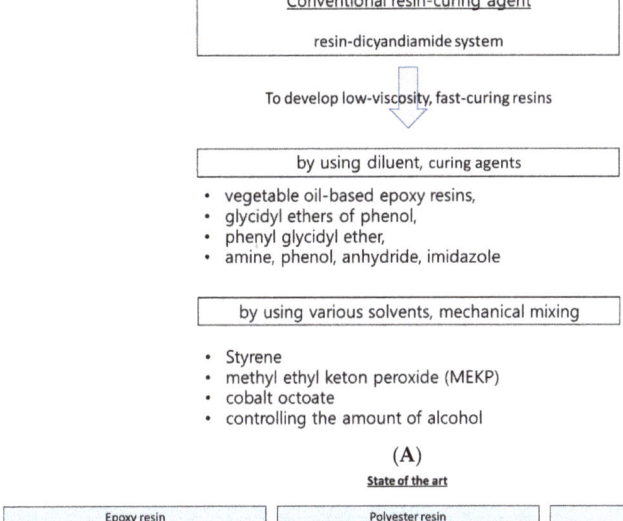

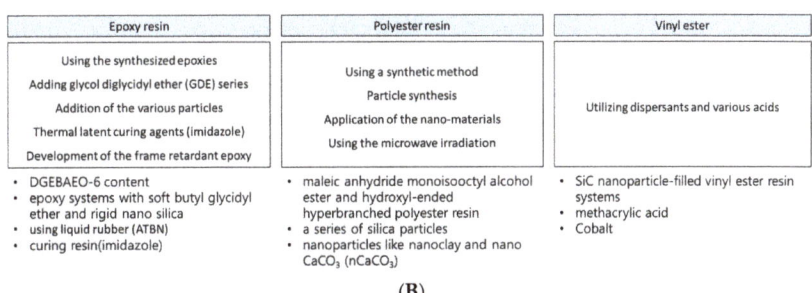

Figure 1. Knowledge Gap: (**A**) Conventional resin-curing system [18–28]. (**B**) State-of-the-art thermoset resin [25–37].

The current state-of-the-art rapidly impregnating thermoplastic resins and their fiber-reinforced composites have been tested with impregnation techniques, material selections, and processing parameters, etc. Several impregnation techniques, such as melt impregnation, powder impregnation, and resin transfer molding, have been explored to achieve the rapid impregnation of thermoplastic resins into fiber reinforcements; they have shown promise in terms of achieving efficient impregnation, reducing cycle times, and enhancing the overall quality of the composite [38,39]. Generally, higher temperatures can lower the viscosity of thermoplastic resins and pressure also helps to drive the resin into the fiber reinforcement, thus enabling better fiber impregnation and complete wetting, as well as the removal of void trapped within the fiber reinforced composites. However, excessively high temperatures may lead to resin degradation, and excessive pressure may lead to fiber deformation or damage. Therefore, optimizing the processing parameters is essential for achieving uniform resin distribution, complete fiber wetting, and strong interfacial bonding between the matrix and reinforcement, leading to the minimization of void content, as well as enhanced mechanical properties [40,41].

2.2. Thermoset Resins

2.2.1. Epoxy

Epoxy resins have been used in the automobile industry since the 1980s, owing to their superior mechanical properties, low shrinkage and creep, and outstanding chemical resistance. The estimated size of the global epoxy resin market was USD 12.5 billion in 2021, and it is anticipated to reach approximately USD 23.4 billion by 2030, with an expected annual growth rate (CAGR) of 7.22% during the forecast period from 2022 to 2030 [18]. A representative commercial epoxy resin is the epoxy-dicyandiamide system, and many strategies have been implemented to develop low-viscosity, fast-curing epoxy resins.

Conventionally, low viscosity can be achieved and controlled by incorporating various diluents, such as epoxy-based reactive diluents, which participate in the polymerization reaction and contribute to the cross-linking network. For example, the preferred viscosity range for the resins used in manufacturing composite materials via liquid molding is generally between 200 and 1000 cP at room temperature for 2~3 h of curing time [42]. Epoxy-based reactive diluents come in various forms, including vegetable oil-based epoxy resins, glycidyl ethers of phenol and paraalkyl substituted phenols, vinylcyclohexane dioxide, the phenyl glycidyl ether, and the trimethylol propane triglycidyl ether [19–24]. In addition, by introducing the catalytic mechanisms wherein epoxy crosslinks with the curing agent, a rapid curing time below 3 h can be obtained with tertiary amines.

Fast-curing epoxy resins can be obtained by adding glycol diglycidyl ether (GDE) series. For example, a low-viscosity acrylate-based epoxy resin (AE)/GDE system was developed by Yang et al., and its rheological behavior is shown in Figure 2A [43]. Seraji et al. [25,44,45] developed a rapid-curing epoxy amine resin with low viscosity, which consists of the diglycidyl ether of bisphenol F, an epoxy phenolic novolac resin, diethyl toluene diamine, and 2-ethyl-4-methylimidazole. The resin system exhibited good thermal and mechanical properties, and superior flame retardancy. Based on these trends, low-viscosity, fast-curing epoxy resins were obtained using the synthesized epoxies. Two resins, the diglycidyl ether of ethoxylated bisphenol-A (BPA) with two and six oxyethylene units (DGEBAEO-2 and DGEBAEO-6), respectively, were synthesized and characterized; the curing exothermic enthalpy decreased with increasing oxyethylene units (Figure 2B) [26]. The viscosities of the blends decreased as the DGEBAEO-6 content increased. In addition, difunctional aromatic epoxy-divinylbenzene dioxide, which was synthesized with epoxidizing divinylbenzene as the catalyst, had a low molecular weight and viscosity, as well as excellent thermal (T_g was approximately 201 °C) and mechanical properties (tensile strength was 131.99 MPa). Wu, Xiankun, et al. [27] and Chen et al. [29] developed a series of epoxy systems with a soft butyl glycidyl ether and rigid nano silica, and a viscosity lower than 600 mPa·s, thus providing an excellent processing performance for the large-scale production of composites in automobile manufacturing. In addition, this system demonstrated improvements in terms of tensile strength and modulus, as well as in elongation at break. Wang et al. [46] reported an epoxy resin-1-(cyanoethyl)-2-ethyl-4-methylimidazol system. The epoxy cured in a few minutes at 120 °C with an acceptable pot life and low water absorption.

The reaction time decreased with the addition of the various particles. Chikhi et al. [30] developed a modified epoxy resin using liquid rubber (ATBN). All reactivity characteristics (gel time, temperature, curing time, and exothermic peaks) decreased. The addition of ATBN led to a reduction in either the glass transition temperature or the stress at break, accompanied by an increase in the elongation at break and the appearance of yielding. Zhang et al. [48] designed a tetrafunctional eugenol-based epoxy resin with a cyclosiloxane structure. Allyl glycidyl ether was selected as the reference compound to generate a silylation epoxy resin. The viscosity of the silicone-containing tetrafunctional epoxy monomers (<0.315 Pa·s) was significantly lower than that of conventional oil-based epoxy resins (14.320 Pa·s) (Figure 2C) [43].

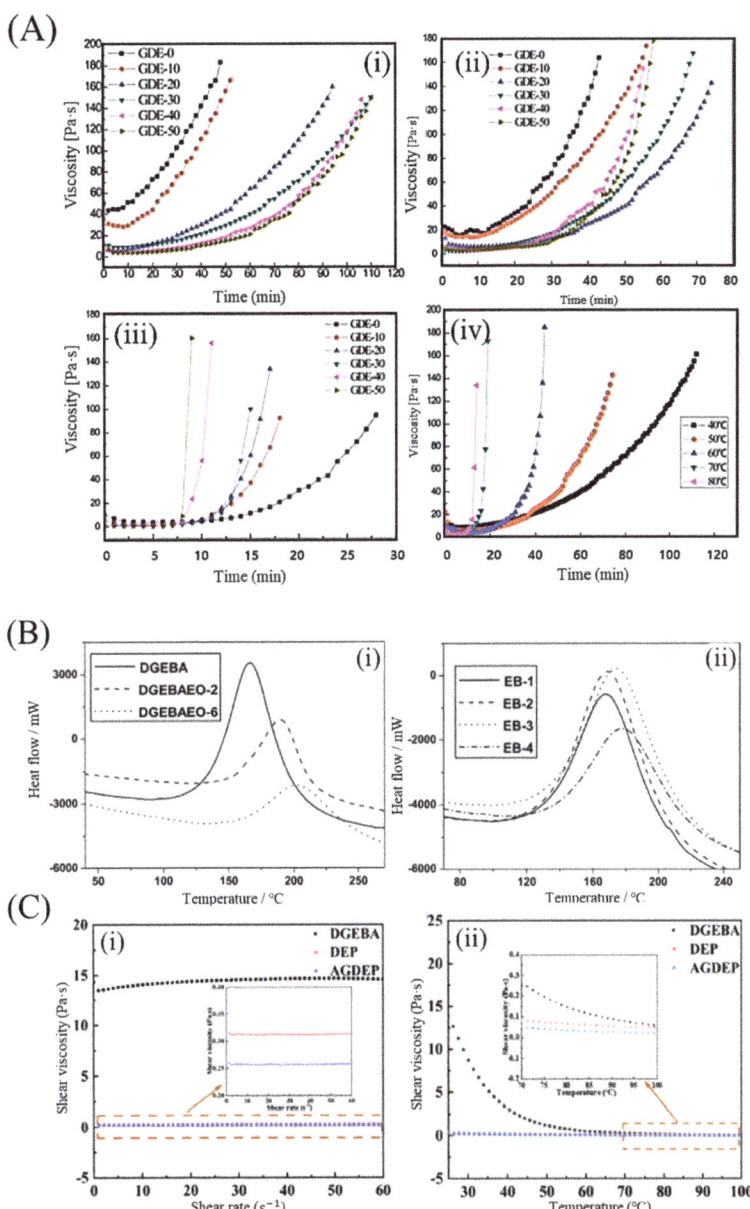

Figure 2. (**A**) Effect of glycol diglycidyl ether (GDE) content on the rheological behavior of the acrylate−based epoxy resin (AE)−40/LPA system: (**i**) 40 °C, (**ii**) 50 °C, (**iii**) 80 °C, (**iv**) 20 phr GDE content. Reproduced with permission [43]. Copyright 2015, John Wiley and Sons. (**B**) DSC curves for (**i**) three neat epoxy resins and (**ii**) DGEBA/DGEBAEO−6 cured using DDM. Reproduced with permission [26]. Copyright 2011, John Wiley and Sons. (**C**) Rheological behaviors of epoxies (without curing agent): (**i**) Change in viscosity, with shear rate from 0.01 to 60 s^{-1} at 25 °C, (**ii**) Change in viscosity with temperature from 25 to 100 °C at a shear rate of 60 s^{-1}. Reproduced with permission [47]. Copyright 2022, John Wiley and Sons.

Moreover, the low viscosity of epoxy resin-based component epoxy systems has recently been obtained for thermal latent curing agents and flame-retardant epoxies (generally below 200 cP at room temperature) [25–27,42]. Thermal latent curing agents of Imidazole are widely employed to fabricate single-component epoxy systems, and they meet the requirements for large-scale industrial production [49–51]. Several phosphorus-modified imidazole derivatives have been developed to combine fast curing rates (below 3 h [25–27] and great flame retardancy characteristics [52–54].

2.2.2. Polyester

Regarding polyester resins, low-viscosity polyester resins can be obtained via particle synthesis. Low-viscosity polyester resins can be applied to produce environmentally friendly coatings, as well as to toughen and reinforce unsaturated polymers. The global market size of unsaturated polyester resins was estimated to be USD 12.2 billion in 2022, and it is expected to grow at an annual growth rate (CAGR) of 7.1% from 2023 to 2030 [55].

Traditionally, low-viscosity polyester resins can be obtained through various methods, including the use of solvents, mechanical mixing methods, etc. For instance, solvents such as styrene, methyl ethyl ketone peroxide (MEKP), and cobalt octoate are typically used. Control and mixing methods involving alcohol are frequently utilized. Nurazzi et al. [56] developed a method to reduce the gel time of unsaturated polyester (UPE) by blending it with methyl ethyl ketone peroxide (MEKP) and various percentages of cobalt. Using this method, the gel time can be reduced by up to 36%.

Recently, alternative approaches have been employed to achieve a lower viscosity (<300 mPa·s) in the compound [57], which facilitates the formation of crosslinking networks. These methods include synthetic techniques, particle synthesis using nanomaterials, microwave irradiation, among others [58,59].

Chen et al. [31] prepared a series of silica particles with different sizes and surface groups through the sol–gel process, using tetraethyl orthosilicate, and they were directly introduced into polyester polyol resins via in situ polymerization. The resulting nanocomposites exhibited lower viscosities than the resins obtained using the blending method. Viscosity increased as the particle concentration increased (Figure 3A). Zhang et al. [28] examined a low-viscosity unsaturated hyperbranched polyester resin (<10,000 cP) using a synthetic method involving a reaction between a maleic anhydride monoisooctyl alcohol ester and a hydroxyl-ended hyperbranched polyester resin prepared from phthalic anhydride and trimethylolpropane. Zhou et al. [57] synthesized a series of unsaturated polyester resins with low viscosities (<300 mPa·s), for a vacuum infusion molding process, by simply controlling the amount of alcohol used in the reactants. Yuan et al. [60] developed a series of low-viscosity transparent UV-curable polyester methacrylate resins, derived from renewable biologically fermented lactic acid (LA), and they reduced the viscosity from 34,620 mPa·s to 160–756 mPa·s by randomly copolymerizing LA and-caprolactone.

The curing time can be reduced using various solvents and applying microwaves. Nasr and Abdel-Azim [33] investigated unsaturated polyester resins, and styrene, methyl ethyl keton peroxide (MEKP), and cobalt octoate were selected as the solvent (monomer), catalyst, and accelerator, respectively. A significant reduction in curing time occurred when the cobalt octoate concentration was increased to 0.02 wt.%. Furthermore, the curing time decreased when the catalyst concentration was increased from zero to 2 wt.%. Mo et al. [32] applied microwave irradiation to the curing of an unsaturated polyester resin with $CaCO_3$ particles, and they showed that microwave irradiation heated the unsaturated polyester resin evenly and rapidly, causing a chain growth reaction which greatly reduced the curing time (Figure 3B). Chirayil et al. [61] prepared nanocellulose-reinforced unsaturated polyester composites via mechanical mixing. The curing time required for gelation in the nanocellulose-filled unsaturated polyester was lower than that for the neat resin, indicating the catalytic action of nanocellulose in the curing reaction (Figure 3C). Kalaee et al. [34] utilized the nanoparticle of $CaCO_3$ ($nCaCO_3$) and found that a decrease in the number of carboxyl groups in the formulation leads to a higher degree of crosslinking.

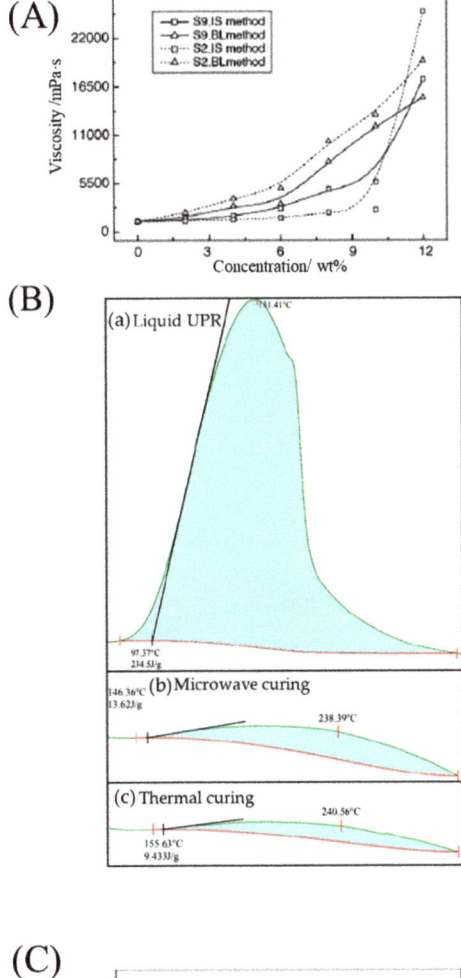

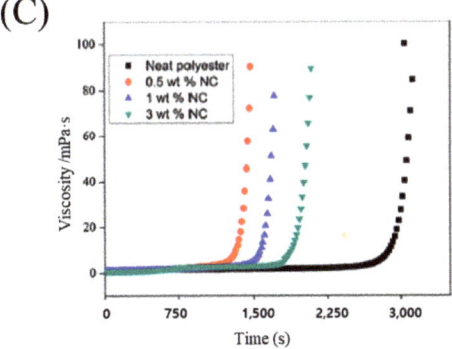

Figure 3. (**A**) Effect of silica content on the viscosity of nanocomposite resins embedded with silica sol S2 or S9. Reproduced with permission [31]. Copyright 2005, Elsevier. (**B**) DSC curves of liquid UPR, and cured samples with microwave curing and thermal curing, respectively. Reproduced with permission [32]. Copyright 2022, MDPI. (**C**) Variation of viscosity over time for NC filled composites. Reproduced with permission [61]. Copyright 2014, Elsevier.

2.2.3. Vinyl Ester

The extensive use of vinyl ester resins as matrix materials in reinforced composites is due to their low viscosity, rapid curing capabilities at room temperature, and cost-effective advantages. Typically, in highly viscous vinyl ester resins, a low-viscosity environment can be achieved by utilizing dispersants and various acids, which effectively reduce their surface activities.

Yong and Hahn [35] conducted a rheological analysis of SiC nanoparticle-filled vinyl ester resin systems using the Bingham, power law, Herschel–Bulkley, and Casson models. The incompatibility between a hydrophilic SiC and a hydrophobic vinyl ester resin can act as the driving force for the formation of SiC aggregates, even when low particle loading occurs (<0.04 volume fraction), resulting in the high viscosity of the resin. The optimum fractional weight percentage of dispersants (wt.% dispersant/wt.% SiC) for dispersion stabilization is 1–3% for particles in the 0.1–3-μm range, and it can be proposed, as follows: the addition of a dispersant at the optimum dosage lowers the viscosity of SiC/vinyl ester suspensions by 50% (Figure 4A).

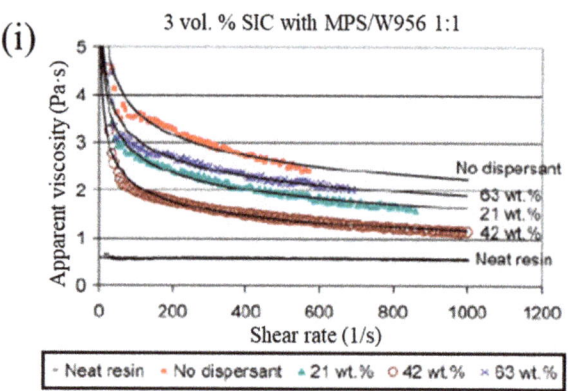

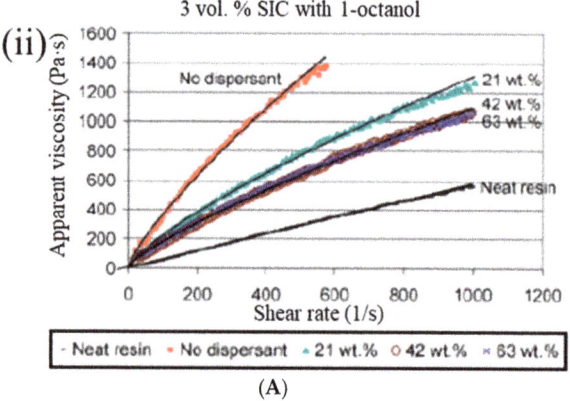

(A)

Figure 4. *Cont.*

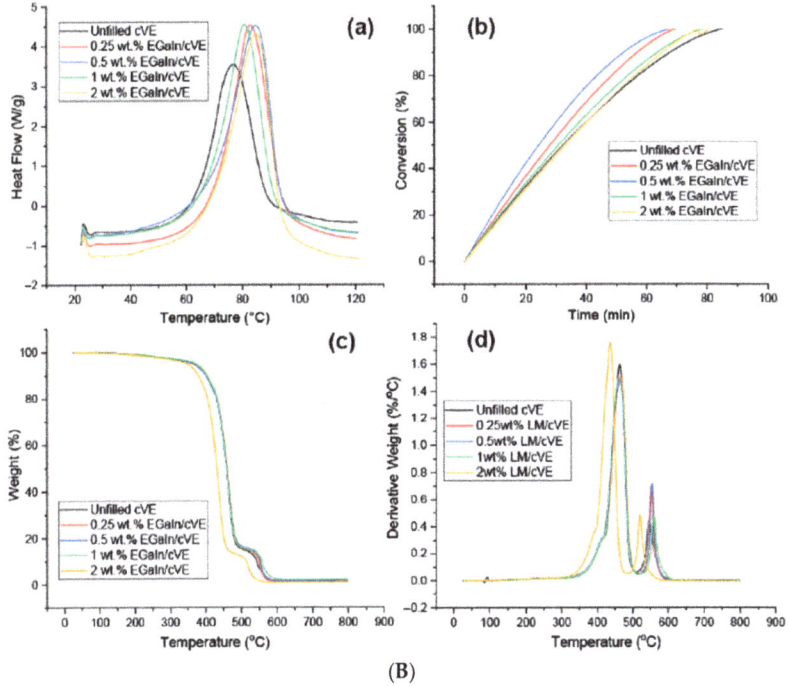

Figure 4. (**A**) (**i**) Viscosity curves of SiC/vinyl ester resin systems with and without MPS/W966. (**ii**) Viscosity curves of SiC/vinyl ester resin systems with and without 1-octanol. Reproduced with permission [35]. Copyright 2006, John Wiley and Sons. (**B**) The DSC graphs (**a**), degree of conversion at 60 °C (**b**), graphs of TGA (**c**), and DTG (**d**) at a heating rate of 5 °C/min for the LM filled and unfilled comonomer vinyl ester composites. Reproduced with permission [62]. Copyright 2022, MDPI.

Gaur et al. [36] obtained the zero-shear viscosity of vinyl ester resins containing styrene (40 wt.%) as the reactive diluent. The curing of vinyl ester resins can be controlled by reacting the epoxy novolac resin with methacrylic acid. They found that the curing and decomposition behavior of vinyl ester resins worsened with an increase in methacrylic acid content (11, 22, 32, 38, and 48 mg KOHg^{-1} solid). The cured product with the lowest acid value was the most thermally stable product. Cook et al. [37] analyzed the gel time and reaction rate of a vinyl ester resin and found that the cobalt species played a dual role in initializing the formation of radicals from MEKP and destroying primary and polymeric radicals. Based on these results, the reaction rate (determined using differential scanning calorimetry, (DSC)) increased and the gel time decreased with increasing concentrations of MEKP. However, cobalt octoate cocatalyst slows the reaction rate, except at very low concentrations. The gel time decreased as MEKP and cobalt octoate concentrations increased. Curing vinyl ester resins with modified silicone-based additives was achieved by Mazali et al. [63]. Silicone-based additives were used to modify the properties of the vinyl ester resin. For the resin cured in the absence of N, N-dimethylaniline, the silicone-based additives acted as retardants of the curing reaction, which is a typical diluent effect, whereas in the presence of this promoter, the reaction enthalpy and rate improved.

The viscosity of the vinyl ester resin could be reduced by increasing the reactive diluent content. Rosu et al. [64] found a linear correlation between the reactive diluent content and the logarithm of viscosity, showing that the presence of reactive diluents accelerated the curing reaction and diminished the gel time. Dang et al. [62] proposed reinforcements for a comonomer vinyl ester (cVE) resin at different weight fractions of up to 2% via a direct polymerization process with a eutectic gallium–indium (EGaIn) alloy and graphene nanoplatelets, showing that sub-micron sized EGaIn (≤1 wt.%) could promote the curing reaction of cVE without changing the curing mechanism (Figure 4B).

2.2.4. Polydicyclopentadiene (p-DCPD)

Dicyclopentadiene (DCPD) is a commercially available monomer that is derived from low-viscosity petrochemicals, making it easy to impregnate into fibers. Due to its impregnation characteristics, its market revenue reached approximately USD 0.86 billion in 2020 and is expected to grow at a CAGR of 5.7% between 2022 and 2030 [65]. Polydicyclopentadiene (PDCPD) is a highly crosslinked polymer formed by the ring-opening metathesis polymerization (ROMP) of its monomer precursor. Exothermic characteristics were observed during the polymerization process because of the relief of the ring strain energy initiated by the transition-metal/alkylidene complexes. Several studies investigated the effects of these catalysts.

Li et al. [66] conducted the ROMP of DCPD using the catalyst systems, WCl6–Et2AlCl and (WCl6–PhCOMe)–Et2AlCl, and their polystyrene-supported counterparts. The acetophenone-modified catalyst system exhibited better catalytic properties than the unmodified system. Moreover, as the polymer yield of ROMP increased, the mechanical properties of notched impact strength (NIS) and the tensile strength (TS) of the synthesized PDCPD increased. Kessler et al. [67] investigated the curing kinetics of PDCPD, prepared via ROMP, with three different concentrations of Grubbs' catalyst using differential scanning calorimetry (Figure 5A). The catalyst concentration had a large effect on the curing kinetics, and the activation energy increased significantly at 30 °C. Yang and Lee [68] investigated the curing kinetics of endo-dicyclopentadiene (DCPD) with two types of Grubbs' catalysts (1st and 2nd generation), using dynamic DSC at different heating rates (Figure 5).

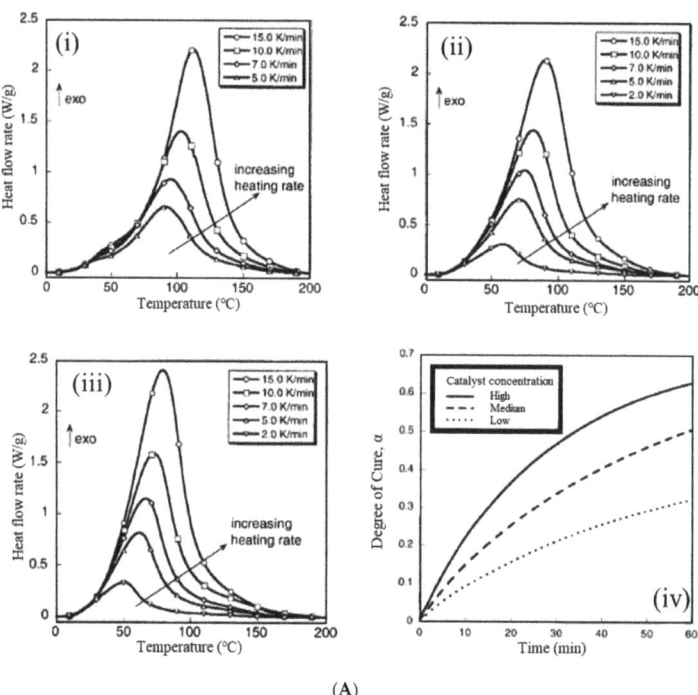

(A)

Figure 5. Cont.

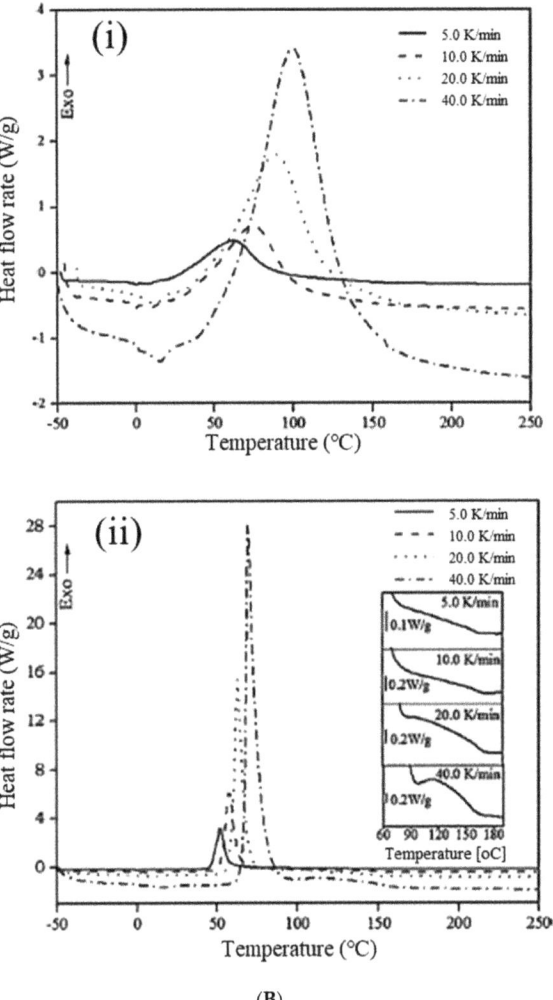

Figure 5. (**A**) The DSC curves for (**i**) low concentration (**ii**) medium-concentration, and (**iii**) high–concentration DCPD and Grubbs' catalyst samples; (**iv**) predictions for isothermal curing at 30 °C based on the model–free iso–conversional method for low, medium, and high catalyst concentrations. Reproduced with permission [67]. Copyright 2002, John Wiley and Sons. (**B**) DSC scans at different heating rates for endo–DCPD with (**i**) 1st generation and (**ii**) 2nd generation Grubbs' catalysts (inset shows the shoulder region). Reproduced with permission [68]. Copyright 2013, Elsevier.

Experimental DSC data obtained at different heating rates were used to evaluate the kinetic parameters of the model-free iso-conversional and model-fitting methods. In the single DSC exotherm of the 1st generation system (Figure 5(Ai)), the appearance of a shoulder above the single exotherm of the 2nd generation system (Figure 5(Aii)) suggests that reaction mechanisms other than ROMP, regarding the norbornene and cyclopentene units, may be involved in this catalyst system. The 2nd generation catalyst system showed a slower initiation rate but a faster polymerization rate compared with the 1st generation.

Yang and Lee [69] also studied two Grubbs' catalysts that exhibited apparent differences in the isothermal curing of endo-dicyclopentadiene (endo-DCPD) via ROMP, using the 1st and 2nd generation Grubbs' catalysts as polymerization initiators. The 2nd generation catalyst was more efficient than the 1st generation catalyst in terms of catalytic activity, as evidenced by the reaction rates and fractional conversions (Figure 6A).

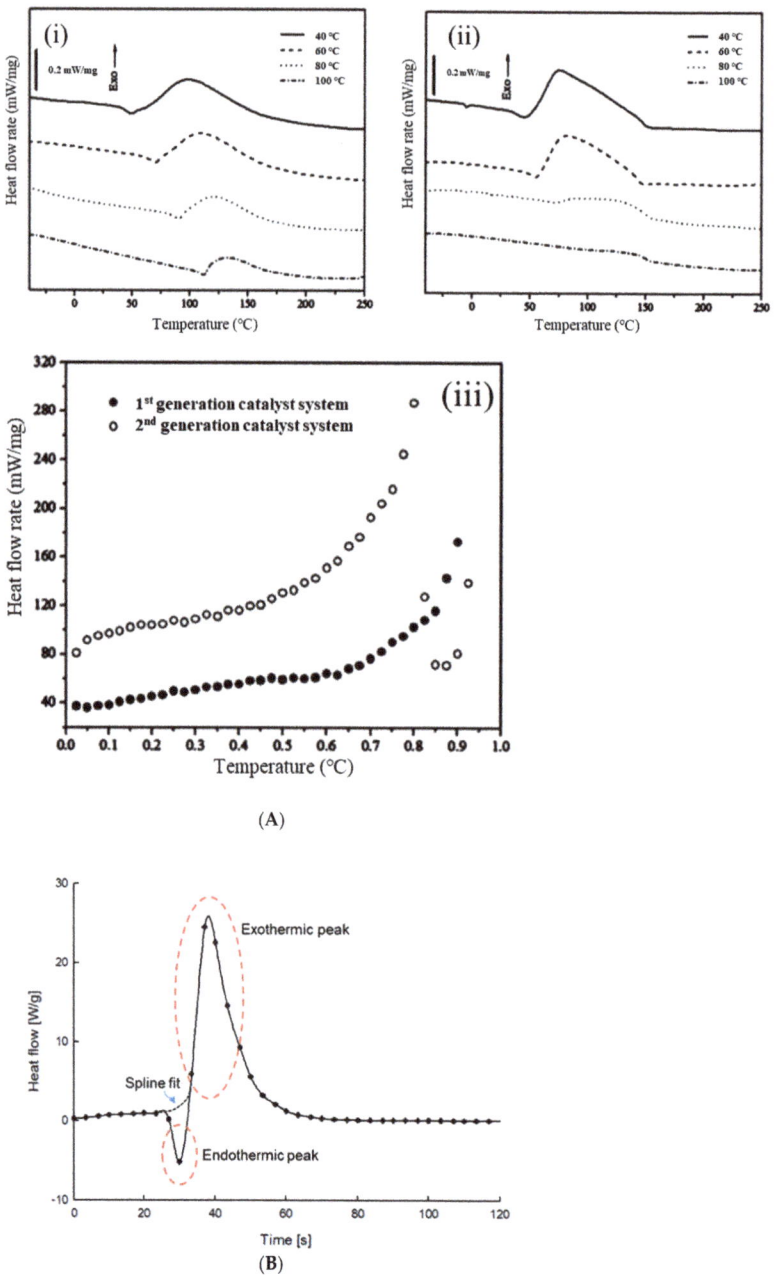

Figure 6. Cont.

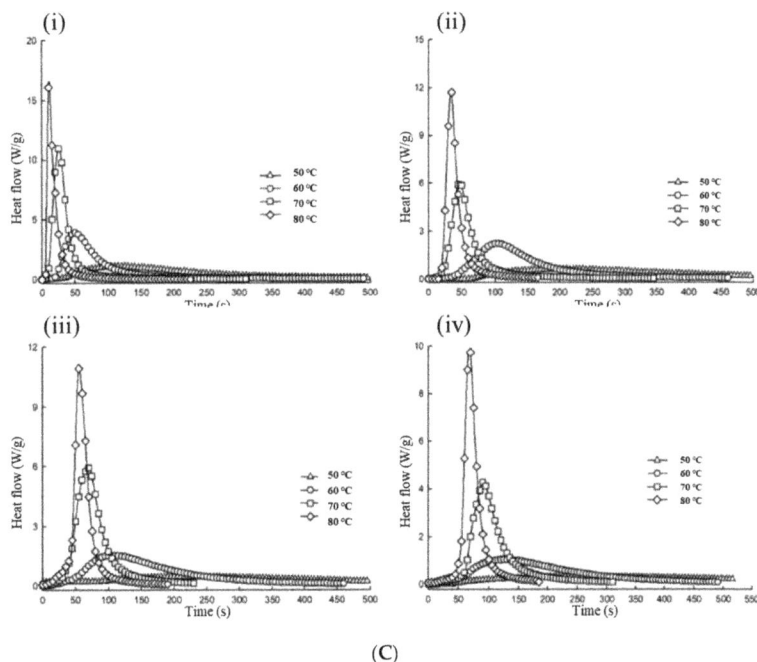

(C)

Figure 6. (**A**) Dynamic DSC scans following the isothermal cure of endo-DCPD with (**i**) the 1st generation and (**ii**) the 2nd generation Grubbs' catalysts. (**iii**) Activation energy as a function of the fractional conversion of endo-DCPD with the 1st and the 2nd generation Grubbs' catalysts. Reproduced with permission [69]. Copyright 2014, American Chemical Society. (**B**) Dynamic DSC curve with endo-DCPD with different amounts of decelerators. Reproduced with permission [70]. Copyright 2019, Elsevier. (**C**) Isothermal DSC profiles of endo-DCPD with different amounts of decelerators, as follows: (**i**) 0.5 wt. mass %, (**ii**) 1.0 wt. mass %, (**iii**) 1.5 wt. mass %, (**iv**) 2.0 wt. mass %. Reproduced with permission [70]. Copyright 2019, Elsevier.

Recent state-of-the-art research on vinyl esters and DCPD has focused on controlling the curing time; this is due to their extremely fast curing times, as shown in Figure 7. Yoo et al. [70] obtained the curing kinetics of endo-DCPD, using isothermal differential scanning calorimetry, by experimentally acquiring kinetic parameters in accordance with model-fitting approaches. Due to the rapid curing of DCPD, a decelerator was included in the manufacturing process. Therefore, the effect of the decelerator was investigated using the curing kinetics of endo-DCPD with different amounts of decelerator solutions, and it was found that the decelerator delayed the reaction and slowed the curing process (Figure 6B,C).

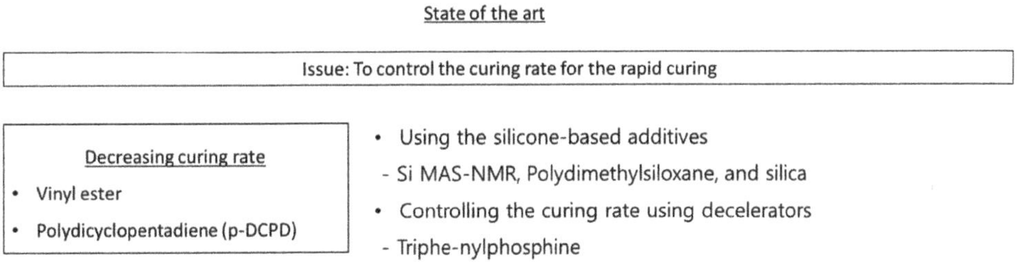

State of the art

Issue: To control the curing rate for the rapid curing

Decreasing curing rate
• Vinyl ester
• Polydicyclopentadiene (p-DCPD)
	• Using the silicone-based additives
	  - Si MAS-NMR, Polydimethylsiloxane, and silica
	• Controlling the curing rate using decelerators
	  - Triphe-nylphosphine

Figure 7. Recent state-of-the-art research on vinyl esters and DCPD [63,70,71].

3. Rapidly Impregnating Thermoplastic Resins and Their Fiber-Reinforced Composites

In recent times, to a certain extent, thermoplastic composites (TPCs) have started to replace thermosetting composites and lightweight metal materials. Worldwide, the market value of TPCs increases every year, from 28 billion U.S. dollars in 2019 to an estimated 36 billion U.S. dollars by 2024; this is because they are very tough, the manufacturing process is faster, they are highly processable and recyclable, they are able to be welded, etc. [72].

Generally, the high melting viscosities of thermoplastic polymers require high processing temperatures and pressures to fully impregnate fibers and reduce defects in products [73]. Subsequently, in situ polymerization methods for fiber-reinforced TPCs have been developed using low-viscosity monomers or oligomeric precursors, such as caprolactam [74–77], laurolactam [78,79], methylmethacrylate (MMA) [80], and cyclic butylene terephthalate (CBT) [81,82], to fabricate fiber-reinforced polyamide 6 (PA6), polyamide 12 (PA12), polymethylmethacrylate (PMMA), and polybutylene terephthalate (PBT) composites, respectively. The global market size for PA 6 is estimated at USD 12.7 billion, and for PA 12, it is estimated at USD 19.43 billion [83]. For the PMMA, the global market size was expected to reach USD 8.33 billion by 2032 and USD 5382 million by 2029 [84]. These monomers (or oligomeric precursors) are polymerized via the addition of catalysts and activators. Table 1 lists several processing parameters and applications of commonly used monomers (or oligomeric precursors) with low viscosities that are suitable for LCM. In this section, we mainly introduce PA6, PA12, PMMA, and PBT thermoplastic composites, and we provide an overview of thermoplastic composites fabricated via in situ polymerization during LCM. Moreover, the effects of reactive processing parameters on the mechanical properties are discussed.

3.1. Polyamide 6 (PA6)

PA6 was synthesized via the anionic ring-opening polymerization of ε-caprolactam, which is a crystalline cyclic amide with a melting temperature of 70 °C, and it is polymerized at 130–170 °C in the presence of a catalyst and activator [85] (Figure 8). PA6 based fiber-reinforced composites can be fabricated within 3–60 min, depending on the type and amount of the catalyst and activator used. Ahmadi et al. [91] suggested that the correct ratio of monomer, catalyst, and activator is a key component in anionic-caprolactam polymerization, and it provides the lowest monomer residue and best properties for the PA6 samples. In addition, polymerization time directly affects the production cycle and cost. Our previous research [76] focused on the effect of polymerization temperature on the degree of polymerization and polymerization time in order to produce perfect products with the shortest molding cycle time. The results showed that the polymerization and crystallization of PA6 occurred simultaneously during heating. As the heating rate increased, the crystallinity decreased, but the degree of polymerization increased. Furthermore, the viscosity of ε-caprolactam varied almost linearly with time in the early stages, whereas it increased exponentially from 20 s after the start of polymerization, indicating the presence of the injection molding cycle time. Ben et al. fabricated glass and carbon fiber hybrid PA6 composites (Figure 9) with A (caprolactam and activator) and B (caprolactam and catalyst) mixtures via vacuum-assisted resin transfer molding (VaRTM) in order to evaluate their mechanical properties when applied to automobile structures [75]. The results showed that the bending, tensile, and compressive strengths of the hybrid-fiber-reinforced PA6 were 594, 315, and 297 MPa, respectively, which were comparable to those of the hybrid, fiber-reinforced fast-curing epoxy (597, 327, and 318 MPa, respectively). However, the flammability of polyamides, which is a key issue in the automobile industry, limits their widespread application. The main challenges include the inhibition of in situ polymerization in the presence of flame retardants and the insolubility of flame retardants due to the filtration of reinforcements such as titanium dioxide, multiwalled carbon nanotubes, phosphorus compounds, etc. [92,93].

Table 1. Processing temperatures and processing times of various monomers.

	Monomer	Processing Temp. (°C)	Processing Time (min)	Application	Ref.
Thermoset	Epoxy (glycol diglycidyl ether (GDE) series diglycidyl ether of bisphenol F, diethyl toluene diamine, and 2-ethyl-4-methylimidazole)	120	180	Aerospace, marine, automobile, construction.	[25–27]
	Polyester (polyethylene terephthalate (PET))	40–80	120	Construction, marine, chemical.	[57]
	Vinyl Ester (vinyl acetate, vinyl propionate, and vinyl laurate)	50–200	<240	Coatings, printed circuit boards, building materials, automotive parts, and fiber-reinforced composites.	[36,37,63]
	p-DCPD (Dicyclopentadiene monomer)	50–90	2	Sporting goods, automotive industries, as well as military and aerospace applications.	[70]
Thermoplastic	PA-6 (ε-caprolactam)	130–170	3–60	Wipers, gears, bearings, and weatherproof coatings.	[85]
	PA-12 (ω-laurolactam)	180–240	with cooling process	Fuel pipes, fuel filters, high pressure oil pipes, gears, brake hoses.	[86,87]
	PMMA (Elium®) (vinyl polymerization of methylmethacrylate (MMA))	80–160 (room temp.~)	low emp.: >900 high temp.: Boiling of monomer	Decorative trims, interior lighting, door entry strips.	[85,88]
	PBT (1,4 butanediol and dimethyltetrephthalate)	180–250	<30	Door handles, bumpers.	[72,89,90]

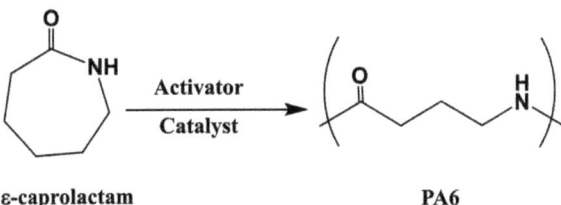

Figure 8. Schematic of the anionic ring open polymerization of PA6.

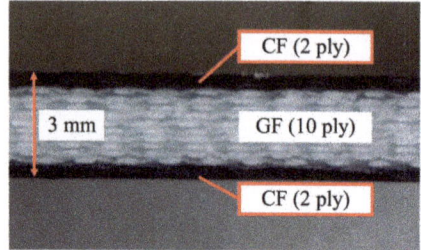
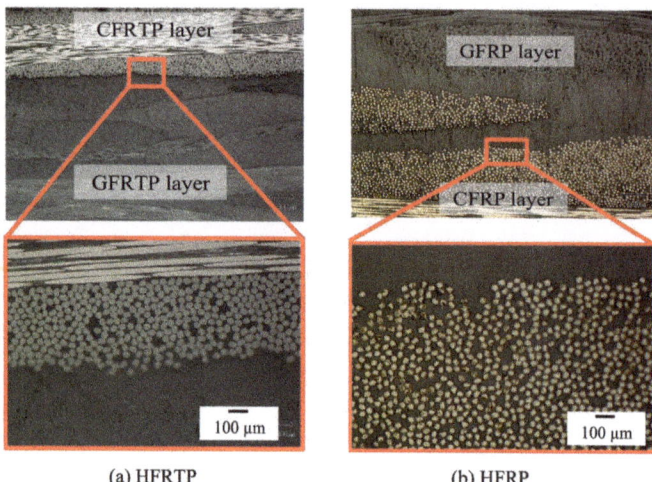

Figure 9. Cross sections of both plates: (**a**) hybrid, fiber-reinforced thermoplastic (HFRTP), (**b**) hybrid, fiber-reinforced plastic (HFRP). Reproduced with permission [75]. Copyright 2015, Elsevier.

In addition, recycling is also an issue for PA6. As a non-degradable plastic, PA6 is extremely challenging to recycle, and it cannot be recycled using traditional methods. Recently, Wursthorn et al. [94] developed lanthanide trisamido catalysts, with which, PA6 can be depolymerized to ε-caprolactam with a high selectivity (more than 95%) and yield (more than 90%), and no solvents or toxic chemicals are used in the whole process. The generated ε-caprolactam can be used as monomers to obtain new PA6, thus, it is feasible to employ this method to recycle PA6 products.

3.2. Polyamide 12 (PA12)

PA12 is called nylon 12, and it is synthesized via the anionic ring-opening polymerization of ω-laurolactam, as shown in Figure 10. ω-laurolactam has a low initial viscosity above its melting point at 153 °C, facilitating the easy and complete impregnation of fibers in the mold. Similar to PA6, PA12-based fiber-reinforced composites can be fabricated using LCM processes, such as thermoplastic resin transfer molding (T-RTM). The desired injection temperature was found to be 170–205 °C, and polymerization started at 180–250 °C, after introducing the catalyst and initiator. Mairtin et al. [79] developed carbon-fiber-reinforced PA12 composites, with a 60% carbon fiber volume fraction, which exhibited high tensile strength (788.3 MPa) and high compression strength (365.7 MPa). It is also reported that the polymerization time is related to polymerization temperature, that is, it takes 8.5 min and 20 min at molding temperatures of 240 °C and 200 °C, respectively.

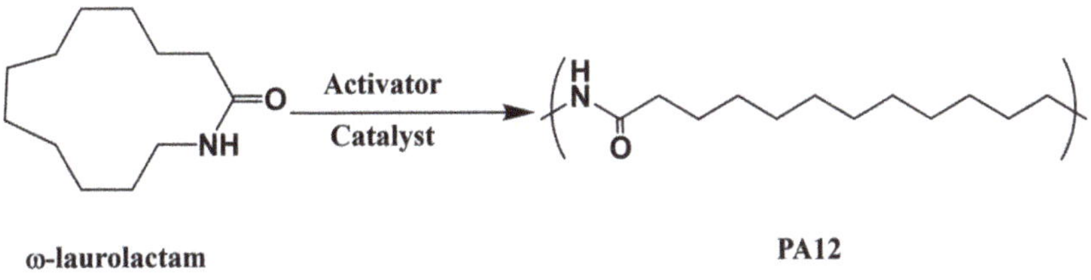

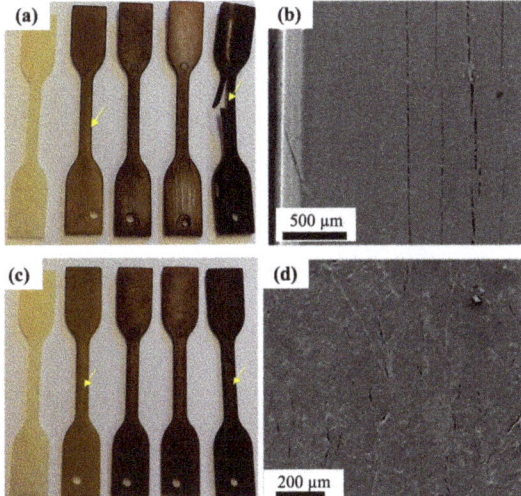

Figure 10. Schematic of the anionic ring open polymerization of PA12.

As listed in Table 1, PA12 is commonly used in fuel filter housing and fuel pipe connectors, which are close to the engine and exposed to fuel and high service temperatures. Therefore, fuel uptake and aging behavior are important factors. Wei et al. [95] found that pure PA12 showed fast and remarkably high fuel uptake when exposed to a mixture of ethanol and gasoline at 120 °C; however, a lower uptake was observed for glass-fiber-reinforced PA12 composites. As shown in Figure 11, the PA12 and glass-fiber-reinforced PA12 composites gradually changed color from white to yellow as the exposure time increased; this is due to the oxidation of PA12, and the cracks in PA12 were larger than those in glass-fiber-reinforced PA12, indicating a suppression effect of glass-fiber on fuel uptake.

Figure 11. Images of (**a**) PA12 and (**c**) glass-fiber-reinforced PA12 composites (from left to right, samples aged for 0, 150, 300, 500, and 700 h). SEM images of the surface of the (**b**) 150 h-aged PA12 sample and (**d**) 150 h-aged glass-fiber-reinforced PA. Reproduced with permission [95]. Copyright 2022, Springer Nature.

In addition, PA12 can be recycled and used in automobiles. It has been reported that an automobile fuel-line clip, produced with recycled PA12 through a selective laser sintering method, provides an 8% reduction in life-cycle global warming potential and life-cycle primary energy demand compared with conventional PA66 [96], thus improving sustainability properties.

3.3. Polymethyl Methacrylate (PMMA) (Elium®)

PMMA is extensively used in the automobile industry to produce various parts and components of vehicles, such as external, rear, and indicator light covers; decorative trims; ambient lighting; door entry strips; and automobile glazing. This is due to its light weight, high scratch resistance, and low stress birefringence. As shown in Figure 12, PMMA was synthesized via the free radical vinyl polymerization of methylmethacrylate (MMA) in the presence of peroxide initiators. The melting temperature of MMA is −48 °C, and it is polymerized at relatively low temperatures (120–160 °C); however, the boiling temperature of MMA is 100 °C, which means that it boils easily and can cause voids in the final products. Moreover, a long cycle time (>900 min) was required to fully polymerize MMA below its boiling temperature.

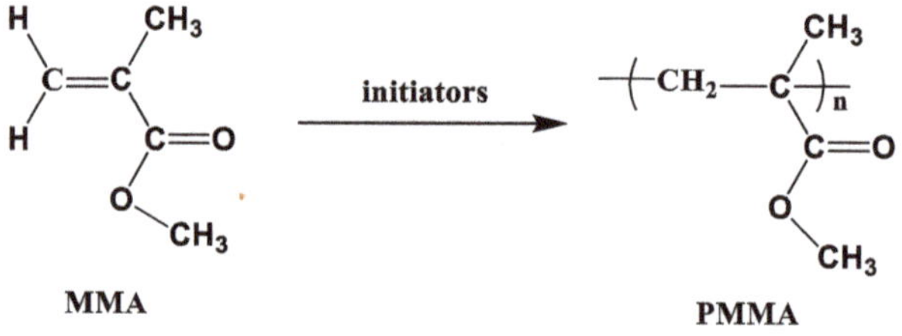

Figure 12. Schematic of the vinyl polymerization of PMMA.

Recently, a novel liquid-reactive MMA has been developed by Arkema, named Elium®. It has low viscosity and low processing temperature (room temperature), and it is used in conjunction with a dibenzoyl peroxide initiator [88]. Elium® can also be used to impregnate fibers via the LCM process, which is the same method used as traditional MMA. Several studies have been conducted to evaluate its mechanical properties. Kazemi et al. [97] studied the dynamic response of carbon-fiber-reinforced Elium® and carbon fiber-reinforced epoxies (Epolam, Sikafloor) using low-velocity impact tests. This study demonstrated the higher plasticity of Elium®-based composites compared with epoxy-based composites, resulting in less structural loss and less absorbed energy, as shown in Figure 13. In addition, many studies have reported on the good mechanical properties of Elium®-based fiber-reinforced composites, such as good toughness, flexural and tensile strength, welding performance, etc. [98–100]. However, Elium® has a much higher shrinkage rate than that of common PMMA due to its fast polymerization, which is a problem to be solved in future studies [101].

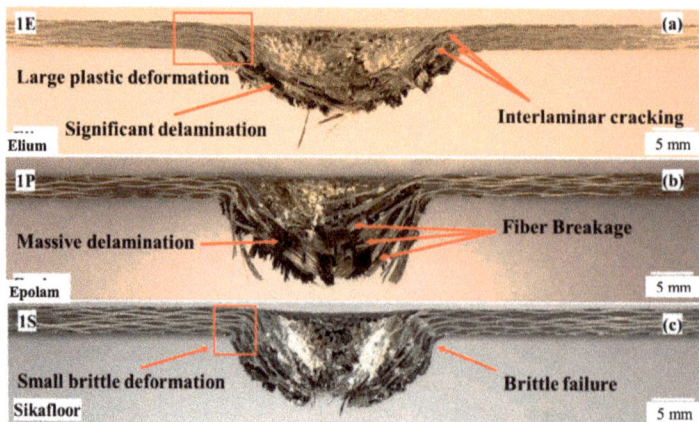

Figure 13. A comparison of cross-sectional observations of (**a**) the carbon-fiber-reinforced Elium® and (**b,c**) thermosetting epoxy (Epolam, Sikafloor) laminates at 20 J. Reproduced with permission [97]. Copyright 2021, Elsevier.

In addition, some recycling technologies for Elium®, such as the mechanical recycling method and chemical recycling method, have already been developed to obtain recycled materials or they have been recovered as monomers [102]. Generally, these recycled materials are reused with virgin materials to enhance mechanical properties, however, recovered monomers could be polymerized to obtain new products. Though a few studies on the characterization and analysis of recycled products have been investigated, more intensive work should be needed to evaluate the life cycle of these recycled products.

3.4. Polybutyleneteraphthalate (PBT)

PBT is widely used in the automobile industry owing to its high stiffness and strength. Indeed, 1,4 butanediol and dimethyltetrephthalate are used as monomers to produce macrocyclic oligomers of CBT with two to seven repeat units [103], and this is followed by the polymerization of semicrystalline PBT in the presence of an initiator (Figure 14). The initial viscosity of CBT is 20 mPa·s at 190 °C, which is suitable for the LCM processing of, for instance, RTM [104]. It is reported that PBT polymerized from CBT via RTM is more brittle than that of conventional PBT due to the high crystallinity of the polymerized PBT [105]. Its toughness could be improved with the addition of nanoparticles [106,107], fibers [108], etc. Baets et al. found that the addition of 0.05–0.1 wt.% of multi-walled carbon nanotubes (MWCNTs) could increase the toughness, stiffness, and strength of PBT composites [109]. They also prepared polycaprolactone-blended CBT/glass-fiber composites to improve the toughness of the composites [109]. Yang et al. [110] found that woven carbon fabric and glass fabric hybrid PBT composites, which are fabricated via a vacuum assisted prepreg process, have a higher impact resistance than that of PBT/carbon fiber (CF) composites, although the presence of fibers may reduce the conversion of CBT. Furthermore, non-isothermal production processes, solvent blending, the addition of plasticizers, and chemical modification can enhance the toughness of CBT composites [81,111,112].

In addition, PBT can be recycled via depolymerization into CBT or monomers (1,4-butanediol and dimethyltetrephthalate) which exhibit properties comparable to those of baseline materials. Cao et al. [113] prepared super tough PBT/MWCNT/epoxidized elastomer composites with excellent mechanical properties for a wide range of PBT applications in the automobile industry.

Despite significant progress, several challenges persist in terms of the rapid impregnation of thermoplastic resins to produce fiber-reinforced composites in the automobile industry. These challenges include achieving uniform resin distribution, controlling fiber wetting, minimizing void content, and maintaining mechanical properties. Recent advancements have focused on addressing these challenges through innovative approaches [41,114]. Furthermore, thermoplastic-based automobile parts are also required to increase automotive plastic reuse, recycling, and recovery, in order to reduce overall automotive plastic waste generation for environmental sustainability. Many studies have reported that plastic parts could be reused or recovered from end-of-life vehicles, and some parts could be recycled via high-vacuum extraction, melt filtration, introducing additives, and so on. However, it is challenging to recycle fiber-reinforced plastic composites or multi-component-blended composites. Recently, a physicochemical recycling method has been developed to recover matrices and fibers which preserve the fibers' lengths [115]. Furthermore, more intensive work on the environmental impacts and life cycle assessments of these recycled products should be investigated [116,117].

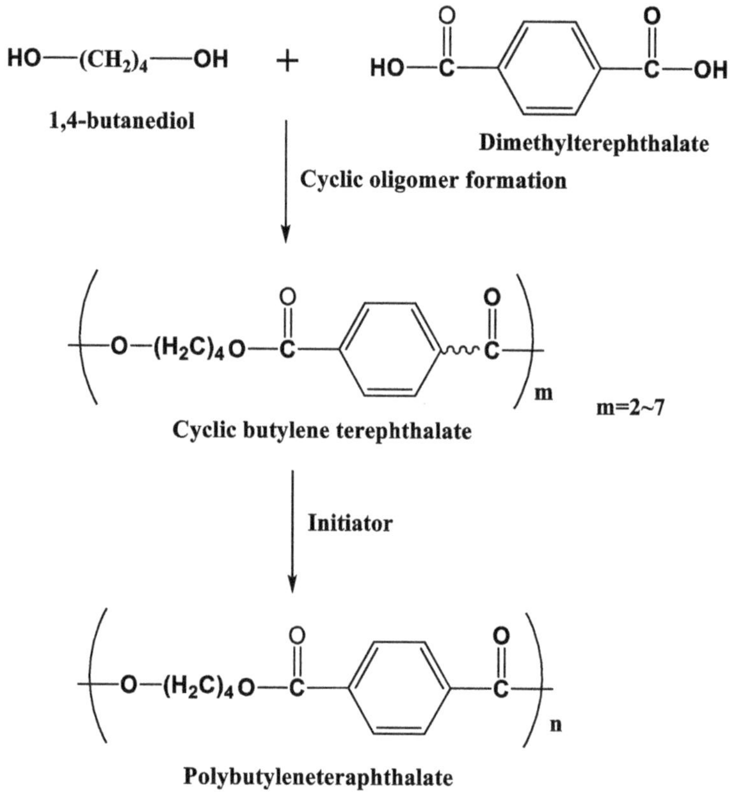

Figure 14. Schematic of the anionic ring open polymerization of PBT.

4. Current Research Gaps and Future Research Outlook

The current research concerning rapidly impregnating resins and the production of fiber-reinforced composites in the automobile industry has identified several research gaps. Addressing these gaps and focusing on future research can lead to advancements and improvements in this field. Here are some of the current research gaps and potential future research directions [118–120].

Enhanced impregnation efficiency: Achieving the uniform impregnation of reinforced fibers with resin is critical for high-quality composites. Research has focused on exploring different impregnation techniques and parameters to minimize voids, ensure uniformity, and enhance interfacial adhesion. Although progress has been made in terms of impregnation techniques, there is a need to further enhance the impregnation efficiency of rapidly impregnating resins, which would enhance the amount of pore space penetrated by the resin [121–124]. Future research should focus on improving resin flow and wetting behavior to achieve the better impregnation of reinforcing fibers. This includes studying the effects of resin viscosity, fiber architecture, and processing parameters on impregnation efficiency.

Optimization of curing processes: A rapid curing process is crucial for the efficient production of fiber-reinforced composites [10,125]. Future research should aim to optimize curing processes by investigating advanced heating methods, optimizing curing temperatures and times, and exploring the use of catalysts or additives to accelerate the curing reaction. Such research will help reduce cycle times and improve the overall productivity of composite manufacturing.

Characterization and optimization of mechanical properties: Understanding and tailoring the mechanical properties of rapidly impregnating resins for specific automotive applications is essential. Future research should focus on exploring the development of new resin formulations that are specifically designed for rapid impregnation; this will involve modifying the viscosity, curing kinetics, or surface tension of the resin to improve its flowability and fiber wetting characteristics. In addition, the resin composition, placement of fiber reinforcement, and processing conditions should be optimized to achieve the desired mechanical properties. This can be achieved through a combination of experimental testing, numerical modeling, and material characterization techniques [126–128].

Durability and long-term performance: Regarding environmental issues, sustainable fibers are of great interest to the automotive industry. However, automotive parts are usually exposed to environmental factors, such as UV radiation, temperature, humidity, and chemical exposure, resulting in poor interfacial properties, water absorption, swelling, etc. [129,130]. Therefore, future research should focus on developing bio-based materials and enhancing the resistance of the composites, as well as their degradation mechanisms [131,132].

Environmental sustainability and recyclability: Given the increasing environmental concerns, future research should focus on developing sustainable and recyclable rapidly impregnating resins [133]. This includes exploring the use of bio-based or recycled materials as resin matrices, investigating recycling techniques for end-of-life composites, and assessing the environmental impact of these materials throughout their lifecycle.

By addressing these research gaps in future research, the automobile industry can benefit from improved rapidly impregnating resins that offer enhanced impregnation efficiency, optimized curing processes and mechanical properties, and improved durability and sustainability.

5. Conclusions and Outlook

The impregnation performance of thermoset resin and thermoplastic resin is crucial for manufacturing composite materials like glass-fiber or carbon-fiber-reinforced polymers. The resins act as a matrix that bind the fibers together, providing strength, stiffness, and durability to the composites. They enable the production of lightweight, yet high-performance, materials that are widely used in aerospace, automotive, and construction industries. One key aspect of impregnating resins is their ability to improve the structural integrity of composite materials. By impregnating fibers or porous structures, the resins enhance the strength, stiffness, and impact resistance of the composite. This is particularly important in industries where lightweight and high-performance materials are sought after, such as in the aerospace and automotive sectors. Therefore, impregnating resins are of significant importance in the polymer resin market due to their ability to enhance the mechanical properties, durability, and protection of materials; this highlights their value and the demand for such specialized resins.

In this paper, rapidly impregnating resins for fiber-reinforced composites are discussed as alternatives to high-performance metal components. An overview of suitable rapidly impregnating resins with low viscosities are introduced, and the differences between thermoset and thermoplastic composites are identified.

Thermoset resins, such as epoxy, polyester, vinyl ester, and DCPD, have excellent dimensional and chemical stabilities and high impact strengths. The epoxy-dicyandiamide system, as a representative commercial epoxy resin, has many strategies which have been implemented to develop low-viscosity, fast-curing epoxy resins, such as the addition of a GDE series and synthesized epoxies. The reaction time decreased with the addition of various particles. The reinforcement of low-viscosity unsaturated polyester resins has also been introduced. Low-viscosity polyester resins can be obtained via particle synthesis. The curing time can be reduced by using various solvents and applying microwaves. Low viscosity, coupled with a rapid curing rate at room temperature, and the relatively low cost of vinyl ester resins, can be obtained using dispersants and various acids to reduce the surface-active properties.

Regarding the thermosetting resins, PA-6, PA-12, PMMA (Elium®), and PBT were introduced, which have high melting viscosities, and they require a high processing temperature and pressure to fully impregnate the fibers and reduce defects in the products. Therefore, in situ polymerization methodologies for fiber-reinforced thermoplastic composites with low viscosities have been developed, and they are suitable for liquid molding processes.

Overall, extensive studies have been conducted on the characterization, analysis, and simulation of rapidly impregnating, resin-based, fiber-reinforced composites. However, the large-scale production of such composites has been rare. Therefore, future research should focus on the large-scale production of composites for the automobile industry, a reduction in their manufacturing time, and an improvement in their performance. In addition, as environmental regulations become stricter, the requirements of automobile materials are also becoming stricter. Some heavy metals and organic substances are banned or restricted to use in automobiles, and automobile parts which cannot be further divided should be merged with homogeneous resins so that they can be recycled more efficiently. Therefore, alternative materials should satisfy the harmlessness to the human body and the environment, and the material itself also needs to satisfy certain performance criteria so that they are comparable to fiber-reinforced or polymer-blended composites. As technologies and industries continue to advance, the importance of rapidly impregnating resins is expected to grow, driven by the need for improved performance, longevity, and reliability of materials and products.

Author Contributions: Conceptualization, M.-X.L.; data curation, H.-L.M. and Y.R.; writing—original draft preparation, M.-X.L., H.-L.M. and S.-K.L.; writing—review and editing, M.-X.L., H.-L.M., S.-K.L., W.Z. and S.-W.C.; visualization, H.-L.M. and Y.R.; supervision, S.-W.C. All authors have read and agreed to the published version of the manuscript.

Funding: This research was supported by the Korea Institute of Energy Technology Evaluation and Planning (KETEP), and the grant was funded by the Korean government (MOTIE) (20213000000020, Development of core equipment and evaluation technology for construction of subsea power grid for offshore wind farm).

Institutional Review Board Statement: Not applicable.

Data Availability Statement: The data presented in this study are available on request from the corresponding author.

Conflicts of Interest: The authors declare no conflict of interest. The funders had no role in the design of the study; in the collection, analyses, or interpretation of data; in the writing of the manuscript; or in the decision to publish the results.

References

1. Annandarajah, C.; Langhorst, A.; Kiziltas, A.; Grewell, D.; Mielewski, D.; Montazami, R. Hybrid cellulose-glass fiber composites for automotive applications. *Materials* **2019**, *12*, 3189. [CrossRef]
2. Warr, C.; Valdoz, J.C.; Bickham, B.P.; Knight, C.J.; Franks, N.A.; Chartrand, N.; Van Ry, P.M.; Christensen, K.A.; Nordin, G.P.; Cook, A.D. Biocompatible PEGDA resin for 3D printing. *ACS Appl. Bio Mater.* **2020**, *3*, 2239–2244. [CrossRef]
3. Moharamzadeh, K.; Ian, M.B.; Richard, V.N. Biocompatibility of resin-based dental materials. *Materials* **2009**, *2*, 514–548. [CrossRef]
4. Guttridge, C.; Shannon, A.; O'Sullivan, A.; O'Sullivan, K.J.; O'Sullivan, L.W. Biocompatible 3D printing resins for medical applications: A review of marketed intended use, biocompatibility certification, and post-processing guidance. *Ann. 3D Print. Med.* **2022**, *5*, 100044. [CrossRef]
5. Nakano, H.; Kato, R.; Kakami, C.; Okamoto, H.; Mamada, K.; Maki, K. Development of biocompatible resins for 3D printing of direct aligners. *J. Photopolym. Sci. Technol.* **2019**, *32*, 209–216. [CrossRef]
6. Thurzo, A.; Kočiš, F.; Novák, B.; Czako, L.; Varga, I. Three-dimensional modeling and 3D printing of biocompatible orthodontic power-arm design with clinical application. *Appl. Sci.* **2021**, *11*, 9693. [CrossRef]
7. Takahashi, J.; Ishikawa, T. Current Japanese activity in CFRTP for industrial application. In Proceedings of the Leuven: International Conference on Textile Composites, Leuven, Belgium, 16–20 September 2013.

8. Matsuo, T.; Uzawa, K.; Orito, Y.; Takahashi, J.; Murayama, H.; Kageyama, K.; Ohsawa, I.; Kanai, M. Investigation about Shear Strength at Welding Area of the Single-Lap Joint and the Scarf Joint for Carbon Fiber Reinforced Thermoplastics. In Proceedings of the American Society for Composites-25th Annual Technical Conference, Dayton, OH, USA, 20–22 September 2010.
9. Ratna, D. *Thermal Properties of Thermosets. Thermosets*; Woodhead Publishing: Cambridge, UK, 2012; pp. 62–91.
10. Zhang, K.; Gu, Y.; Zhang, Z. Effect of rapid curing process on the properties of carbon fiber/epoxy composite fabricated using vacuum assisted resin infusion molding. *Mater. Des. (1980–2015)* **2014**, *54*, 624–631. [CrossRef]
11. Odom, M.G.; Sweeney, C.B.; Parviz, D.; Sill, L.P.; Saed, M.A.; Green, M.J. Rapid curing and additive manufacturing of thermoset systems using scanning microwave heating of carbon nanotube/epoxy composites. *Carbon* **2017**, *120*, 447–453. [CrossRef]
12. Gan, H.; Seraji, S.M.; Zhang, J.; Swan, S.R.; Issazadeh, S.; Varley, R.J. Synthesis of a phosphorus-silicone modifier imparting excellent flame retardancy and improved mechanical properties to a rapid cure epoxy. *React. Funct. Polym.* **2020**, *157*, 104743. [CrossRef]
13. Reichanadter, A.; Bank, D.; Mansson, J.A.E. A novel rapid cure epoxy resin with internal mold release. *Polym. Eng. Sci.* **2021**, *61*, 1819–1828. [CrossRef]
14. Boros, R.; Sibikin, I.; Ageyeva, T.; Kovács, J.G. Development and validation of a test mold for thermoplastic resin transfer molding of reactive PA-6. *Polymers* **2020**, *12*, 976. [CrossRef] [PubMed]
15. Ota, Y.; Yamamoto, T. Improved metal-resin adhesion via electroplating-induced polymer particle adsorption. *Surf. Coat. Technol.* **2020**, *388*, 125591. [CrossRef]
16. Willicombe, K.; Elkington, M.; Hamerton, I.; Ward, C. Development of novel transportation shells for the rapid, automated manufacture of automotive composite parts. *Procedia Manuf.* **2020**, *51*, 818–825. [CrossRef]
17. Zin, M.H.; Razzi, M.F.; Othman, S.; Liew, K.; Abdan, K.; Mazlan, N. *A Review on the Fabrication Method of Bio-Sourced Hybrid Composites for Aerospace and Automotive Applications*; IOP Conference Series: Materials Science and Engineering; IOP Publishing: Bristol, UK, 2016; Volume 152, No. 1.
18. Kalita, D.J.; Tarnavchyk, I.; Kalita, H.; Chisholm, B.J.; Webster, D.C. Novel bio-based epoxy resins from eugenol derived copolymers as an alternative to DGEBA resin. *Prog. Org. Coat.* **2023**, *178*, 107471. [CrossRef]
19. Lakho, D.A.; Yao, Z.; Cho, K.; Ishaq, M.; Wang, Y. Study of the curing kinetics toward development of fast-curing epoxy resins. *Polym.-Plast. Technol. Eng.* **2017**, *56*, 161–170. [CrossRef]
20. Johansson, K.; Bergman, T.; Johansson, M. Hyperbranched aliphatic polyesters and reactive diluents in thermally cured coil coatings. *ACS Appl. Mater. Interfaces* **2009**, *1*, 211–217. [CrossRef]
21. Núñez-Regueira, L.; Villanueva, M.; Fraga-Rivas, I. Effect of a reactive diluent on the curing and dynamomechanical properties of an epoxy-diamine system. *J. Therm. Anal. Calorim.* **2006**, *86*, 463–468. [CrossRef]
22. Villanueva, M.; Fraga, I.; Rodríguez-Añón, J.; Proupín-Castiñeiras, J. Study of the influence of a reactive diluent on the rheological properties of an epoxy-diamine system. *J. Therm. Anal. Calorim.* **2009**, *98*, 521–525. [CrossRef]
23. Costa, M.L.; Rezende, M.C.; Pardini, L.C. Effect on the reactive diluent PGE on the cure ki-netics of an epoxy resin used in structural composites. *Química Nova* **2000**, *23*, 320–325. [CrossRef]
24. Cicala, G.; Recca, G.; Carciotto, S.; Restuccia, C.L. Development of epoxy/hyperbranched blends for resin transfer molding and vacuum assisted resin transfer molding applications: Effect of a reactive diluent. *Polym. Eng. Sci.* **2009**, *49*, 577–584. [CrossRef]
25. Seraji, S.M.; Gan, H.; Issazadeh, S.; Varley, R.J. Investigation of the dual polymerization of rapid curing organophosphorous modified epoxy/amine resins and subsequent flame retardancy. *Macromol. Chem. Phys.* **2021**, *222*, 2000342. [CrossRef]
26. Yang, X.; Huang, W.; Yu, Y. Epoxy toughening using low viscosity liquid diglycidyl ether of ethoxylated bisphenol-A. *J. Appl. Polym. Sci.* **2012**, *123*, 1913–1921. [CrossRef]
27. Wu, X.; Xu, C.A.; Lu, M.; Zheng, X.; Zhan, Y.; Chen, B.; Wang, K.; Meng, H. Preparation and characterization of a low viscosity epoxy resin derived from m-divinylbenzene. *High Perform. Polym.* **2023**, *35*, 153–165. [CrossRef]
28. Zhang, D.; Wang, J.; Li, T.; Zhang, A.; Jia, D. Synthesis and Characterization of a Novel Low-Viscosity Unsaturated Hyperbranched Polyester Resin. *Chem. Eng. Technol.* **2011**, *34*, 119–126. [CrossRef]
29. Chen, H.; Lian, Q.; Xu, W.; Hou, X.; Li, Y.; Wang, Z.; An, D.; Liu, Y. Insights into the synergistic mechanism of reactive aliphatic soft chains and nano-silica on toughening epoxy resins with improved mechanical properties and low viscosity. *J. Appl. Polym. Sci.* **2021**, *138*, 50484. [CrossRef]
30. Chikhi, N.; Fellahi, S.; Bakar, M. Modification of epoxy resin using reactive liquid (ATBN) rubber. *Eur. Polym. J.* **2002**, *38*, 251–264. [CrossRef]
31. Chen, Y.; Zhou, S.; Chen, G.; Wu, L. Preparation and characterization of polyester/silica nanocomposite resins. *Prog. Org. Coat.* **2005**, *54*, 120–126. [CrossRef]
32. Mo, Q.; Huang, Y.; Ma, L.; Lai, W.; Zheng, Y.; Li, Y.; Xu, M.; Huang, Z. Study on Microwave Curing of Unsaturated Polyester Resin and Its Composites Containing Calcium Carbonate. *Polymers* **2022**, *14*, 2598. [CrossRef]
33. Nasr, E.S.; Abdel-Azim, A.A. The effect of curing conditions on the physical and mechanical properties of styrenated polyester. *Polym. Adv. Technol.* **1992**, *3*, 407–411. [CrossRef]
34. Kalaee, M.; Akhlaghi, S.; Nouri, A.; Mazinani, S.; Mortezaei, M.; Afshari, M.; Mostafanezhad, D.; Allahbakhsh, A.; Dehaghi, H.A.; Amirsadri, A.; et al. Effect of nano-sized calcium carbonate on cure kinetics and properties of polyester/epoxy blend powder coatings. *Prog. Org. Coat.* **2011**, *71*, 173–180. [CrossRef]
35. Yong, V.; Hahn, H.T. Rheology of silicon carbide/vinyl ester nanocomposites. *J. Appl. Polym. Sci.* **2006**, *102*, 4365–4371. [CrossRef]

36. Gaur, B.; and Rai, J.S.P. Rheological behavior of vinyl ester resin. *Polym.-Plast. Technol. Eng.* **2006**, *45*, 197–203. [CrossRef]
37. Cook, W.D.; Simon, G.P.; Burchill, P.J.; Lau, M.; Fitch, T.J. Curing kinetics and thermal properties of vinyl ester resins. *J. Appl. Polym. Sci.* **1997**, *64*, 769–781. [CrossRef]
38. Xue, Y.; Zhao, H.; Zhang, Y.; Gao, Z.; Zhai, D.; Li, Q.; Zhao, G. Design and multi-objective optimization of the bumper beams prepared in long glass fiber-reinforced polypropylene. *Polym. Compos.* **2021**, *42*, 2933–2947. [CrossRef]
39. Yang, B.; Lu, L.; Liu, X.; Xie, Y.; Li, J.; Tang, Y. Uniaxial tensile and impact investigation of carbon-fabric/polycarbonate composites with different weave tow widths. *Mater. Des.* **2017**, *131*, 470–480. [CrossRef]
40. Kovács, Z.; Pomázi, Á.; Toldy, A. The flame retardancy of polyamide 6—Prepared by in situ polymerisation of ε-caprolactam—For T-RTM applications. *Polym. Degrad. Stab.* **2022**, *195*, 109797. [CrossRef]
41. Henning, F.; Kärger, L.; Dörr, D.; Schirmaier, F.J.; Seuffert, J.; Bernath, A. Fast processing and continuous simulation of automotive structural composite components. *Compos. Sci. Technol.* **2019**, *171*, 261–279. [CrossRef]
42. Zhang, Y.; Thakur, V.K.; Li, Y.; Garrison, T.F.; Gao, Z.; Gu, J.; Kessler, M.R. Soybean-oil-based thermosetting resins with methacrylated vanillyl alcohol as bio-based, low-viscosity comonomer. *Macromol. Mater. Eng.* **2018**, *303*, 1700278. [CrossRef]
43. Yang, Y.; Zhao, Y.F.; Wang, J.Y.; Zhao, C.; Tong, L.; Liu, X.Y.; Zhang, J.-H.; Zhan, M.-S. Structure and properties of a low viscosity acrylate based flexible epoxy resin. *J. Appl. Polym. Sci.* **2016**, *133*, 42959. [CrossRef]
44. Seraji, S.M.; Gan, H.; Swan, S.R.; Varley, R.J. Phosphazene as an effective flame retardant for rapid curing epoxy resins. *React. Funct. Polym.* **2021**, *164*, 104910. [CrossRef]
45. Seraji, S.M.; Gan, H.; Le, N.D.; Zhang, J.; Varley, R.J. The effect of DOPO concentration and epoxy amine stoichiometry on the rheological, thermal, mechanical and fire-retardant properties of crosslinked networks. *Polym. Int.* **2022**, *71*, 1320–1329. [CrossRef]
46. Wang, H.; Wang, H.; Zhou, G. Synthesis of rosin-based imidoamine-type curing agents and curing behavior with epoxy resin. *Polym. Int.* **2011**, *60*, 557–563. [CrossRef]
47. Zhang, P.; Yao, T.; Xue, K.; Meng, X.; Zhang, J.; Liu, L. Low-dielectric constant and viscosity tetrafunctional bio-based epoxy resin containing cyclic siloxane blocks. *J. Appl. Polym. Sci.* **2022**, *139*, 52176. [CrossRef]
48. Zhang, Z.; Pinnaratip, R.; Ong, K.G.; Lee, B.P. Correlating the mass and mechanical property changes during the degradation of PEG-based adhesive. *J. Appl. Polym. Sci.* **2020**, *137*, 48451. [CrossRef] [PubMed]
49. Ueyama, J.; Ogawa, R.; Ota, K.; Mori, Y.; Tsuge, A.; Endo, T. Rapid Curing System of a Cyanate Ester Resin/Epoxy Resin with a Thermal Latent Polymeric Hardener Based on a Phenol–Amine Salt. *ACS Appl. Polym. Mater.* **2021**, *4*, 84–90. [CrossRef]
50. Yang, B.; Mao, Y.; Zhang, Y.; Bian, G.; Zhang, L.; Wei, Y.; Jiang, Q.; Qiu, Y.; Liu, W. A novel liquid imidazole- copper (II) complex as a thermal latent curing agent for epoxy resins. *Polymer* **2019**, *178*, 121586. [CrossRef]
51. Huo, S.; Yang, S.; Wang, J.; Cheng, J.; Zhang, Q.; Hu, Y.; Ding, G.; Zhang, Q.; Song, P.; Wang, H. A liquid phosphaphenanthrene-derived imidazole for improved flame retardancy and smoke suppression of epoxy resin. *ACS Appl. Polym. Mater.* **2020**, *2*, 3566–3575. [CrossRef]
52. Xu, Y.J.; Chen, L.; Rao, W.H.; Qi, M.; Guo, D.M.; Liao, W.; Wang, Y.Z. Latent curing epoxy system with excellent thermal stability, flame re-tardance and dielectric property. *Chem. Eng. J.* **2018**, *347*, 223–232. [CrossRef]
53. Huo, S.; Yang, S.; Wang, J.; Cheng, J.; Zhang, Q.; Hu, Y.; Ding, G.; Zhang, Q.; Song, P. A liquid phosphorus-containing imidazole derivative as flame-retardant curing agent for epoxy resin with enhanced thermal latency, mechanical, and flame-retardant performances. *J. Hazard. Mater.* **2020**, *386*, 121984. [CrossRef]
54. Xu, Y.J.; Wang, J.; Tan, Y.; Qi, M.; Chen, L.; Wang, Y.Z. A novel and feasible approach for one-pack flame-retardant epoxy resin with long pot life and fast curing. *Chem. Eng. J.* **2018**, *337*, 30–39. [CrossRef]
55. Koottatep, T. *Non-Recyclable Plastics: Management Practices and Implications*; WA Publishing: London, UK, 2023.
56. Nurazzi, N.M.; Khalina, A.; Sapuan, S.M.; Laila, A.D.; Rahmah, M. Curing behaviour of unsaturated polyester resin and interfacial shear stress of sugar palm fibre. *Jour-Nal Mech. Eng. Sci.* **2017**, *11*, 2650–2664. [CrossRef]
57. Zhou, X.; Zhou, R.; Zuo, J.; Tu, S.; Yin, Y.; Ye, L. Unsaturated polyester resins with low-viscosity: Preparation and mechanical properties enhancement by isophorone diisocyanate (IPDI) modification. *Mater. Res. Express* **2019**, *6*, 115305. [CrossRef]
58. Hazarika, A.; Deka, B.K.; Kim, D.; Kong, K.; Park, Y.B.; Park, H.W. Microwave-synthesized freestanding iron-carbon nanotubes on polyester composites of woven Kevlar fibre and silver nanoparticle-decorated graphene. *Sci. Rep.* **2017**, *7*, 40386. [CrossRef] [PubMed]
59. Piszko, P.; Kryszak, B.; Piszko, A.; Szustakiewicz, K. Brief review on poly (glycerol sebacate) as an emerging polyester in biomedical application: Struc-ture, properties and modifications. *Polym. Med* **2021**, *51*, 43–50. [CrossRef]
60. Yuan, Z.; Liu, Q.; Pan, X.; Wang, J.; Jin, M.; Li, J. Preparation and properties of star-shaped UV-curable polyester methacrylate resins with Low viscosity derived from renewable resources. *Prog. Org. Coat.* **2021**, *157*, 106324. [CrossRef]
61. Chirayil, C.J.; Mathew, L.; Hassan, P.A.; Mozetic, M.; Thomas, S. Rheological behaviour of nanocellulose reinforced unsaturated polyester nanocomposites. *Int. J. Biol. Macromol.* **2014**, *69*, 274–281. [CrossRef]
62. Dang, T.K.M.; Nikzad, M.; Arablouei, R.; Masood, S.; Bui, D.K.; Truong, V.K.; Sbarski, I. Thermomechanical Properties and Fracture Toughness Improvement of Thermosetting Vinyl Ester Using Liquid Metal and Graphene Nanoplatelets. *Polymers* **2022**, *14*, 5397. [CrossRef]
63. Mazali, C.A.I.; Felisberti, M.I. Vinyl ester resin modified with silicone-based additives: III. Curing kinetics. *Eur. Polym. J.* **2009**, *45*, 2222–2233. [CrossRef]
64. Rosu, L.; Cascaval, C.N.; Rosu, D. Curing of vinyl ester resins. *Rheol. Behaviour. J. Optoelectron. Adv. Mater.* **2006**, *8*, 690.

65. Grand View Research. Plastic Market Size, Share & Trends Analysis Report by Product (PE, PP, PU, PVC, PET, Polystyrene, ABS, PBT, PPO, Epoxy Polymers, LCP, PC, Polyamide), by Application, by Region, and Segment Forecasts, 2020–2027. In *Market Analysis Report*; Grand View Research: San Francisco, CA, USA, 2020.
66. Li, H.; Wang, Z.; He, B. Ring-opening metathesis polymerization of dicyclopentadiene by tungsten catalysts supported on polystyrene. *J. Mol. Catal. A Chem.* **1999**, *147*, 83–88. [CrossRef]
67. Kessler, M.R.; White, S.R. Cure kinetics of the ring-opening metathesis polymerization of dicyclopentadiene. *J. Polym. Sci. Part A Polym. Chem.* **2002**, *40*, 2373–2383. [CrossRef]
68. Yang, G.; Lee, J.K. Effect of Grubbs' catalysts on cure kinetics of endo-dicyclopentadiene. *Thermochim. Acta* **2013**, *566*, 105–111. [CrossRef]
69. Yang, G.; Lee, J.K. Curing kinetics and mechanical properties of endo-dicyclopentadiene synthesized using different Grubbs' catalysts. *Ind. Eng. Chem. Res.* **2014**, *53*, 3001–3011. [CrossRef]
70. Yoo, H.M.; Jeon, J.H.; Li, M.X.; Lee, W.I.; Choi, S.W. Analysis of curing behavior of endo-dicyclopentadiene using different amounts of decelerator solution. *Compos. Part B Eng.* **2019**, *161*, 439–454. [CrossRef]
71. Sideridou, I.D.; Achilias, D.S.; Karava, O. Reactivity of benzoyl peroxide/amine system as an initiator for the free radical polymerization of dental and orthopaedic dimethacrylate monomers: Effect of the amine and mono-mer chemical structure. *Macromolecules* **2006**, *39*, 2072–2080. [CrossRef]
72. Krawczak, P. Recycling of thermoset structural composites: Would textile technology bring a high added-value solution? *Express Polym. Lett.* **2011**, *5*, 238. [CrossRef]
73. Akca, E.; Gürsel, A. A review on superalloys and IN718 nickel-based INCONEL superalloy. *Period. Eng. Nat. Sci.* **2015**, *3*. [CrossRef]
74. Ben, G.; Hirabayashi, A.; Sakata, K.; Nakamura, K.; Hirayama, N. Evaluation of new GFRTP and CFRTP using epsilon caprolactam as matrix fabricated with VaRTM. *Sci. Eng. Compos. Mater.* **2015**, *22*, 633–641. [CrossRef]
75. Ben, G.; Sakata, K. Fast fabrication method and evaluation of performance of hybrid FRTPs for applying them to automotive structural members. *Compos. Struct.* **2015**, *133*, 1160–1167. [CrossRef]
76. Li, M.X.; Lee, D.; Lee, G.H.; Kim, S.M.; Ben, G.; Lee, W.I.; Choi, S.W. Effect of temperature on the mechanical properties and polymerization kinetics of polyamide-6 composites. *Polymers* **2020**, *12*, 1133. [CrossRef]
77. Li, M.X.; Mo, H.L.; Ren, Y.; Choi, S.W. Carbon Fiber-Reinforced Polyamide 6 Composites Formed by In Situ Polymerization—Experimental and Numerical Analysis of the Influence of Polymerization Temperature. *Coatings* **2022**, *12*, 947. [CrossRef]
78. Zingraff, L.; Michaud, V.; Bourban, P.E.; Månson, J.A. Resin transfer moulding of anionically polymerised polyamide 12. *Compos. Part A Appl. Sci. Manuf.* **2005**, *36*, 1675–1686. [CrossRef]
79. Máirtín, P.Ó.; McDonnell, P.; Connor, M.T.; Eder, R.; Brádaigh, C.Ó. Process investigation of a liquid PA-12/carbon fibre moulding system. *Compos. Part A Appl. Sci. Manuf.* **2001**, *32*, 915–923. [CrossRef]
80. Banerjee, M.; Sain, S.; Mukhopadhyay, A.; Sengupta, S.; Kar, T.; Ray, D. Surface treatment of cellulose fibers with methyl-methacrylate for enhanced properties of in situ polymerized PMMA/cellulose composites. *J. Appl. Polym. Sci.* **2014**, *131*, 32. [CrossRef]
81. Abt, T.; Sanchez-Soto, M. A review of the recent advances in cyclic butylene terephthalate technology and its composites. *Crit. Rev. Solid State Mater. Sci.* **2017**, *42*, 173–217. [CrossRef]
82. Yan, C.; Liu, L.; Zhu, Y.; Xu, H.; Liu, D. Properties of polymerized cyclic butylene terephthalate and its composites via ring-opening polymerization. *J. Thermoplast. Compos. Mater.* **2018**, *31*, 181–201. [CrossRef]
83. Pervaiz, M.; Faruq, M.; Jawaid, M.; Sain, M. Polyamides: Developments and applications towards next-generation engineered plastics. *Curr. Org. Synth.* **2017**, *14*, 146–155. [CrossRef]
84. Global Market Study on Methyl Methacrylate (MMA)—Increasing Application in PMMA Production to Account for Significant Revenue Generation Opportunities. Available online: http://www.persistencemarketresearch.com/market-research/methyl-methacrylate-market.asp (accessed on 17 September 2023).
85. Van Rijswijk, K.; Bersee, H.E.N. Reactive processing of textile fiber-reinforced thermoplastic composites–An overview. *Compos. Part A Appl. Sci. Manuf.* **2007**, *38*, 666–681. [CrossRef]
86. Rosso, P.; Friedrich, K.; Wollny, A.; Mülhaupt, R. A novel polyamide 12 polymerization system and its use for a LCM-process to produce CFRP. *J. Thermoplast. Compos. Mater.* **2005**, *18*, 77–90. [CrossRef]
87. Wei, Y.; Cui, M.; Ye, Z.; Guo, Q. Environmental challenges from the increasing medical waste since SARS outbreak. *J. Clean. Prod.* **2021**, *291*, 125246. [CrossRef]
88. Bodaghi, M.; Park, C.H.; Krawczak, P. Reactive processing of acrylic-based thermoplastic composites: A mini-review. *Front. Mater.* **2022**, *9*, 931338. [CrossRef]
89. Tran, N.-T.; Pham, N.T.-H. Investigation of the effect of polycarbonate rate on mechanical properties of polybutylene terephthalate/polycarbonate blends. *Int. J. Polym. Sci.* **2021**, *2021*, 7635048. [CrossRef]
90. Davoodi, M.M.; Sapuan, S.M.; Ahmad, D.; Aidy, A.; Khalina, A.; Jonoobi, M. Effect of polybutylene terephthalate (PBT) on impact property improvement of hybrid kenaf/glass epoxy composite. *Mater. Lett.* **2012**, *67*, 5–7. [CrossRef]
91. Ahmadi, S.; Morshedian, J.; Hashemi, S.A.; Carreau, P.J.; Leelapornpisit, W. Novel anionic polymerization of ε-caprolactam towards polyamide 6 containing nanofibrils. *Iran. Polym. J.* **2010**, *19*, 229–240.

92. Dabees, S.; Osman, T.; Kamel, B.M. Mechanical, thermal, and flammability properties of polyamide-6 reinforced with a combination of carbon nanotubes and titanium dioxide for under-the-hood applications. *J. Thermoplast. Compos. Mater.* **2023**, *36*, 1545–1575. [CrossRef]
93. Aydogan, Y.; Atabek Savas, L.; Erdem, A.; Hacioglu, F.; Dogan, M. Performance evaluation of various phosphorus compounds on the flammability properties of short carbon fiber-reinforced polyamide 6 composites. *Fire Mater.* **2022**, *47*, 837–847. [CrossRef]
94. Wursthorn, L.; Beckett, K.; Rothbaum, J.O.; Cywar, R.M.; Lincoln, C.; Kratish, Y.; Marks, T.J. Selective Lanthanide-Organic Catalyzed Depolymerization of Nylon-6 to ε-Caprolactam. *Angew. Chem. Int. Ed.* **2023**, *62*, e202212543. [CrossRef]
95. Wei, X.F.; Kallio, K.J.; Olsson, R.T.; Hedenqvist, M.S. Performance of glass fiber reinforced polyamide composites exposed to bioethanol fuel at high temperature. *npj Mater. Degrad.* **2022**, *6*, 69. [CrossRef]
96. He, D.; Kim, H.C.; De Kleine, R.; Soo, V.K.; Kiziltas, A.; Compston, P.; Doolan, M. Life cycle energy and greenhouse gas emissions implications of polyamide 12 recycling from selective laser sintering for an injection-molded automotive component. *J. Ind. Ecol.* **2022**, *26*, 1378–1388. [CrossRef]
97. Kazemi, M.E.; Shanmugam, L.; Dadashi, A.; Shakouri, M.; Lu, D.; Du, Z.; Hu, Y.; Wang, J.; Zhang, W.; Yang, L.; et al. Investigating the roles of fiber, resin, and stacking sequence on the low-velocity impact response of novel hybrid thermoplastic composites. *Compos. Part B Eng.* **2021**, *207*, 108554. [CrossRef]
98. Bhudolia, S.K.; Gohel, G.; Fai, L.K.; Barsotti Jr, R.J. Fatigue response of ultrasonically welded carbon/Elium® thermoplastic composites. *Mater. Lett.* **2020**, *264*, 127362. [CrossRef]
99. Bhudolia, S.K.; Gohel, G.; Leong, K.F.; Barsotti Jr, R.J. Investigation on ultrasonic welding attributes of novel carbon/elium® composites. *Materials* **2020**, *13*, 1117. [CrossRef] [PubMed]
100. Obande, W.; Brádaigh, C.M.Ó.; Ray, D. Continuous fibre-reinforced thermoplastic acrylic-matrix composites prepared by liquid resin infusion–A review. *Compos. Part B Eng.* **2021**, *215*, 108771. [CrossRef]
101. Obande, W.; Mamalis, D.; Ray, D.; Yang, L.; Brádaigh, C.M.Ó. Mechanical and thermomechanical characterisation of vacuum-infused thermoplastic-and thermoset-based composites. *Mater. Des.* **2019**, *175*, 107828. [CrossRef]
102. Allagui, S.; El Mahi, A.; Rebiere, J.L.; Beyaoui, M.; Bouguecha, A.; Haddar, M. Effect of recycling cycles on the mechanical and damping properties of flax fibre reinforced elium composite: Experimental and numerical studies. *J. Renew. Mater.* **2021**, *9*, 695. [CrossRef]
103. Brunelle, D.J.; Bradt, J.E.; Serth-Guzzo, J.; Takekoshi, T.; Evans, T.L.; Pearce, E.J.; Wilson, P.R. Semicrystalline polymers via ring-opening polymerization: Preparation and polymerization of alkylene phthalate cyclic oligomers. *Macromolecules* **1998**, *31*, 4782–4790. [CrossRef] [PubMed]
104. Ishak, Z.A.M.; Gatos, K.G.; Karger-Kocsis, J. On the in-situ polymerization of cyclic butylene terephthalate oligomers: DSC and rheological studies. *Polym. Eng. Sci.* **2006**, *46*, 743–750. [CrossRef]
105. Parton, H.; Baets, J.; Lipnik, P.; Goderis, B.; Devaux, J.; Verpoest, I. Properties of poly (butylene terephthatlate) polymerized from cyclic oligomers and its composites. *Polymer* **2005**, *46*, 9871–9880. [CrossRef]
106. Baets, J.; Godara, A.; Devaux, J.; Verpoest, I. Toughening of polymerized cyclic butylene terephthalate with carbon nanotubes for use in composites. *Compos. Part A Appl. Sci. Manuf.* **2008**, *39*, 1756–1761. [CrossRef]
107. Broza, G.; Kwiatkowska, M.; Rosłaniec, Z.; Schulte, K. Processing and assessment of poly (butylene terephthalate) nanocomposites reinforced with oxidized single wall carbon nanotubes. *Polymer* **2005**, *46*, 5860–5867. [CrossRef]
108. Swolfs, Y.; Gorbatikh, L.; Verpoest, I. Fibre hybridisation in polymer composites: A review. *Compos. Part A Appl. Sci. Manuf.* **2014**, *67*, 181–200. [CrossRef]
109. Baets, J.; Godara, A.; Devaux, J.; Verpoest, I. Toughening of isothermally polymerized cyclic butylene terephthalate for use in composites. *Polym. Degrad. Stab.* **2010**, *95*, 346–352. [CrossRef]
110. Yang, B.; Wang, Z.; Zhou, L.; Zhang, J.; Liang, W. Experimental and numerical investigation of interply hybrid composites based on woven fabrics and PCBT resin subjected to low-velocity impact. *Compos. Struct.* **2015**, *132*, 464–476. [CrossRef]
111. Abt, T.; Sánchez-Soto, M.; de Ilarduya, A.M. Toughening of in situ polymerized cyclic butylene terephthalate by addition of tetrahydrofuran. *Polym. Int.* **2011**, *60*, 549–556. [CrossRef]
112. Baets, J.; Devaux, J.; Verpoest, I. Toughening of basalt fiber-reinforced composites with a cyclic butylene terephthalate matrix by a nonisothermal production method. *Adv. Polym. Technol.* **2010**, *29*, 70–79. [CrossRef]
113. Cao, Y.; Xu, P.; Wu, B.; Hoch, M.; Lemstra, P.J.; Yang, W.; Dong, W.; Du, M.; Liu, T.; Ma, P. High-performance and functional PBT/EVMG/CNTs nanocomposites from recycled sources by in situ multistep reaction-induced interfacial control. *Compos. Sci. Technol.* **2020**, *190*, 108043. [CrossRef]
114. Emmenegger, C.; Norman, D. The challenges of automation in the automobile: Commentary on Hancock (2019) Some pitfalls in the promises of automated and autonomous vehicles. *Ergonomics* **2019**, *62*, 512–513. [CrossRef]
115. Gérard, P. Sustainable Management of Manufacturing Wastes and End-Of-Life Parts of Novel Fully Recyclable Thermoplastic Composites. In Proceedings of the SFIP International Congress "Plastics Industry and Environment", Douai, France, 18–19 May 2022.
116. Ravina, M.; Bianco, I.; Ruffino, B.; Minardi, M.; Panepinto, D.; Zanetti, M. Hard-to-recycle plastics in the automotive sector: Economic, environmental and technical analyses of possible actions. *J. Clean. Prod.* **2023**, *394*, 136227. [CrossRef]
117. Miller, L.; Soulliere, K.; Sawyer-Beaulieu, S.; Tseng, S.; Tam, E. Challenges and alternatives to plastics recycling in the automotive sector. *Materials* **2014**, *7*, 5883–5902. [CrossRef]

118. Zhang, W.; Xu, J. Advanced lightweight materials for Automobiles: A review. *Mater. Des.* **2022**, *221*, 110994. [CrossRef]
119. Roy, K.; Debnath, S.C.; Pongwisuthiruchte, A.; Potiyaraj, P. Recent advances of natural fibers based green rubber composites: Properties, current status, and future perspectives. *J. Appl. Polym. Sci.* **2021**, *138*, 50866. [CrossRef]
120. Hu, X.B.; Cui, X.Y.; Liang, Z.M.; Li, G.Y. The performance prediction and optimization of the fiber-reinforced composite structure with uncertain param-eters. *Compos. Struct.* **2017**, *164*, 207–218. [CrossRef]
121. Lee, K.H.; Choe, C.R.; Yoon, B.I. Resin impregnation efficiency of carbon-carbon composites. *J. Mater. Sci. Lett.* **1993**, *12*, 199–200. [CrossRef]
122. Pugnetti, M.; Zhou, Y.; Biedermann, A.R. Ferrofluid impregnation efficiency and its spatial variability in natural and synthetic porous media: Implications for magnetic pore fabric studies. *Transp. Porous Media* **2022**, *144*, 367–400. [CrossRef]
123. Kobayashi, S.; Tsukada, T.; Morimoto, T. Resin impregnation behavior in carbon fiber reinforced polyamide 6 composite: Effects of yarn thickness, fabric lamination and sizing agent. *Compos. Part A Appl. Sci. Manuf.* **2017**, *101*, 283–289. [CrossRef]
124. Valente, M.; Rossitti, I.; Sambucci, M. Different Production Processes for Thermoplastic Composite Materials: Sustainability versus Mechanical Properties and Processes Parameter. *Polymers* **2023**, *15*, 242. [CrossRef]
125. Herring, M.L.; Fox, B.L. The effect of a rapid curing process on the surface finish of a carbon fibre epoxy composite. *Compos. Part B Eng.* **2011**, *42*, 1035–1043. [CrossRef]
126. Kazano, S.; Osada, T.; Kobayashi, S.; Goto, K. Experimental and analytical investigation on resin impregnation behavior in continuous carbon fiber reinforced thermoplastic polyimide composites. *Mech. Adv. Mater. Mod. Process.* **2018**, *4*, 6. [CrossRef]
127. Zhao, C.; Yang, B.; Wang, S.; Ma, C.; Wang, S.; Bi, F. Three-dimensional numerical simulation of meso-scale-void formation during the mold-filling process of LCM. *Appl. Compos. Mater.* **2019**, *26*, 1121–1137.
128. Tzetzis, D.; Hogg, P.J. The influence of surface morphology on the interfacial adhesion and fracture behavior of vacuum infused carbon fiber reinforced polymeric repairs. *Polym. Compos.* **2008**, *29*, 92–108. [CrossRef]
129. Maiti, S.; Islam, M.R.; Uddin, M.A.; Afroj, S.; Eichhorn, S.J.; Karim, N. Sustainable fiber-reinforced composites: A Review. *Adv. Sustain. Syst.* **2022**, *6*, 2200258.
130. Huang, S.; Fu, Q.; Yan, L.; Kasal, B. Characterization of interfacial properties between fibre and polymer matrix in composite materials–A critical review. *J. Mater. Res. Technol.* **2021**, *13*, 1441–1484.
131. Rodrigues, I.; Mata, T.M.; Martins, A.A. Environmental analysis of a bio-based coating material for automobile interiors. *J. Clean. Prod.* **2022**, *367*, 133011.
132. Sharma, H.; Kumar, A.; Rana, S.; Sahoo, N.G.; Jamil, M.; Kumar, R.; Sharma, S.; Li, C.; Kumar, A.; Eldin, S.M.; et al. Critical review on advancements on the fiber-reinforced composites: Role of fiber/matrix modification on the performance of the fibrous composites. *J. Mater. Res. Technol.* **2023**, *26*, 2975–3002.
133. Zhang, J.; Chevali, V.S.; Wang, H.; Wang, C.H. Current status of carbon fibre and carbon fibre composites recycling. *Compos. Part B Eng.* **2020**, *193*, 108053.

Disclaimer/Publisher's Note: The statements, opinions and data contained in all publications are solely those of the individual author(s) and contributor(s) and not of MDPI and/or the editor(s). MDPI and/or the editor(s) disclaim responsibility for any injury to people or property resulting from any ideas, methods, instructions or products referred to in the content.

Article

Cashew Nutshells: A Promising Filler for 3D Printing Filaments

María José Paternina Reyes [1], Jimy Unfried Silgado [1], Juan Felipe Santa Marín [2,†], Henry Alonso Colorado Lopera [3] and Luis Armando Espitia Sanjuán [1,*]

1. Engineering, Science and Technology Research Group, Mechanical Engineering Department, University of Córdoba, Cr. 6 No. 76-103, Montería 230002, Córdoba, Colombia; mpaterninareyes34@correo.unicordoba.edu.co (M.J.P.R.); jimyunfried@correo.unicordoba.edu.co (J.U.S.)
2. Grupo de Investigación Materiales Avanzados y Energía—MATyER, Instituto Tecnológico Metropolitano, Medellín 050012, Antioquia, Colombia; jfsanta@unal.edu.co
3. CCComposites Laboratory, University of Antioquia, Calle 67 No. 53-108, Medellín 050010, Antioquia, Colombia; henry.colorado@udea.edu.co
* Correspondence: luisespitia@correo.unicordoba.edu.co
† Current address: Tribology and Surfaces Group, Universidad Nacional de Colombia, Medellín 050034, Antioquia, Colombia.

Citation: Paternina Reyes, M.J.; Unfried Silgado, J.; Santa Marín, J.F.; Colorado Lopera, H.A.; Espitia Sanjuán, L.A. Cashew Nutshells: A Promising Filler for 3D Printing Filaments. *Polymers* 2023, *15*, 4347. https://doi.org/10.3390/polym15224347

Academic Editors: Patricia Krawczak, Chung Hae Park and Abderrahmane Ayadi

Received: 20 September 2023
Revised: 23 October 2023
Accepted: 25 October 2023
Published: 7 November 2023

Copyright: © 2023 by the authors. Licensee MDPI, Basel, Switzerland. This article is an open access article distributed under the terms and conditions of the Creative Commons Attribution (CC BY) license (https://creativecommons.org/licenses/by/4.0/).

Abstract: Cashew nutshells from the northern region of Colombia were prepared to assess their potential use as a filler in polymer matrix filaments for 3D printing. After drying and grinding processes, cashew nutshells were characterized using scanning electron microscopy (SEM), attenuated total reflectance Fourier-transform infrared (ATR-FTIR), and thermogravimetric analyses (TGA). Three different filaments were fabricated from polylactic acid pellets and cashew nutshell particles at 0.5, 1.0, and 2.0 weight percentages using a single-screw extruder. Subsequently, single-filament tensile tests were carried out on them. SEM images showed rough and porous particles composed of an arrangement of cellulose microfibrils embedded in a hemicellulose and lignin matrix, the typical microstructure reported for natural fibers. These characteristics observed in the particles are favorable for improving filler–matrix adhesion in polymer matrix composites. In addition, their low density of 0.337 g/cm^3 makes them attractive for lightweight applications. ATR-FTIR spectra exhibited specific functional groups attributed to hemicellulose, cellulose, and lignin, as well as a possible transformation to crystalline cellulose during drying treatment. According to TGA analyses, the thermal stability of cashew nutshell particles is around 320 °C. The three polylactic acid–cashew nutshell particle filaments prepared in this work showed higher tensile strength and elongation at break when compared to polylactic acid filament. The characteristics displayed by these cashew nutshell particles make them a promising filler for 3D printing filaments.

Keywords: natural fiber-reinforced polymers; cashew nutshell particles; 3D printing filament

1. Introduction

A remarkable augmentation in the use of natural fibers as reinforcement or dispersed phase in polymer matrix composites to replace pure polymers or traditional reinforcements has attracted a great deal of attention in the past years; however, these bio-composites exhibit large variability in their properties. The electrical, mechanical, and rheological properties of date palm waste-derived biochar–polypropylene matrix composites were investigated [1]. The authors reported that the increase in biochar content decreased tensile strength and ductility, but increased the modulus of elasticity, the electrical conductivity, and the storage moduli of composites. The effect of fiber content and fiber treatment on the mechanical behavior of banana fiber–polyethylene matrix composites was assessed [2]. In all cases, the 20 weight percentage fiber addition increased tensile strength, modulus of elasticity, and elongation at break. However, both 40 and 60 weight percentages, as a function of treatment, increased and decreased these properties, a fact attributed to the different chemical composition (lignin and cellulose contents) and the degree of crystallinity produced on the fibers by each treatment. Other researchers prepared bio-composites by compounding unmodified and acylated cork powder with polylactic acid (PLA), and polycaprolactone (PCL) matrices at 1, 5, 10, 20, and 30 weight percentages of powder [3]. The results indicated that unmodified and acylated cork contents equal or lower to 10 mass percentage slightly affected the modulus of elasticity of both composites, whereas a considerable decrease in this property was observed for 20 and 30 mass percentages. In contrast, tensile strength decreased with the augment on cork content, regardless of treatment and polymer matrix. The weathering performance of bio-composites has been evaluated by some researchers. The effect of moisture on the mechanical properties of PLA matrix composites reinforced with ramie, flax, and cotton fibers at 10, 20, 30, 40, and 50 mass percentages was reported [4]. Independently of moisture and fiber content, the fiber addition improved the flexural properties of composites; however, impact behavior decreased with increasing fiber content. Furthermore, the weathering resistance of bio-composites produced by the addition of microcrystalline cellulose and nutshell fibers into a high-density polyethylene matrix was investigated [5]. A weathering test was used to simulate outdoor degrading factors for 672 h. It was reported that bio-composites with nutshell fibers were less affected by weathering exposure. The thermal degradation and fire resistance of unsaturated polyester (UP) and unsaturated polyester with acrylic acid (Modar) matrices reinforced with different natural fibers (including jute, flax, and sisal at a 30 volume percentage) was compared to glass fiber composites [6]. According to the results, Modar matrix composites were more resistant to temperature than UP matrix composites. Nevertheless, the fire risk was similar between them. Among the fibers, flax fibers are the most adequate to be used. They showed the best thermal resistance due to their low lignin content, a long time to ignition, and a long period before reaching the flashover. Glass fiber composites showed more flame resistance than the bio-composites, but exhibited higher emissions of CO and CO_2. Other disadvantages reported for synthetic fibers include a limited recycling capability that increases their carbon footprint due to further CO_2 emissions, unsustainable production processes, the high energy involved in their manufacturing processes, and their nonbiodegradability [7,8]. Composites constituted by both synthetic and natural fibers and polymer matrix have also been reported. Hemp as continuous fiber reinforcement and palm shell and coconut shell powders as particle reinforcements were used to obtain hybrid epoxy matrix composites [9]. The researchers showed that the inclusion of an optimum amount of coconut shell and palm shell powders improved tensile strength, failure strain, flexural strength, resistance to shock, and hardness. Consequently, these composites may be used in structural applications. Thus, the addition of fibers and particles obtained from agroindustrial residues on polymer matrix composites may either increase or decrease properties. This dispersed behavior exhibited by natural fibers is associated with growing conditions and plant maturity at harvest, poor fiber extracting and processing methods, and differences in farming practices, among others. Despite this, natural fibers are low-cost, biodegradable, originate from renewable resources, exhibit low density and have other characteristics that make them attractive for lightweight applications. Particularly, weak interfacial bonding emerges as a common issue in natural fiber–polymer matrix composites, promoting variability in mechanical properties and low load transfers from matrix to fibers [7]. Some characteristics of natural fibers are considered relevant to increase interfacial bonding, such as rough and porous surfaces, high specific surface area, chemical composition (the content of hemicellulose, cellulose, and lignin), and crystallinity [1,2,5,7]. Efforts towards enhancing natural fiber–polymer matrix adhesion are mandatory to overcome these adversities.

The region of Córdoba in Colombia mainly bases its economy on agriculture. Cashew plantation is one of the crops grown in the region. Besides the nut, a rich-phenol liquid known as cashew nutshell liquid (CNSL) is extracted from cashew nutshells. This liquid provides essential components used in the fabrication of sustainable and green products, such as resins, rubbers, biodiesel, and insecticides [10,11]. Other derivative and value-added products obtained from cashew nutshell, cashew apple, and cashew biomass can be found in the literature [12–17]. Unfortunately, Córdoba's farmers extract only the nuts, and the cashew nutshell residues are not used. Occasionally, these residues are improperly disposed of in open fields, causing environmental damage associated with the pH of the anacardic acid from the cashew nutshells [18]. Moreover, these residues are available at low or no cost because they are considered waste material, and fillers fabricated from cashew nutshells are not commercially available in the market. Therefore, a feasible alternative to convert these agroindustrial residues into valuable filaments for 3D printing arises, promoting economic and environmental benefits. Although research has been oriented toward the isolation and application of cellulose, CNSL, anacardic acid [16,19–23], and biochar production from cashew nutshells [24,25], few studies have been developed on evaluating their behavior in polymer matrix composites [26], and no work was found on their use on filaments for 3D printing.

Additive manufacturing techniques such as 3D and 4D printing comprise a promissory set of material processes and allow the fabrication of shape-memory materials [27], biocompatible materials [28], blending materials [29], filler powder-reinforced materials [30], and continuous or discontinuous fiber-reinforced composite materials [31], among others. Many high-value characteristics such as sustainability, assembling complexity, and flexibility are present in these manufacturing processes [32,33]. Particularly, the development of filaments containing cashew nutshell particles is not only a novel approach for the development of bio-composites but also allows the possibility of combining the biodegradability of natural fibers and PLA with the precise component fabrication achieved by 3D printing according to customer requirements [34,35]. In this work, cashew nutshell particles were prepared and their potential use as filler for filaments for 3D printing was assessed in terms of surface characteristics, apparent density, thermal stability, and content of hemicellulose, cellulose, and lignin. Furthermore, single-filament tensile tests were carried out on filaments fabricated with polylactic acid and three different mass percentages of cashew nutshell particles.

2. Materials and Methods

Cashew nutshells were kindly provided by Asopromarsab, a producer organization located in Chinú, Córdoba, Colombia, as shown in Figure 1. Cashew nutshells were received after manual removal of the nut. No chemical, physical, or any other treatment was carried out on them. From this point on, this original condition is referenced as the as-received condition.

Figure 1. Localization of Asopromarsab and cashew nutshells used in this work.

Cashew nutshells were dried at 250 °C for 30 min in a muffle furnace and air-cooled to room temperature. Shore D hardness was measured according to the ASTM D2240 standard [36], and afterwards, the nutshells were submitted to a grinding process.

2.1. Apparent Density and Morphology Analysis

Apparent density was calculated using a predefined volume container of 123.45 cm^3, a scale of 0.001 g, and the following equation:

$$\rho_a = \frac{m}{v} \quad (1)$$

where m is the mass of the particles deposited in the container (g), v is the volume of the particles including internal pores (cm^3), and ρ_a is the apparent density (g/cm^3). In addition, the morphology, surface characteristics, and microstructure of the ground material were assessed by scanning electron microscopy (SEM).

2.2. Bromatological Analyses

Bromatological analysis was developed on the dried-cashew nutshells. Mass percentages of dry matter, ashes, acid detergent fiber (ADF), neutral detergent fiber (NDF), and cellulose were determined according to standard methods proposed by AOAC International [37]. Equations (2) and (3) were used to calculate the mass percentages of lignin and hemicellulose:

$$\text{Lignin} = \text{ADF} - \text{Cellulose} \quad (2)$$

$$\text{Hemicellulose} = \text{NDF} - \text{ADF} \quad (3)$$

2.3. ATR-FTIR and TGA Analyses

Attenuated total reflectance Fourier-transform infrared (ATR-FTIR) spectra were collected using a resolution of 4 cm^{-1} from 4000 to 500 cm^{-1}. Thermogravimetric analyses (TGA) were carried out from 24 °C to 900 °C using a heating rate of 5 °C/min in a nitrogen atmosphere. The first derivative of the TGA curve (the dW/dT curve) was also plotted to determine inflection points useful for in-depth interpretation of the thermal behavior of cashew nutshell particles. In both analyses, the as-received condition was used for comparison purposes.

2.4. Preparation of 3D Filaments

Three different filaments were fabricated from polylactic acid (PLA) pellets and cashew nutshell particles at 0.5, 1.0, and 2.0 weight percentages (wt%) using a single-screw extruder. The maximum value of 2.0 wt% of particles was selected to avoid nozzle clogging during the 3D printing process. The filaments were continuously produced at 190 °C, with an extrusion speed of 7 RPM and a nozzle diameter of 1 mm. PLA filament for comparison purposes was also fabricated. The nominal properties of the PLA pellets used in this work are displayed in Table 1.

Table 1. Nominal properties of the PLA pellets used in this work.

Properties	Values
Modulus of elasticity	3450 MPa
Tensile strength	63 MPa
Tensile strain at tensile strength	4%
Tensile stress at break	44 MPa
Tensile strain at break	10%
Melting temperature	>155 °C
Density	1.25 g/cm^3

2.5. Tensile Tests

Single-filament tensile tests were carried out to measure tensile strength and elongation at break percentage. The ends of a 100 mm-length single filament were glued on a 30 × 110 mm cardboard. The tests were conducted using a testing speed of 5 mm/min and a gauge length of 50 mm. The cross-sectional area of the filaments was calculated using the mean diameter determined by digital image analysis. Five tests were performed on each filament.

3. Results and Discussion

Figure 2 shows the cashew nutshell particles produced after the drying and grinding processes. Apparent density and Shore hardness were 0.337 g/cm^3 and 39.67 ± 12.80 D, respectively. These low values agree with those reported for other natural fibers, and are desirable for bio-composite manufacturing due to the opportunity to produce lightweight and biodegradable components [38,39].

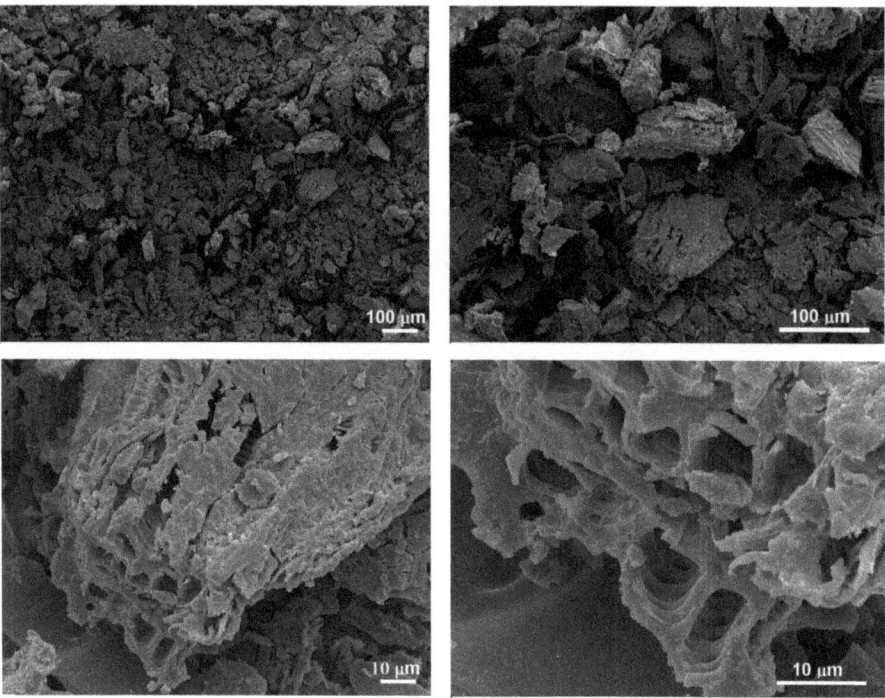

Figure 2. Morphology of cashew nutshell particles after drying and grinding processes. Scanning electron microscopy images.

Figure 2 shows angular particles with a rough and porous surface and heterogeneous particle sizes ranging from ~7 to ~150 μm. Irregular and porous particles and even surfaces may enhance the filler matrix bonding in composites due to higher mechanical interlocking [7]. The influence of fillers on the properties of polymer matrix composites has been evaluated by some researchers [9]. They reported that the rigid and 3D irregular shapes of coconut shell particles provided mechanical support and were suitable for distributing stresses. The composites evidenced a higher ultimate strength of 22.9 ± 2.49 MPa; however, as the particles are nonporous, the mechanical interlocking was lower, resulting in lower failure strain values. In contrast, polymer matrix composites with palm shell particles showed extensive interlocking due to their porosity, promoting better failure strain and higher ductility. They also reported an increase in composite hardness due to the augmentation in particle percentage. Other researchers recommend enhancements on natural particle properties such as porosity and surface functionalization by physical or chemical methods. These characteristics might improve filler–matrix interaction and result in superior composite properties [1].

The characteristic plant cell wall microstructure is observable in cashew nutshell particles: an arrangement of cellulose microfibrils embedded in a hemicellulose and lignin matrix [8]. The major components in natural fibers include cellulose, hemicellulose, and lignin, while minor components include volatiles, ash, and lipid [40]. Essentially, cellulose microfibrils act as a skeletal component, hemicellulose surrounds them, and lignin provides binding and protection.

Table 2 summarizes the chemical composition of dried-cashew nutshells. Cellulose, hemicellulose, and lignin contents agree with those reported for several natural fibers [7].

Table 2. Chemical composition of cashew nutshells dried at 250°.

Component	Mass Percentage	Method
Dry matter	97.39	AOAC 930.39
Ashes	2.01	AOAC 942.05
ADF	88.88	AOAC 973.18
NDF	89.58	AOAC 2002.04
Lignin	24.31	Equation (1)
Cellulose	64.57	AOAC 973.18
Hemicellulose	0.70	Equation (2)

Particularly, cellulose percentages in dried-cashew nutshell particles are comparable with hemp, jute, flax, and sisal fibers [6,7]. Cellulose is a strong, linear, and unbranched molecule composed of β-1,4-linked d-glucose polysaccharide units arranged into crystalline and amorphous regions. It constitutes about 9 to 80 wt% of lignocellulosic biomass [40]. Some benefits of cellulose in polymer matrix composites can be found in the literature. Storage modulus and thermal stability increased with the addition of cellulosic reinforcement in PLA composites [41]. Crystalline cellulose ensures a strong reinforcement in polymer matrix and improves tensile and flexural strength of polymer composites [42]. The strength of fibrillar cellulose combined with its economic advantages presents an opportunity to develop lighter and stronger composites [43].

Figure 3 shows the ATR-FTIR spectra for the cashew nutshells in the as-received condition and dried at 250 °C.

In both spectra, the typical wave numbers of cellulose, hemicellulose, and lignin previously reported for natural fibers are observed [2,44–46]. The valleys at 1374 cm^{-1} and 752 cm^{-1} evidence the hemicellulose and anacardic acid degradation, respectively, during the drying process. Wave numbers located at 3420 cm^{-1} and 1615 cm^{-1} are attributed to humidity absorption; nevertheless, the former has been assigned to crystalline cellulose [45,46]. It is possible that during the drying treatment, crystalline cellulose was formed. Many studies have been carried out in obtaining microcrystalline cellulose by alkaline or acid treatments. For instance, a conversion to crystalline cellulose from cellulose I in banana rachis treated with an 18 wt% KOH solution was reported [46]. The FTIR spectra showed two bands located at 3487 cm^{-1} and 3442 cm^{-1} associated with intramolecular hydrogen bonding in cellulose II due to changes in the hydrogen bonding system. This transition was also confirmed by X-ray diffraction and electron diffraction coupled with transmission electron microscopy. However, transformation to crystalline cellulose involving temperature treatments was not found in the literature. Further research must be done to understand this phenomenon. Table 3 shows the indexation carried out in this work.

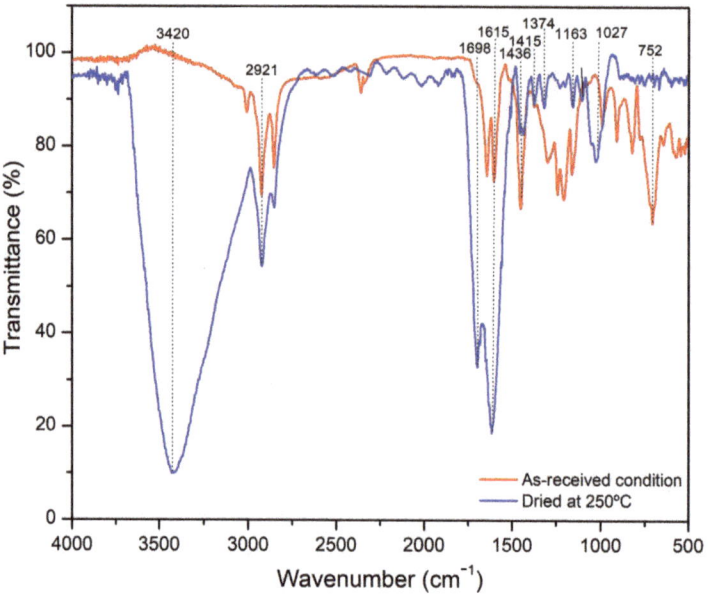

Figure 3. ATR-FTIR spectra for the cashew nutshells in the as-received condition and dried at 250 °C.

Table 3. ATR-FTIR indexation for the cashew nutshells in the as-received condition and dried at 250 °C.

Wave Number (cm^{-1})	Functional Group	Components	Observation
3420	–OH	Humidity absorption, possible cellulose ii formation	Hydrogen bonding
2921	C–H	Cellulose, hemicellulose, lignin	Aliphatic group
1698	C=O	Hemicellulose, lignin	Ester and acetyl groups of polysaccharides
1615	–(H)C=O	Cellulose, hemicellulose, lignin, humidity absorption	Carboxyl ions
1436	–CH	Lignin	Stretching on the aldehyde group
1415	–(Ar)C=C	Lignin	Stretching on aromatic fractions
1374	C–O	Hemicellulose, lignin	Acetyl group
1163	C=O	Lignin	Symmetric stretching of lignin
1027	C–O–C	Lignin, cellulose	Secondary alcohols, aliphatic ethers, cellulose monomer bonds
752	C–H	Lignin, anacardic acid	Bending vibrations on the aromatic group

The thermal behavior of cashew nutshells in the as-received condition and dried at 250 °C is shown in Figure 4.

The as-received condition shows four significant changes around 250 °C, 320 °C, 350 °C, and 430 °C. This behavior is consistent with the analyses described by other researchers [19,20,24,47,48], and can be associated with the following stages.

Stage I (24–200 °C): Release of the moisture absorbed between 24 °C and 120 °C [22], and decarboxylation reaction of anacardic acid around 177 °C [20,47]. This stage shows an 8 wt% reduction.

Stage II (200–375 °C): The weight reduction at this stage is attributed to three factors: 1. hemicellulose degradation from ~220 °C to ~300 °C [24,47], 2. CNSL decomposition at ~275 °C [47,48], and 3. the beginning of cellulose degradation at ~280 °C [17,22,35]. This stage exhibited a 32.4 wt% reduction.

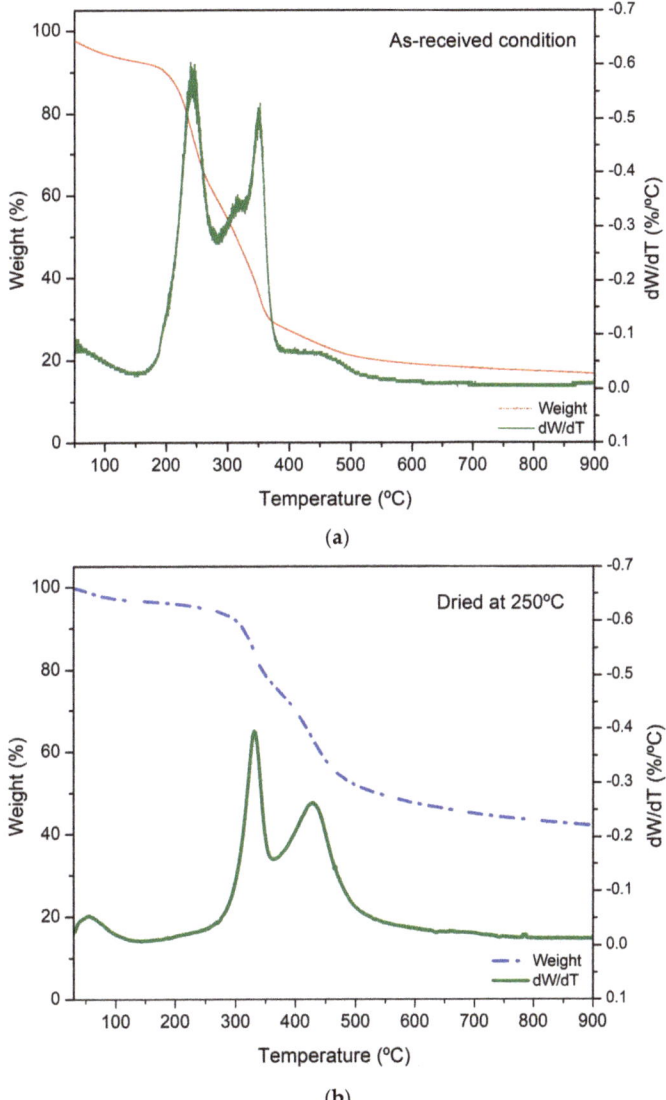

Figure 4. TGA and dW/dT curves of cashew nutshells in (**a**) as-received condition and (**b**) dried at 250 °C.

Stage III (375–600 °C): Cellulose degradation carries on until 450 °C [19,47], and around this temperature, lignin degradation occurs [47]. From this point on, the weight percentage stabilized due to the decomposition of the material to residual char. This stage exhibited a higher weight percentage reduction of 40.66%. In contrast, on dried-cashew nutshells, changes at ~320 °C and ~430 °C only were identified, corresponding to cellulose and lignin degradation, respectively. Both moisture release and anacardic acid decarboxylation reaction occurred during the drying process at 250 °C. It is worth mentioning that ABS and PLA are the most used polymers with natural fiber fillers and their printing temperature range is from 130 °C to 250 °C [35]. These temperatures are lower than the degradation temperature of cellulose; therefore, the advantages and characteristics of dried-cashew nutshell particles discussed in this work might remain during filament manufacturing and the 3D printing process.

Figure 5 shows the PLA–cashew nutshell particle filaments fabricated in this work. Broadly speaking, the filaments exhibited an even and regular surface and were free of perceptible protuberance, voids, or cross-sectional area changes. Very few particles were observed at the surface: most of them were homogeneously dispersed throughout the filament volume. The filament diameter was 1.05 ± 0.02 mm, 1.08 ± 0.01 mm, and 1.28 ± 0.01 mm for 0.5, 1.0, and 2.0 wt% of particles, respectively.

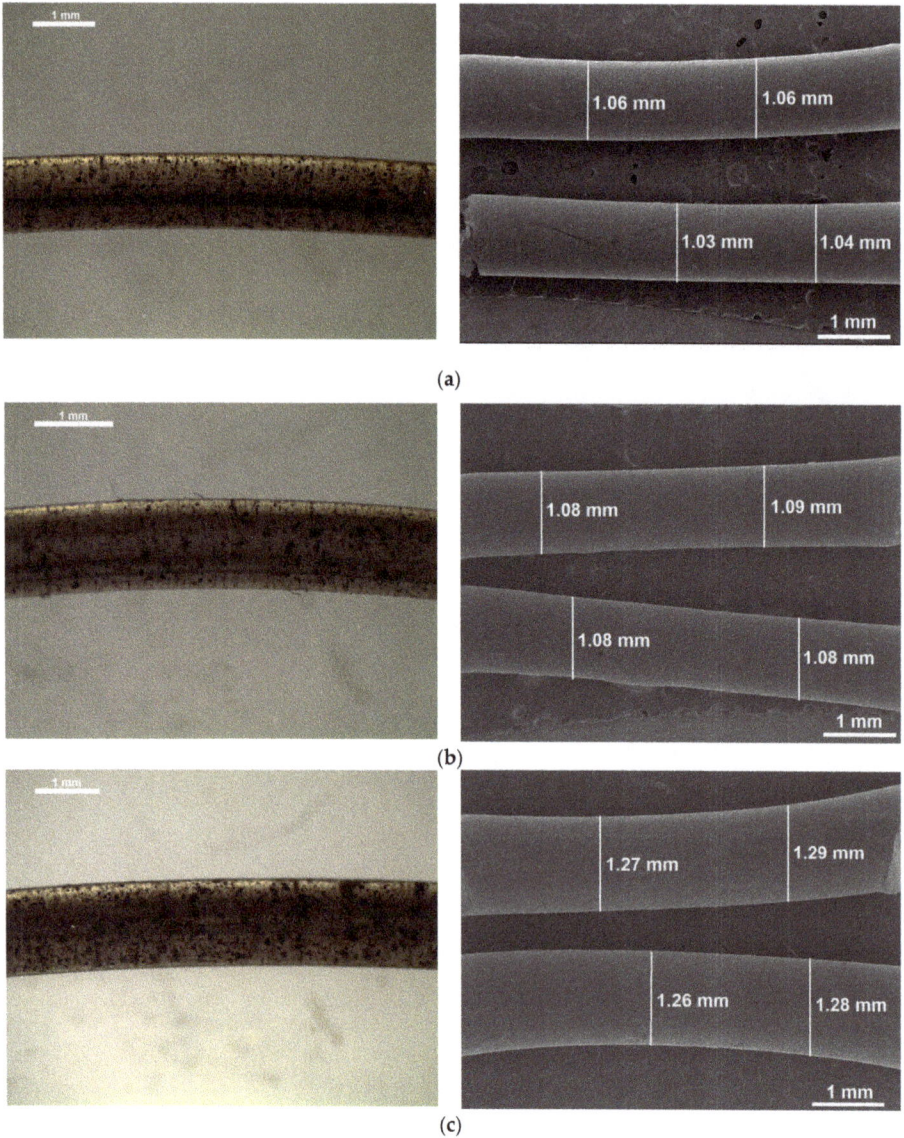

Figure 5. PLA–cashew nutshell particle filaments, (**a**) 0.5 wt% of particles, (**b**) 1.0 wt% of particles and (**c**) 2.0 wt% of particles. Left side: stereo-optical microscopy images. Right side: scanning electron microscopy images.

Figure 6 shows the tensile strength as a function of elongation percentage, while Table 4 summarizes the tensile strength and elongation at break of PLA filaments and PLA–cashew nutshell particle filaments.

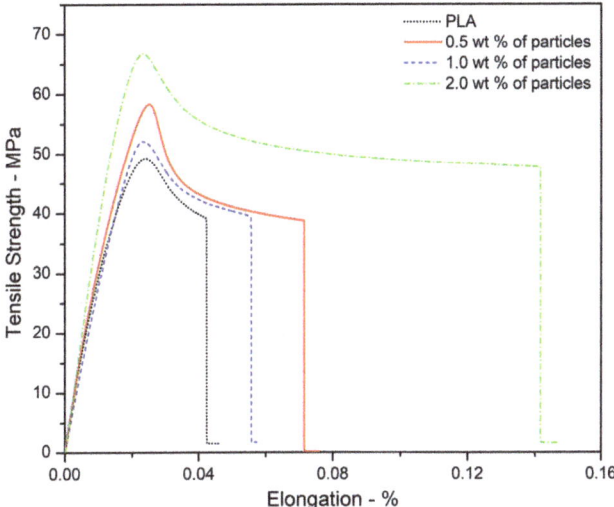

Figure 6. Tensile strength as a function of elongation percentage of PLA filaments and PLA–cashew nutshell particle filaments.

Table 4. Tensile strength and elongation at break of PLA filaments and PLA–cashew nutshell particle filaments.

Filament	Tensile Strength MPa	Elongation at Break %
PLA	47.30 ± 6.72	4.2 ± 0.19
0.5 wt% of particles	58.28 ± 5.0	7.8 ± 0.07
1.0 wt% of particles	52.03 ± 5.17	5.6 ± 0.05
2.0 wt% of particles	64.20 ± 2.28	14.1 ± 0.08

Cashew nutshell particle addition increased both tensile strength and elongation at break in every percentage when compared to PLA filament. The highest values of these properties were exhibited by the 2.0 wt% particle filaments. Particularly, elongation at break percentage increased ~3.36 times, evidencing a significant improvement in ductility, which might reduce the brittle response of PLA when submitted to impact conditions. Furthermore, the tensile strength showed by PLA–cashew nutshell particle filaments is greater than those reported for kenaf fiber–PLA filaments [49] and similar to crab shell–PLA filaments [50]. However, the highest elongations at break shown by those filaments were ~3.6 and ~4.2% respectively, values extremely low in comparison to the 14.1 ± 0.08% shown by 2.0 wt% PLA–cashew nutshell particle filaments fabricated in this research. Regarding other fillers in PLA matrix, wood reduces the modulus of elasticity and tensile strength of filaments by about 50%, and metallic particles including Fe, Cu, Al, and bronze have a negligible or negative impact on tensile strength, elongation at break, and flexural properties. In contrast, carbon nanotubes increase tensile strength and modulus of elasticity by around 50% and 60% [51]. However, a cost–benefit analysis should be carried out to determine whether this remarkable increase in mechanical properties is worth it.

The above characteristics for the PLA–cashew nutshell particle-reinforced composites show the feasibility of this new composite material as an alternative for 3D printing filaments. Among these characteristics are low density, surface characteristics, thermal stability, cellulose content, and homogeneous particle dispersion along the PLA matrix. In addition to these characteristics, other advantages are as follows. First, cashew nutshell is a waste material, not properly disposed of or used in many countries, and thus its use for 3D printing filaments is good for the environment and for the economy as well (as a new process can generate employment in the region and perhaps reduce costs depending on the application). Second, cashew nutshell particles significantly reinforced PLA with the particle loading used, as shown in Figure 6. This is very important for filaments, since particles can deteriorate the composite properties, particularly under tensile applications.

A complete characterization of bio-composites fabricated with these filaments, including printing ability, microstructural features, and mechanical properties, will be presented in a further work.

4. Conclusions

Cashew nutshell particles were produced by drying and grinding processes. Their potential use as filler for filaments for 3D printing was assessed in terms of surface characteristics, apparent density, thermal stability, and content of hemicellulose, cellulose, and lignin. Three different filaments were fabricated from polylactic acid pellets and cashew nutshell particles at 0.5, 1.0, and 2.0 wt%, and then single-filament tensile tests were carried out on them. The main conclusions can be summarized as follows.

Drying and grinding processes produced irregular, rough, and porous cashew nutshell particles with a Shore hardness of 39.67 ± 12.80D, an apparent density of 0.337 g/cm^3, and size particles ranging from ~7 to ~150 μm. The surface characteristics of the particles are desirable because they provide a greater adhesion force due to mechanical interlocking at the interface of natural fiber and polymer matrices. The hardness and low apparent density of the particles encourage their use in the fabrication of light and biodegradable components.

The thermal stability of the cashew nutshell particles was around 320 °C; therefore, their characteristics might remain during filament production and subsequently 3D printing of polymer matrix composites.

Cashew nutshell particles exhibited the characteristic microstructure observed in natural fibers; an arrangement of cellulose microfibrils embedded in a hemicellulose and lignin matrix. The mass percentages of cellulose, hemicellulose, and lignin were 64.57, 0.70 and 24.31, respectively. Cellulose within the structure may enhance the mechanical strength and thermal stability of cashew nutshell particle–polymer matrix filaments. A possible transformation to crystalline cellulose might occur during the drying process at 250 °C. Additional research is encouraged to gain a better understanding of this phenomenon.

The filaments fabricated with polylactic acid and cashew nutshell particles are free of perceptible defects and exhibit a homogeneous distribution of particles along them. In addition, these filaments showed higher tensile strength and elongation at break in comparison to polylactic acid filament. The 2.0 wt% polylactic acid–cashew nutshell particle filaments showed the highest tensile strength and elongation at break, with 64.20 ± 2.28 MPa and 14.1 ± 0.08%.

These results suggest that the cashew nutshell particles prepared in this work are promising as natural fillers for polylactic acid matrix filaments for 3D printing.

5. Patents

A national utility patent application has been submitted to the Superintendency of Industry and Commerce (SIC) in Colombia: NC2023/0005743.

Author Contributions: M.J.P.R.: investigation, writing—original draft. J.U.S.: conceptualization, methodology, writing—review and editing, funding acquisition. J.F.S.M.: investigation, writing—review and editing. H.A.C.L.: methodology, writing—review and editing. L.A.E.S.: conceptualization, methodology, investigation, supervision, writing—review and editing. All authors have read and agreed to the published version of the manuscript.

Funding: This research was funded by the University of Cordoba—Colombia, funding number FI-06-19.

Institutional Review Board Statement: Not applicable.

Data Availability Statement: Data used in this work are confidential.

Acknowledgments: The authors thank Asopromarsab for providing the cashew nutshells used in this work.

Conflicts of Interest: The authors declare no conflict of interest.

References

1. Poulose, A.M.; Elnour, A.Y.; Anis, A.; Shaikh, H.; Al-Zahrani, S.; George, J.; Al-Wabel, M.I.; Usman, A.R.; Ok, Y.S.; Tsang, D.C.; et al. Date palm biochar-polymer composites: An investigation of electrical, mechanical, thermal and rheological characteristics. *Sci. Total. Environ.* **2018**, *619–620*, 311–318. [CrossRef]
2. Ibrahim, M.M.; Dufresne, A.; El-Zawawy, W.K.; Agblevor, F.A. Banana fibers and microfibrils as lignocellulosic reinforcements in polymer composites. *Carbohydr. Polym.* **2010**, *81*, 811–819. [CrossRef]
3. Vilela, C.; Sousa, A.F.; Freire, C.S.; Silvestre, A.J.; Neto, C.P. Novel sustainable composites prepared from cork residues and biopolymers. *Biomass-Bioenergy* **2013**, *55*, 148–155. [CrossRef]
4. van den Oever, M.; Beck, B.; Müssig, J. Agrofibre reinforced poly(lactic acid) composites: Effect of moisture on degradation and mechanical properties. *Compos. Part A Appl. Sci. Manuf.* **2010**, *41*, 1628–1635. [CrossRef]
5. Torun, S.B.; Tomak, E.D.; Cavdar, A.D.; Mengeloglu, F. Characterization of weathered MCC/nutshell reinforced composites. *Polym. Test.* **2021**, *101*, 107290. [CrossRef]
6. Manfredi, L.B.; Rodríguez, E.S.; Wladyka-Przybylak, M.; Vázquez, A. Thermal degradation and fire resistance of unsaturated polyester, modified acrylic resins and their composites with natural fibres. *Polym. Degrad. Stab.* **2006**, *91*, 255–261. [CrossRef]
7. Burton, K.; Hazael, R.; Critchley, R.; Bloodworth-Race, S. Lignocellulosic Natural Fibers in Polymer Composite Materials: Benefits, Challenges and Applications. In *Encyclopedia of Materials: Plastics and Polymers*; Hashmi, M.S.J., Ed.; Elsevier: Amsterdam, The Netherlands, 2022; pp. 353–369, ISBN 9780128232910. [CrossRef]
8. Gbadeyan, O.J.; Sarp, A.; Glen, B.; Sithole, B. Nanofiller/Natural Fiber Filled Polymer Hybrid Composite: A Review. *J. Eng. Sci. Technol. Rev.* **2021**, *14*, 61–74. [CrossRef]
9. Jani, S.P.; Kumar, A.S.; Khan, M.A.; Sajith, S.; Saravanan, A. Influence of Natural Filler on Mechanical Properties of Hemp/Kevlar Hybrid Green Composite and Analysis of Change in Material Behavior Using Acoustic Emission. *J. Nat. Fibers* **2021**, *18*, 1580–1591. [CrossRef]
10. Ike, D.C.; Ibezim-Ezeani, M.U.; Akaranta, O. Cashew nutshell liquid and its derivatives in oil field applications: An update. *Green Chem. Lett. Rev.* **2021**, *14*, 620–633. [CrossRef]
11. Kyei, S.K.; Eke, W.I.; Nagre, R.D.; Mensah, I.; Akaranta, O. A comprehensive review on waste valorization of cashew nutshell liquid: Sustainable development and industrial applications. *Clean. Waste Syst.* **2023**, *6*, 100116. [CrossRef]
12. Zafeer, M.K.; Bhat, K.S. Valorisation of agro-waste cashew nut husk (Testa) for different value-added products. *Sustain. Chem. Clim. Action* **2023**, *2*, 100014. [CrossRef]
13. Shenoy, D.; Pai, A.; Vikas, R.; Neeraja, H.; Deeksha, J.; Nayak, C.; Rao, C.V. A study on bioethanol production from cashew apple pulp and coffee pulp waste. *Biomass-Bioenergy* **2011**, *35*, 4107–4111. [CrossRef]
14. Das, P.; Sreelatha, T.; Ganesh, A. Bio oil from pyrolysis of cashew nut shell-characterisation and related properties. *Biomass-Bioenergy* **2004**, *27*, 265–275. [CrossRef]
15. Serpa, J.d.F.; Silva, J.d.S.; Reis, C.L.B.; Micoli, L.; e Silva, L.M.A.; Canuto, K.M.; de Macedo, A.C.; Rocha, M.V.P. Extraction and characterization of lignins from cashew apple bagasse obtained by different treatments. *Biomass-Bioenergy* **2020**, *141*, 105728. [CrossRef]
16. Anilkumar, P. (Ed.) *Cashew Nut Shell Liquid: A Goldfield for Functional Materials*; Springer International Publishing: Berlin/Heidelberg, Germany, 2017; ISBN 978-3-319-47454-0. [CrossRef]
17. Sharma, P.; Gaur, V.K.; Sirohi, R.; Larroche, C.; Kim, S.H.; Pandey, A. Valorization of cashew nut processing residues for industrial applications. *Ind. Crop. Prod.* **2020**, *152*, 112550. [CrossRef]
18. Sakinah, N.; Djoefrie, H.M.H.B.; Hariyadi; Manohara, D. Utilization of Cashew Nut Shell as Organic Fertilizer and Fungicide. [IPB University Bogor Indonesia]. In *MT Agriculture*; Montana Agricultural Statistics: Helena, MT, USA, 2013; Available online: https://repository.ipb.ac.id/handle/123456789/67037 (accessed on 19 September 2023).
19. Bamgbola, A.; Adeyemi, O.; Olubomehin, O.; Akinlabi, A.; Sojinu, O.; Iwuchukwu, P. Isolation and characterization of cellulose from cashew (*Anacardium occidentale* L.) nut shells. *Curr. Res. Green Sustain. Chem.* **2020**, *3*, 100032. [CrossRef]
20. Tyman, J.H.P.; Johnson, R.A.; Muir, M.; Rokhgar, R. The extraction of natural cashew nut-shell liquid from the cashew nut (*Anacardium occidentale*). *J. Am. Oil Chem. Soc.* **1989**, *66*, 553–557. [CrossRef]
21. Nambela, L.; Haule, L.V.; Mgani, Q.A. Anacardic acid isolated from cashew nut shells liquid: A potential precursor for the synthesis of anthraquinone dyes. *Clean. Chem. Eng.* **2022**, *3*, 100056. [CrossRef]

22. Kasemsiri, P.; Hiziroglu, S.; Rimdusit, S. Properties of wood polymer composites from eastern redcedar particles reinforced with benzoxazine resin/cashew nut shell liquid copolymer. *Compos. Part A Appl. Sci. Manuf.* **2011**, *42*, 1454–1462. [CrossRef]
23. Agag, T.; An, S.Y.; Ishida, H. 1,3-bis(benzoxazine) from cashew nut shell oil and diaminodiphenyl methane and its composites with wood flour. *J. Appl. Polym. Sci.* **2013**, *127*, 2710–2714. [CrossRef]
24. Moreira, R.; Orsini, R.d.R.; Vaz, J.M.; Penteado, J.C.; Spinacé, E.V. Production of Biochar, Bio-Oil and Synthesis Gas from Cashew Nut Shell by Slow Pyrolysis. *Waste Biomass-Valorization* **2017**, *8*, 217–224. [CrossRef]
25. Kaur, R.; Kumar, V.T.; Krishna, B.B.; Bhaskar, T. Characterization of slow pyrolysis products from three different cashew wastes. *Bioresour. Technol.* **2023**, *376*, 128859. [CrossRef]
26. Sundarakannan, R.; Arumugaprabu, V.; Manikandan, V.; Vigneshwaran, S. Mechanical property analysis of biochar derived from cashew nut shell waste reinforced polymer matrix. *Mater. Res. Express* **2019**, *6*, 125349. [CrossRef]
27. Rahmatabadi, D.; Soltanmohammadi, K.; Pahlavani, M.; Aberoumand, M.; Soleyman, E.; Ghasemi, I.; Baghani, M. Shape memory performance assessment of FDM 3D printed PLA-TPU composites by Box-Behnken response surface methodology. *Int. J. Adv. Manuf. Technol.* **2023**, *127*, 935–950. [CrossRef]
28. Rahmatabadi, D.; Aberoumand, M.; Soltanmohammadi, K.; Soleyman, E.; Ghasemi, I.; Baniassadi, M.; Abrinia, K.; Bodaghi, M.; Baghani, M. Toughening PVC with Biocompatible PCL Softeners for Supreme Mechanical Properties, Morphology, Shape Memory Effects, and FFF Printability. *Macromol. Mater. Eng.* **2023**, *308*, 2300114. [CrossRef]
29. Soleyman, E.; Aberoumand, M.; Soltanmohammadi, K.; Rahmatabadi, D.; Ghasemi, I.; Baniassadi, M.; Abrinia, K.; Baghani, M. 4D printing of PET-G via FDM including tailormade excess third shape. *Manuf. Lett.* **2022**, *33*, 1–4. [CrossRef]
30. Saroia, J.; Wang, Y.; Wei, Q.; Lei, M.; Li, X.; Guo, Y.; Zhang, K. A review on 3D printed matrix polymer composites: Its potential and future challenges. *Int. J. Adv. Manuf. Technol.* **2020**, *106*, 1695–1721. [CrossRef]
31. Periyasamy, R.; Hemanth Kumar, M.; Rangappa, S.M.; Siengchin, S. A comprehensive review on natural fillers reinforced polymer composites using fused deposition modeling. *Polym. Compos.* **2023**, *44*, 3715–3747. [CrossRef]
32. Saran, O.S.; Reddy, A.P.; Chaturya, L.; Kumar, M.P. 3D printing of composite materials: A short review. *Mater. Today Proc.* **2022**, *64*, 615–619. [CrossRef]
33. Ribeiro, I.; Matos, F.; Jacinto, C.; Salman, H.; Cardeal, G.; Carvalho, H.; Godina, R.; Peças, P. Framework for Life Cycle Sustainability Assessment of Additive Manufacturing. *Sustainability* **2020**, *12*, 929. [CrossRef]
34. Bi, X.; Huang, R. 3D printing of natural fiber and composites: A state-of-the-art review. *Mater. Des.* **2022**, *222*, 111065. [CrossRef]
35. Ahmed, W.; Alnajjar, F.; Zaneldin, E.; Al-Marzouqi, A.H.; Gochoo, M.; Khalid, S. Implementing FDM 3D Printing Strategies Using Natural Fibers to Produce Biomass Composite. *Materials* **2020**, *13*, 4065. [CrossRef] [PubMed]
36. ASTM D2240-15; Standard Test Method for Rubber Property—Durometer Hardness, Book of Standards Volume: 09.01, Developed by Sub-Committee: D11.10. ASTM International: West Conshohocken, PA, USA, 2021; p. 13. [CrossRef]
37. Hall, M.B.; Mertens, D.R. Comparison of alternative neutral detergent fiber methods to the AOAC definitive method. *J. Dairy Sci.* **2023**, *106*, 5364–5378. [CrossRef] [PubMed]
38. Arockiam, A.J.; Subramanian, K.; Padmanabhan, R.; Selvaraj, R.; Bagal, D.K.; Rajesh, S. A review on PLA with different fillers used as a filament in 3D printing. *Mater. Today Proc.* **2022**, *50*, 2057–2064. [CrossRef]
39. Okele, A.I.; Mamza, P.A.P.; Nkeoye, P.; Marut, A.J. Mechanical properties of cashew nut shell powder (CSNP) filled natural rubber volcanizate. *Chem. Technol. Indian J. (CTIJ)* **2016**, *11*, 36–42.
40. Ilham, Z. Chapter 3—Biomass Classification and Characterization for Conversion to Biofuels. In *Nor Adilla Rashidi, Value-Chain of Biofuels*; Yusup, S., Ed.; Elsevier: Amsterdam, The Netherlands, 2022; pp. 69–87, ISBN 9780128243886. [CrossRef]
41. Mathew, A.P.; Oksman, K.; Sain, M. Mechanical properties of biodegradable composites from poly lactic acid (PLA) and microcrystalline cellulose (MCC). *J. Appl. Polym. Sci.* **2005**, *97*, 2014–2025. [CrossRef]
42. Boran, S.; Kiziltas, A.; Gardner, D.J. Characterization of Ultrafine Cellulose-filled High-Density Polyethylene Composites Prepared using Different Compounding Methods. *BioResources* **2016**, *11*, 8178–8199. [CrossRef]
43. Bharimalla, A.K.; Deshmukh, S.P.; Patil, P.G.; Vigneshwaran, N. Energy Efficient Manufacturing of Nanocellulose by Chemo- and Bio-Mechanical Processes: A Review. *World J. Nano Sci. Eng.* **2015**, *5*, 204–212. [CrossRef]
44. Biswas, S.; Rahaman, T.; Gupta, P.; Mitra, R.; Dutta, S.; Kharlyngdoh, E.; Guha, S.; Ganguly, J.; Pal, A.; Das, M. Cellulose and lignin profiling in seven, economically important bamboo species of India by anatomical, biochemical, FTIR spectroscopy and thermogravimetric analysis. *Biomass-Bioenergy* **2022**, *158*, 106362. [CrossRef]
45. Taharuddin, N.H.; Jumaidin, R.; Mansor, M.R.; Yusof, F.A.M.; Alamjuri, R.H. Characterization of Potential Cellulose from Hylocereus Polyrhizus (Dragon Fruit) peel: A Study on Physicochemical and Thermal Properties. *J. Renew. Mater.* **2023**, *11*, 131–145. [CrossRef]
46. Zuluaga, R.; Putaux, J.L.; Cruz, J.; Vélez, J.; Mondragon, I.; Gañán, P. Cellulose microfibrils from banana rachis: Effect of alkaline treatments on structural and morphological features. *Carbohydr. Polym.* **2009**, *76*, 51–59. [CrossRef]
47. Melzer, M.; Blin, J.; Bensakhria, A.; Valette, J.; Broust, F. Pyrolysis of extractive rich agroindustrial residues. *J. Anal. Appl. Pyrolysis* **2013**, *104*, 448–460. [CrossRef]
48. Ábrego, J.; Plaza, D.; Luño, F.; Atienza-Martínez, M.; Gea, G. Pyrolysis of cashew nutshells: Characterization of products and energy balance. *Energy* **2018**, *158*, 72–80. [CrossRef]
49. Lau, H.Y.; Hussin, M.S.; Hamat, S.; Abdul, M.S.; Ibrahim, M.; Zakaria, H. Effect of kenaf fiber loading on the tensile properties of 3D printing PLA filament. *Mater. Today Proc.* **2023**. [CrossRef]

50. Palaniyappan, S.; Sivakumar, N.K. Development of crab shell particle reinforced polylactic acid filaments for 3D printing application. *Mater. Lett.* **2023**, *341*, 134257. [CrossRef]
51. Angelopoulos, P.M.; Samouhos, M.; Taxiarchou, M. Functional fillers in composite filaments for fused filament fabrication; a review. *Mater. Today Proc.* **2021**, *37*, 4031–4043. [CrossRef]

Disclaimer/Publisher's Note: The statements, opinions and data contained in all publications are solely those of the individual author(s) and contributor(s) and not of MDPI and/or the editor(s). MDPI and/or the editor(s) disclaim responsibility for any injury to people or property resulting from any ideas, methods, instructions or products referred to in the content.

Article

Effect of the Chemical Properties of Silane Coupling Agents on Interfacial Bonding Strength with Thermoplastics in the Resizing of Recycled Carbon Fibers

Hyunkyung Lee [1,2], Minsu Kim [1], Gyungha Kim [1,*] and Daeup Kim [1,*]

[1] Carbon & Light Materials Application Group, Korea Institute of Industrial Technology, Bucheon 14449, Republic of Korea; dori9424@gmail.com (H.L.); mskim85@kitech.re.kr (M.K.)
[2] Department of Carbon Material Fiber Engineering, Chonbuk National University, Jeonju 54896, Republic of Korea
* Correspondence: gyungha@kitech.re.kr (G.K.); dukim@kitech.re.kr (D.K.)

Citation: Lee, H.; Kim, M.; Kim, G.; Kim, D. Effect of the Chemical Properties of Silane Coupling Agents on Interfacial Bonding Strength with Thermoplastics in the Resizing of Recycled Carbon Fibers. *Polymers* 2023, 15, 4273. https://doi.org/10.3390/polym15214273

Academic Editors: Patricia Krawczak, Chung Hae Park and Abderrahmane Ayadi

Received: 15 September 2023
Revised: 24 October 2023
Accepted: 24 October 2023
Published: 30 October 2023

Copyright: © 2023 by the authors. Licensee MDPI, Basel, Switzerland. This article is an open access article distributed under the terms and conditions of the Creative Commons Attribution (CC BY) license (https:// creativecommons.org/licenses/by/ 4.0/).

Abstract: Upcycling recycled carbon fibers recovered from waste carbon composites can reduce the price of carbon fibers while improving disposal-related environmental problems. This study assessed and characterized recycled carbon fibers subjected to sizing treatment using N-(2-aminoethyl)-3-aminopropyltrimethoxysilane (APS) chemically coordinated with polyamide 6 (PA6) and polypropylene (PP) resins. Sizing treatment with 1 wt.% APS for 10 s yielded O=C-O on the surface of the carbon fiber, and the -SiOH in the APS underwent a dehydration–condensation reaction that converted O=C-O (lactone groups) into bonds of C-O (hydroxyl groups) and C=O (carbonyl groups). The effects of C-O and C=O on the interfacial bonding force increased to a maximum, resulting in an oxygen-to-carbon ratio (O/C) of 0.26. The polar/surface energy ratio showed the highest value of 32.29% at 10 s, and the interfacial bonding force showed the maximum value of 32 MPa at 10 s, which is about 15% better than that of commercial carbon fiber (PA6-based condition). In 10 s resizing treatments with 0.5 wt.% 3-methacryloxypropyltrimethoxysilane (MPS), C-O, C=O, and O=C-O underwent a dehydration–condensation reaction with -SiOH, which broke the bonds between carbon and oxygen and introduced a methacrylate group ($H_2C=C(CH_3)CO_2H$), resulting in a significant increase in C-O and C=O, with an O/C of 0.51. The polar/surface free energy ratio was about 38% at 10 s, with the interfacial bonding force increasing to 27% compared to commercial carbon fiber (PP-based conditions). MPS exhibited a superior interfacial shear strength improvement, two times higher than that of APS, with excellent coordination with PP resin and commercial carbon fiber, although the interfacial bonding strength of the PP resin was significantly lower.

Keywords: carbon fiber; resizing; silane coupling agent; thermoplastic; interfacial shear strength; mechanism; oxygen functional group

1. Introduction

Carbon fibers are lightweight materials with low density, high specific strength, heat resistance, and excellent thermal and electrical conductivity, and their applications are expected to expand not only in the aerospace industry but also in all industries in the future [1–9]. However, due to the high price of carbon fiber and the expensive manufacturing process, it is only used for expensive parts, such as aerospace, shipbuilding, and sporting goods, and it is difficult to expand its application to fields such as general commercial vehicles due to its high price [10,11]. In addition, carbon composites currently in use are made of thermosetting resins, which are difficult to recycle, and most are disposed of by landfill and incineration, causing environmental pollution [12]. To reduce the price of carbon fibers and solve environmental pollution problems, upcycling technology to recover waste carbon composites for recycling is absolutely necessary [13–16].

The surface properties of carbon fibers greatly affect the mechanical properties of carbon composites, and surface treatment and sizing are essential for upcycling recycled carbon fibers recovered from previously used carbon composites to achieve properties equivalent to commercial carbon fiber without degradation [17]. Sizing treatment is a simple process that protects the surface by coating the carbon fiber with an interfacial binder, while offering a stable interface by improving the chemical bonding force with the resin to yield better chemical and mechanical properties than those obtained by surface treatment of general carbon fibers [18–20]. Sizing treatments include coating with organic polymers and coating with metal oxides, which are inorganic molecules, and converting them into metal crystals to form a protective film [19,20]. The sizing treatment of carbon fibers with poly(phthalazinone ether ketone) was previously shown to result in C-N and C=N bonds present in the phthalazine ring, which improved the thermal stability, and the surface energy was enhanced by the increase in C=O bonds [21].

The coating of basalt fibers with an amino–silane coupling agent reportedly enhanced the interfacial bonding of basalt fibers and PA66 by the non-polar CH_2 chains and polar amino groups of the silane coupling agent. As the number of CH_2 chains increased, the chain entanglement between Si molecules and PA66 improved the interfacial bonding [22]. By coating the basalt felt (BF) surface with a nickel-based metal-organic framework (Ni-MOF), the papers reported that the weak interfacial bond with the epoxy resin was improved by the self-lubricating behavior of Ni-MOF during friction, increasing the interfacial bond strength by about 15.19%. [23]. Changes in the chemical properties of carbon fibers sized with E51 epoxy resin and the curing agent DDS, analyzed using X-ray photoelectron spectroscopy (XPS), showed enhanced epoxide bond formation and interfacial shear strength (IFSS), which were not seen in untreated carbon fibers [5]. Further, sizing treatment with 4,4'-diphenylmethane diisocyanate can increase the content of oxygen functional groups via the chemical reaction of carbonyl, carboxyl, and -NCO groups on the carbon fiber surface and can improve the wettability between carbon fiber and resin [24].

Carbon fibers treated with poly(amidoamine) can improve the mechanical properties of carbon composites by forming covalent bonds between amino groups and epoxy resin [25], and carbon fibers sized with carboxylic polyphenylene sulfide (PPS-COOH) have been shown to increase interfacial bonding with polyphenylene sulfide resin by eliminating C-N bonds, forming new C-S bonds, and increasing the content of C=O [26]. According to research so far, most studies have been conducted on the mechanical and chemical properties according to the sizing type and processing conditions of commercial carbon fiber and resin; however, the mechanism of surface chemical structure changes during resizing treatment using recycled carbon fibers has not been clarified.

In this study, to improve the interfacial bonding between recycled carbon fibers and resins for the purpose of upcycling recycled carbon fibers, recycled carbon fibers were desized and then resized using silane coupling agents that are chemically compatible with thermoplastic PA6 and PP resins. This study aimed to identify the optimal conditions under which recycled carbon fibers with resizing treatment have the same physical and chemical properties as commercial carbon fiber and to investigate the effects and mechanisms of PA6 and PP silane coupling agents on the interfacial bonding force between recycled carbon fibers and thermoplastic resins.

2. Materials and Methods
2.1. Materials

This study used recycled carbon fibers recovered from hydrogen tanks, and their physical properties were compared with Toray's commercial carbon fiber, as shown in Table 1. In accordance with the ASTM D3822 standard [27], tensile tests were conducted at a tensile speed of 5 mm/min, and the average value was calculated for the results of >20 tests per condition. The silane coupling agents used in the resizing process were N-(2-aminoethyl)-3-aminopropyltrimethoxy silane (KBM-602, 99.9%, Shin Etsu, Tokyo, Japan, hereinafter referred to as APS), which has good chemical harmony with PA6 (Figure 1a), and 3-methacryloxypropyltrimethoxysilane (KBM-503, 99.9%, Shin Etsu, Tokyo, Japan), hereinafter referred to as MPS), which has good chemical harmony with PP (Figure 1b). The chemical structures of the sizing agents used in this study are shown in Figure 1.

Table 1. Properties of carbon fiber in this study.

Type	Commercial CF	Recycled CF
Tensile strength (Gpa)	4.49	3.45
Modulus (Gpa)	261	256
Elongation (%)	2.62	2.08
Density (g/cm^3)	1.80	1.80

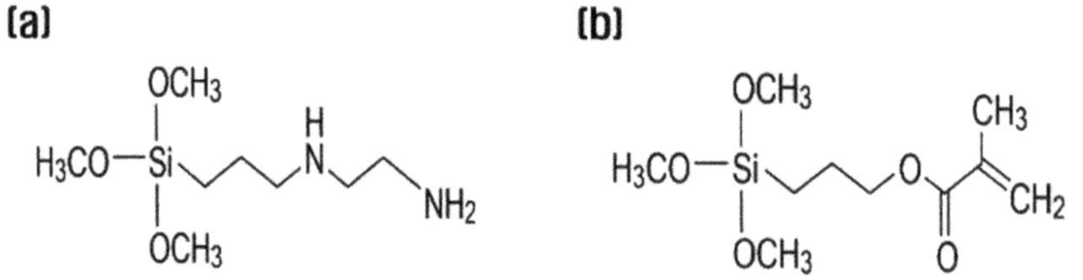

Figure 1. Chemical formula of (**a**) N-(2-aminoethyl)-3-aminopropyltrimethoxysilane and (**b**) 3-methacryloxypropyltrimethoxysilane.

2.2. Experimental Methods

To desize recycled carbon fibers, they were treated in acetone (99.5%, Daejung Chemical, Siheung-si, Republic of Korea) at 60 °C for 30 min to completely remove the chemical components remaining on the surface of the carbon fibers when they were recovered and separated from the waste carbon composite. Surface treatment was then performed by immersion in nitric acid (60.0%, Samchun Pure Chemical, Seoul, Republic of Korea) at 100 °C for 1 h, and the carbon fiber without surface treatment and resizing was labeled "untreated." Recycled carbon fiber with nitric acid surface treatment was subjected to resizing treatment, and the sizing agent was prepared by adding 0.5–2 wt.% silane coupling agent to ethanol (99.5%, Daejung Chemical, Siheung-si, Republic of Korea) and distilled water. The recycled carbon fibers with nitric acid surface treatment were immersed in the sizing agent for 3–15 s for resizing and then dried at 120 °C for 2 h.

2.3. Characteristic Analysis

The recycled carbon fiber with the resizing treatment was analyzed for the amount of sizing agent coated on the surface of the recycled carbon fiber using thermogravimetric analysis (TGA, WATERS (TA Instruments, New Castle, DE, USA) (Discovery SDT 650)), and the sample was heated up to 1000 °C at a ramping rate of 10 °C/min in a nitrogen atmosphere to analyze the change in thermal weight. IFSS (Interfacial shear strength tester, ST-1000) was performed based on ASTM C1557 [28] to analyze the interfacial bonding strength between carbon fiber and resin; the interfacial shear strength was evaluated through the pull-out method by depositing 200 μm of carbon fiber into the resin and pulling it out at a speed of 0.1 mm/min, and the average value was used for 25 tests per test condition.

XPS from Nexsa (Thermo Fisher Scientific Inc., Whaltman, MA, USA) was used to investigate the changes in the chemical functional groups on the surface of recycled carbon fibers following resizing treatment. The specimens were irradiated with monochromatic Al Kα (1486.6 eV), and high-resolution spectra were obtained at a pass energy of 50 eV and a beam size of 400 μm. In addition, to analyze the surface energy changes, the contact angle of each condition was measured using the Wilhelmy Plate Method based on ASTM D1331-

20 [29], in which the carbon fiber was dropped into hydrophilic water and hydrophobic diiodomethane (99.9%, Sigma–Aldrich Co., Llc., St. Louis, MO, USA) at a constant injection rate of 6 mm/min. The contact angle measurements were evaluated five times per condition, and the surface energy was calculated from the contact angle obtained from the angle of immersion of the carbon fiber in the sample and its exit angle.

3. Results and Discussion

3.1. Thermal Properties of Recycled Carbon Fibers

The amount of sizing agent bound to the surface of the recycled carbon fiber was evaluated using TGA to assess changes in thermogravimetric weight. Figure 2 shows the TGA graph as a function of the sizing agent concentration and the resizing treatment time. To select the optimal concentration, the treatment time was fixed at 10 s, and the resizing was performed according to the change in concentration from 0.5 to 2 wt.% (Figure 2a–d). The weight loss of the PA6-based APS sizing agent was about 0.48% at a concentration of 0.5 wt.%, about 1.05% when resizing was performed at 1 wt.% and about 2.23% for a concentration of 2 wt.%. The weight loss of the PP-based MPS sizing agent was about 1.08% at a concentration of 0.5 wt.% and about 2.27% at a concentration of 1 wt.%, which was significantly higher than that of APS. Regarding silane coupling agents, the best properties are reportedly obtained when the carbon fibers are coated with 1% of the agent. When sizing agents with a concentration of 1 wt.% or more are used, the concentration of silane-based substances that exhibit stiff properties should be gradually increased to minimize the impact on the properties [24]. Therefore, in this study, the concentration of 1 wt.% for APS and 0.5 wt.% for MPS was optimally fixed as a coating condition of 1% from the TGA results, and the changes with treatment time were evaluated.

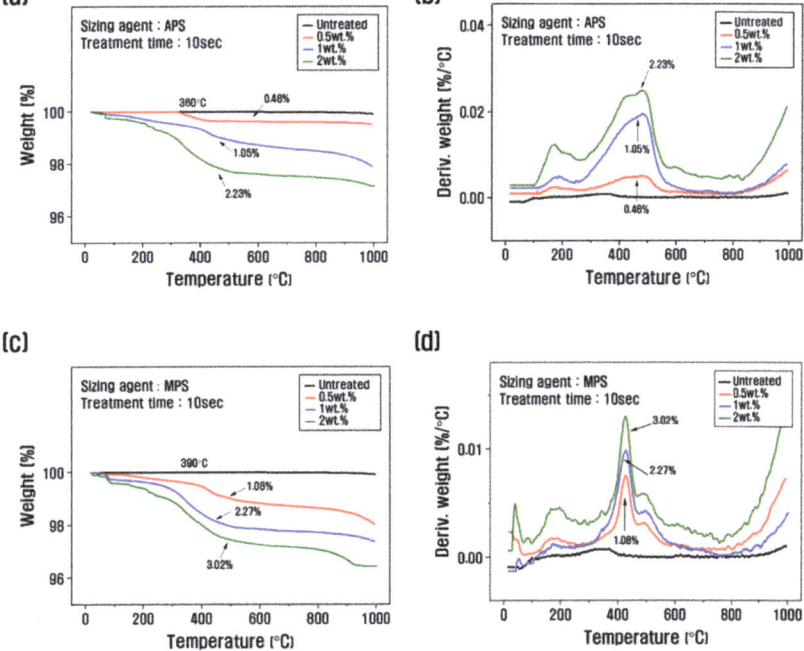

Figure 2. *Cont.*

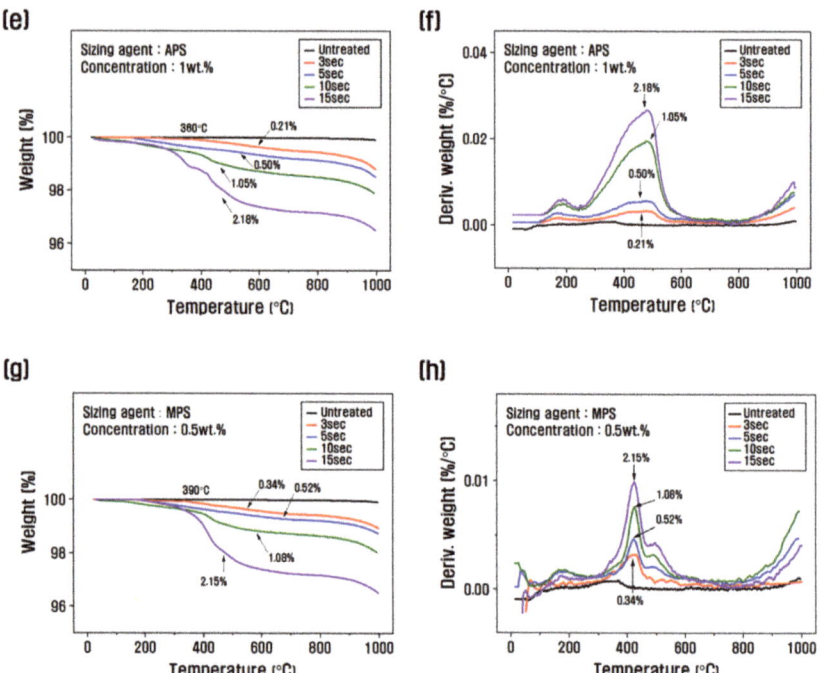

Figure 2. Thermogravimetric analysis and derivative thermogravimetry of recycled carbon fibers according to the concentration and treatment time of the silane coupling agent: (**a,b,e,f**) rCF/APS, (**c,d,g,h**) rCF/MPS.

When fixing the APS concentration at 1 wt.% and observing the weight loss as a function of treatment time, the weight change was about 0.21% at 3 s, about 0.5% at 5 s, about 1.05% at 10 s, and about 2.18% at 15 s, and the weight loss gradually increased (Figure 2e,f). When the MPS concentration was fixed at 0.5 wt.% and the weight change with treatment time was analyzed, the weight change was about 0.34% and about 0.52% for the resizing treatment for 3 s and 5 s, respectively, about 1.08% at 10 s, and about 2.15% at 15 s (Figure 2g,h). This is believed to be due to the larger molecular chain and molecular weight of MPS compared to APS, which results in a faster coating of the recycled carbon fiber. According to the TGA graph, the weight loss in the range of 100–200 °C was caused by the evaporation of water present in the recycled carbon fiber, and the weight change after 300 °C was due to the removal of water molecules in the -SiOH condensation reaction on the surface of the recycled carbon fiber and the thermal degradation of the silane coupling agent [30,31].

Other studies have reported that when more than 1% sizing treatment is applied to the surface of carbon fibers, fiber-to-fiber bonding and agglomeration are observed. This phenomenon causes the deterioration of the mechanical properties of carbon composites, due to the difficulty of the resin in penetrating between the carbon fibers when mixed to produce carbon composites, and becomes more severe as the coating amount of the sizing agent increases [21,32,33]. In this study, TGA results showed that a 1% sizing agent coating on the surface of recycled carbon fibers was the optimal condition. The PA6 sizing agent was optimal at a concentration of 1 wt.% and a treatment time of 10 s, whereas the PP sizing agent was optimal at a concentration of 0.5 wt.% and a treatment time of 10 s.

3.2. Mechanical Properties of Recycled Carbon Fibers

The interfacial shear strength was evaluated as a function of treatment time at the optimum concentration, and the results are shown in Figure 3. The recycled carbon fibers treated with PA6-based APS sizing agent were compared with PA6 resin, and the recycled carbon fibers treated with PP-based MPS sizing agent were compared with PP resin. For the recycled carbon fibers treated with APS at a concentration of 1 wt.%, the interfacial shear strength increased with increasing treatment time, reaching a maximum of 32 MPa at 10 s, and began to decrease at 15 s (Figure 3a). For MPS at a concentration of 0.5 wt.%, the interfacial shear strength gradually increased until the treatment time of 10 s, and decreased by about 11% at 15 s, compared to 10 s.

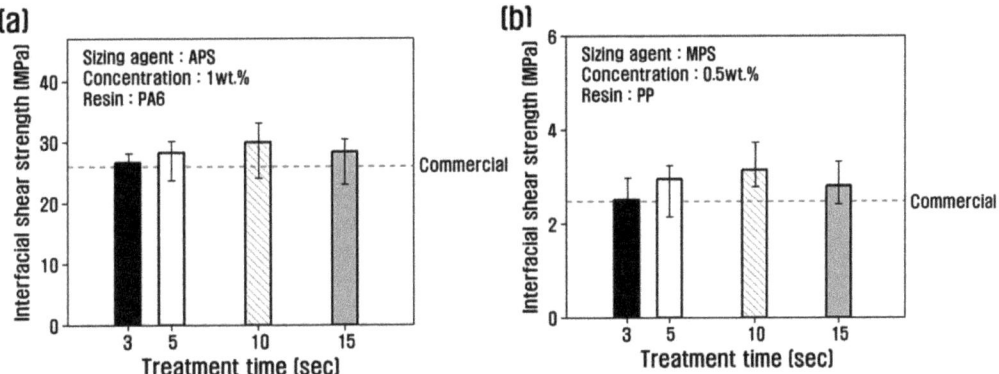

Figure 3. Interfacial shear stress for (**a**) rCF/APS and (**b**) rCF/MPS by single fiber pull-out testing.

Na Sun et al. reported that when the amount of sizing agent coated is low, the microscopic grooves present on the surface during carbon fiber manufacturing are not completely filled, which leads to pores at the interface of PA6 resin and carbon fiber during composite formation and reduces the interfacial bonding force [33]. The sizing agent-coated layer can improve interfacial bonding through chemical bonding and intermolecular attraction by chain entanglement, thereby preventing cracks from occurring at the interface with the resin [26]. A study reported that the interfacial shear strength was improved due to better impregnation, a rough surface, and high surface free energy between carbon fiber and resin [21].

This study confirmed that the interfacial shear strength was improved, compared to commercial carbon fiber, by resizing treatment under optimal conditions. When treated with APS for 10 s, the interfacial shear strength increased by about 15% compared to commercial carbon fiber at a concentration of 1 wt.%, and it increased by about 27% when treated with MPS at a concentration of 0.5 wt.% for 10 s. This is attributed to the improvement of the interfacial bonding force through a more active chemical reaction between the carbon fiber and the resin with the use of a sizing agent which has better chemical coordination with each resin than the epoxy sizing agent coated on commercial carbon fiber.

The rate of increase in the interfacial bonding force of carbon fibers treated with a non-polar PP-based MPS sizing agent, which has a significantly lower interfacial bonding force with carbon fiber and resin than the PA6-based APS sizing agent, was higher. This is attributed to the optimal surface treatment in this study and the resizing treatment with MPS, which has a good chemical bond with PP resin and contributes to an improved interfacial bonding force with carbon fiber than using epoxy, a sizing agent used on the surface of commercial carbon fiber. However, the interfacial shear strength decreased when the concentration of APS was 1 wt.% for more than 10 s and when MPS was 0.5 wt.% for more than 10 s. Previous studies have indicated that increased resizing treatment time can result in an uneven coating layer on the carbon fiber surface when the sizing agent is 1% or more, resulting in sizing agent agglomeration and the breakdown of the interfacial bonding in the sizing layer due to van der Waals interaction between sizing agent molecules, which results in a decrease in the interfacial shear strength [34].

3.3. Chemical Properties of Recycled Carbon Fibers

Figure 4 shows the C1s and O1s XPS spectra from the analysis of the chemical state of the recycled carbon fiber surface according to the resizing conditions. The oxygen-to-carbon ratio (O/C) for judging the degree of composition change and oxygen content increase is summarized in Table 2. With a 1 wt.% PA6-based APS sizing agent, the content of carbon and oxygen decreased, and the content of nitrogen and silicon increased, compared to the untreated control. The carbon content decreased until 10 s and increased at 15 s; oxygen showed the highest value at 10 s, and the nitrogen and silicon content increased continuously as the treatment time increased. By contrast, with the 0.5 wt.% PP-based MPS sizing agent, the amount of carbon decreased as the treatment time increased, and the amount of oxygen and silicon increased. The ratio of O/C, which indicates the degree of activity of the carbon fiber surface, showed the highest value of 0.26 at 10 s with APS treatment, and the recycled carbon fiber treated with MPS gradually increased as treatment time increased. The optimum value of 0.51 at 10 s was about twice that of the APS. Other studies have reported that the surface activity of carbon fibers is enhanced when the O/C is higher than 0.26 [11]. As shown in Table 2, the O/C of commercial carbon fiber is 0.28; thus, the recycled carbon fibers in this study are considered to have sufficiently introduced oxygen functional groups. Further, the reason for the significantly higher O/C of MPS compared to APS is thought to be the increased introduction of oxygen functional groups due to the large amount of oxygen contained in MPS.

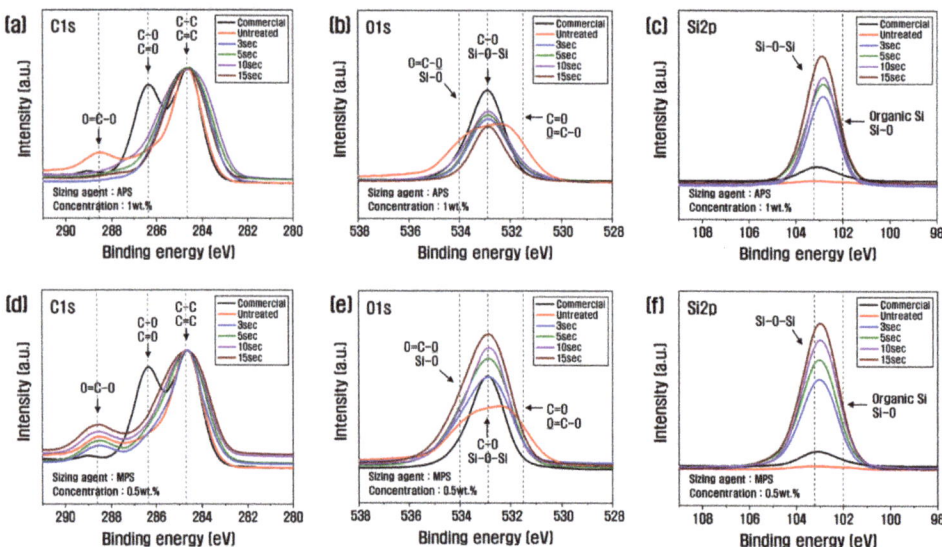

Figure 4. C1s and O1s XPS spectra of recycled carbon fibers covered with (**a**–**c**) APS and (**d**–**f**) MPS.

According to the C1s spectra, the O=C-O (lactone group), which was present in the untreated carbon fiber without sizing treatment, decreased significantly as time increased, while the C-O (hydroxyl group) and C=O (carbonyl group) increased up to 10 s of treatment time and decreased at 15 s (Figure 4a, Table 3). This is because, up to 10 s, the -SiOH (silanol group) present in the APS reacted with the O=C-O on the surface of the recycled carbon fiber in a dehydration–condensation reaction, causing the O=C-O bond to decrease and the O connected to the O=C-O to bond with another carbon, resulting in an increase in C-O and C=O; however, beyond 10 s, the hydrolysis of APS is actively occurring, resulting in a decrease in O=C-O, C-O, and C=O [34]. However, after the resizing treatment with MPS, O=C-O, C-O, and C=O increased continuously with the increasing treatment time (Figure 4d, Table 3). This is possibly due to the -SiOH in MPS that reacted with the O=C-O in the recycled carbon fiber, breaking the bonds and increasing the amount of C-O and C=O. MPS contains a large amount of methacrylate groups, which are rich in oxygen; thus, O=C-O, C-O, and C=O tended to increase gradually in treatments with MPS. According to the O1s spectra, C-O increased up to 10 s of treatment time but decreased at 15 s as a result of resizing with APS (Figure 4b). This means that O=C-O reacted with -SiOH contained in APS to become C-O and C=O, and C-O increased up to 10 s of treatment time. After 10 s, APS actively reacted with both O=C-O and C-O, and C-O decreased.

Table 2. Surface element composition of recycled carbon fibers according to sizing treatment conditions.

Treatment Condition		Elemental Composition (at. %)				O/C
Sizing Agent	Time (s)	Carbon	Oxygen	Nitrogen	Silicon	
Commercial CF		76.31	21.31	0.75	1.63	0.28
Untreated		73.91	22.81	2.73	0.55	0.31
APS (1 wt.%)	3	63.79	13.86	13.44	8.91	0.22
	5	62.71	14.28	13.73	9.28	0.23
	10	60.61	15.59	14.11	9.69	0.26
	15	62.54	12.56	14.77	10.13	0.20
MPS (0.5 wt.%)	3	64.08	27.51	0.27	8.14	0.43
	5	60.64	29.42	0.32	9.62	0.49
	10	58.55	29.80	0.30	11.35	0.51
	15	55.20	31.28	0.28	13.24	0.57

Table 3. Functional group according to sizing agent and treatment time by XPS.

Treatment Condition		C1s (at. %)			
Sizing Agent	Time (s)	C-C, C=C	C-O, C=O	C-N	O=C-O
Commercial CF		71.09	26.86	0.98	1.07
Untreated		74.00	8.12	6.28	11.59
APS (1 wt.%)	3	60.75	20.04	11.84	7.31
	5	60.13	20.83	12.91	6.13
	10	60.06	22.19	13.74	4.01
	15	63.19	19.95	14.27	2.59
MPS (0.5 wt.%)	3	76.05	10.53	5.27	8.15
	5	71.40	14.36	4.70	9.54
	10	62.43	22.47	3.70	11.40
	15	58.58	24.13	3.21	14.08

Further, the C-O of recycled carbon fiber treated with MPS gradually increased until the treatment time of 15 s (Figure 4e). This is because MPS has a large amount of oxygen. Thus, C-O steadily increased as the processing time increased. According to the Si2p spectra, the Si-O-Si of APS tended to increase continuously as the treatment time increased, which this study attributed to the binding of -SiOH contained in the sizing agent during the resizing process to remove H_2O. MPS showed the same trend, with Si-O-Si gradually increasing with treatment time, which is thought to be due to the active reaction between -SiOH (Figure 4c,f).

Previous studies have reported that after sizing treatment with DMHM (N-(4′4-diaminodiphenyl methane)-2-hydroxypropyl methacrylate), vinyl functional groups (-CH=CH$_2$) are introduced to the carbon fiber surface, increasing the radial width of the C=C peak, and O=C-N bonds are formed by the reaction of COOH (carboxyl group) and NH$_2$ (amino group) at the carbon fiber surface [35,36]. Furthermore, sizing with polydopamine can increase the amount of carbon and nitrogen, and C-N bonds have been shown to be generated through the spontaneous oxidative polymerization of dopamine [37]. When treated with poly(phthalazinone ether ketone), bonds such as C-N and C=N appeared due to the formation of phthalazine rings [21], and when an amino–silane coupling agent is used, the content of carbon increases as the length of the chain increases; the Si content also increases as the Si-O-Si bond increases, but the content of Si is thought to decrease because excessively long chains cover the Si located inside [21]. Vinyl ester treated with the R806 sizing agent contains many C-O bonds, which is expected because vinyl ester is the product of an unsaturated monoacid reaction with epoxy [38]. Further, with MR13006, a sizing agent with fewer C-O and more O-C=O than R806, it was difficult to distribute the agent uniformly on the carbon fiber, and it was reported that C-O has better compatibility with the carbon fiber surface than O=C-O [38].

In this study, C-O and C=O increased due to the dehydration–condensation reaction of O=C-O and -SiOH in the sizing agent on the surface of recycled carbon fiber, up to the optimum concentration of 1 wt.% and the treatment time of 10 s for APS, and Si-O-Si increased slightly due to the bonding of -SiOH on the surface of carbon fiber. When the treatment time was more than 10 s, C-O and C=O decreased due to the combination of APS, and Si-O-Si increased significantly due to the active reaction of -SiOH on the surface of the recycled carbon fiber. In the case of MPS, at the optimum concentration of 0.5 wt.%, C-O and C=O continuously increased with increasing treatment time due to the methacrylate group present at the end of MPS, and Si-O-Si increased significantly due to the reaction of -SiOH on the surface of recycled carbon fiber. Plausibly, these oxygen functional groups increase the surface energy of the recycled carbon fiber and improve the interfacial bonding.

To investigate the changes in surface free energy of recycled carbon fibers due to the resizing process, the contact angle was measured, and the values were substituted into the following Equation (1) [39] to calculate the polarized and non-polarized surface free energy, which is shown in Figure 5.

$$\frac{\gamma_L(1+\cos\theta)}{2(\gamma_L^D)^{\frac{1}{2}}} = \left(\gamma_S^P\right)^{\frac{1}{2}} \times \left(\frac{\gamma_L^P}{\gamma_L^D}\right)^{\frac{1}{2}} + \left(\gamma_S^D\right)^{\frac{1}{2}} \tag{1}$$

With APS at 1 wt.%, the contact angle decreased up to 10 s of treatment time, reaching the lowest value of about 41.36°, and then increased slightly at 15 s. The contact angle of carbon fiber treated with MPS at a concentration of 0.5 wt.% decreased until 10 s of treatment time. By contrast, the carbon fiber treated with MPS at a concentration of 0.5 wt.% showed a decreasing trend up to 10 s and almost no change beyond 10 s. From the contact angle results, the surface energy as a function of the resizing treatment time showed that the surface energy of the recycled carbon fiber treated with 1 wt.% APS increased up to 10 s of treatment time and decreased beyond 10 s. The increase was due to the polar surface area. The increase up to 10 s was due to the increase in polar surface energy, while the non-polar surface energy did not change. The polarity/surface energy ratio showed the highest value of 32.29% at 10 s, which is about 2.6 times higher than that of commercial carbon fiber, and decreased at treatment times above 10 s, compared to 10 s. It is possible that when APS is coated on recycled carbon fibers at 1% or more, it breaks the oxygen functional groups and actively binds the APS, reducing the amount of oxygen and thus reducing the polar surface energy.

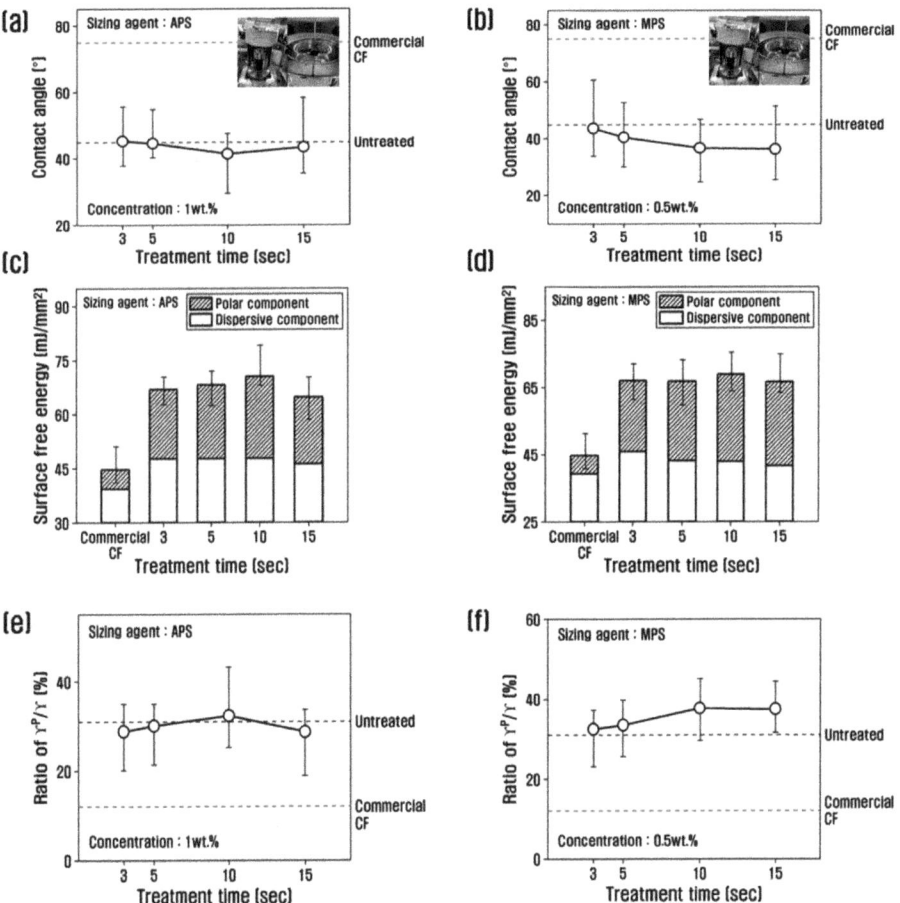

Figure 5. Variation of contact angle, surface free energy, and γ^p/γ of recycled carbon fibers treated with (**a,c,e**) APS and (**b,d,f**) MPS.

When treated with MPS, the polar surface energy was the highest at 10 s; the change was insignificant after 10 s and the polar/surface free energy ratio was about 38% at 10 s, which is about 3 times higher than that of commercial carbon fiber. This is believed to be due to the large amount of oxygen contained in MPS, which greatly enhances polar surface energy. In previous studies, the polar surface free energy was shown to increase due to -NH$_2$ after treatment with poly(amidoamine), and it increased continuously with increasing concentration [34]. Sizing agents can also increase the amount of oxygen functional groups, such as C=O, in carbon fibers to improve the polar surface energy [18].

3.4. Mechanism of Functional Group Change during Resizing Treatment

To investigate the effect and mechanism of PA6 and PP silane coupling agents on the interfacial bonding force between recycled carbon fibers and thermoplastics, surface treatment with nitric acid was performed under optimal conditions, followed by sizing treatment. Based on the results of analyzing the mechanical and chemical properties of recycled carbon fibers according to the concentration of the sizing agent and treatment time, the chemical structure and functional group mechanisms of recycled carbon fibers are shown in Figure 6.

(a) rCF + 1wt.% APS

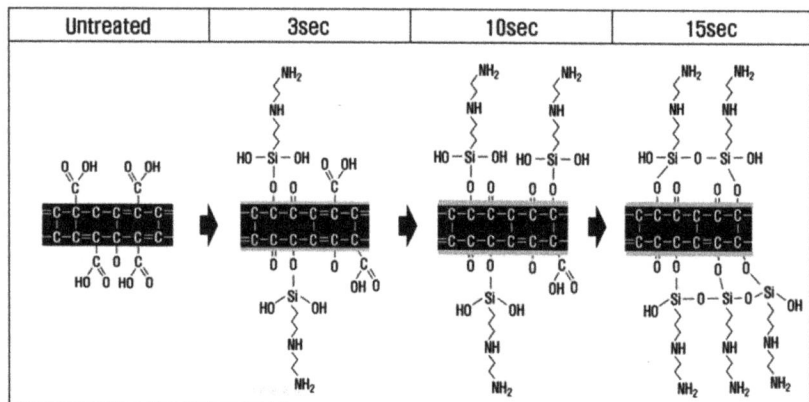

(b) rCF + 0.5wt.% MPS

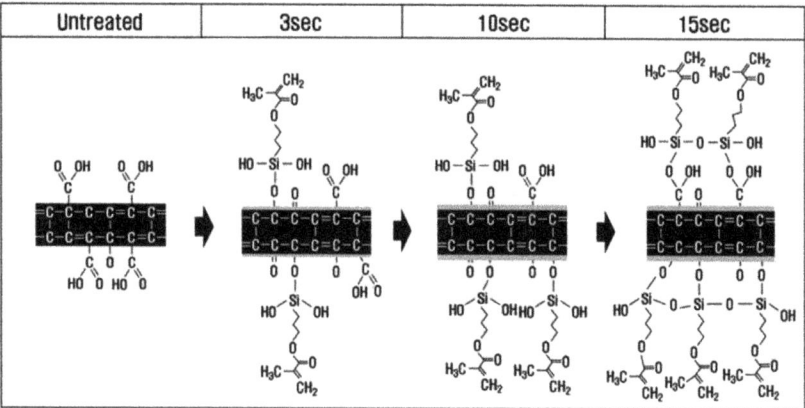

Figure 6. Schematic of the chemical reaction of carbon fibers according to sizing treatment time. (a) rCF/APS; (b) rCF/MPS.

With a PA6-based 1 wt.% APS sizing agent, H_2O was removed by a dehydration–condensation reaction between O=C-O present on the surface of recycled carbon fiber and -SiOH contained in APS at a treatment time of 3 s (Equations (2) and (3)). At this time, O=C-O was converted into C-O and C=O by breaking the bonds between carbon and oxygen, and it seems that O=C-O gradually decreased and C-O and C=O increased slightly. In general, sizing agents react with water to undergo hydrolysis, in which -OCH$_3$ (methyl groups) become -OH (hydroxyl groups), and undergo a dehydration–condensation reaction with the oxygen functional groups of recycled carbon fibers [40,41]. In this study, H_2O was removed, and O=C-O was converted to C-O and C=O through a dehydration–condensation reaction between O=C-O present on the carbon fiber surface and -SiOH contained in APS. At 10 s, similar to the trend at 3 s, O=C-O decreased, and C-O and C=O continued to increase, suggesting that -SiOH mainly reacted with O=C-O at this time. Thus, this study can conclude that at a treatment time of 10 s, C-O and C=O increased to the maximum, and the interfacial bonding force was maximized due to the optimal oxygen to carbon ratio (O/C). When the treatment time was 15 s or more, O=C-O, C-O, and C=O were all reduced due to the dehydration–condensation reaction with the -SiOH contained in the sizing agent, which broke the bond between carbon and oxygen. Si-O-Si was greatly increased by reacting with each other and combining with the -SiOH contained in the sizing agent during the resizing treatment. As a result, O/C decreased due to the decrease in C-O and C=O, and the interfacial binding force decreased somewhat more than at the treatment time of 10 s.

In the case of the resizing treatment with 0.5 wt.% of PP-based MPS sizing agent, H_2O was removed by a dehydration–condensation reaction of O=C-O on the carbon fiber surface with the -SiOH contained in the sizing agent at a treatment time of 3 s. At this time, the bond between the carbon and oxygen of O=C-O was broken, and the methacrylate group (H_2C=C(CH_3)CO_2H) containing a large amount of oxygen from the MPS was introduced into the broken bond, and O=C-O, C-O, and C=O increased (Equations (2)–(4)). At a treatment time 10 s, similar to the trend at 3 s, O=C-O and the -SiOH contained in the sizing agent underwent a dehydration–condensation reaction. The bonds of O=C-O were broken, and the methacrylate group was continuously and significantly introduced into the broken bonds, and O=C-O, C-O, and C=O continued to increase. For up to 10 s of treatment time, SiOH was judged to have reacted mainly with O=C-O. As a result, at 10 s of treatment time, the optimum O/C was obtained due to an increase in O=C-O, C-O, and C=O, which represents the maximum interfacial bonding force. At 15 s or more, not only O=C-O but also C-O and C=O present on the surface of the carbon fiber underwent a dehydration–condensation reaction with the -SiOH contained in the sizing agent, breaking the bond between carbon and oxygen but greatly introducing methacrylate groups containing a large amount of oxygen, resulting in a significant increase in C-O, C=O, and Si-O-Si. In addition, at a treatment time of 15 s or more, the -SiOH contained in the sizing agent reacted and bonded with each other, resulting in an increase in Si-O-Si. As a result, C-O and C=O continued to increase at 15 s rather than at 10 s, but the interfacial bonding force decreased within the error range. Thus, this study judged that optimal sizing was achieved at the shortest time of 10 s.

$$O = C - O \downarrow + - SiOH$$

$$-SiOH + -SiOH \rightarrow$$

$$\nearrow C = O \uparrow + C - O - Si + H_2O \quad (2)$$
$$\searrow C - O \uparrow + C - O - Si + H_2O \quad (3)$$
$$Si - O - Si + H_2O \quad (4)$$

4. Conclusions

In this study, the thermal, mechanical, and chemical properties of recycled carbon fibers were analyzed according to the sizing agent concentration and treatment time after desizing, the optimal conditions were derived, and the chemical changes and functional group mechanisms according to the desizing treatment conditions were identified.

1. In the case of resizing with the PA6-based APS sizing agent, at a concentration of 1 wt.% of the sizing agent and a treatment time of 10 s, the O=C-O present on the surface of the carbon fiber and the -SiOH contained in the APS underwent a dehydration–condensation reaction, and the O=C-O was converted into the bonds of C-O and C=O, the C-O and C=O increased, and from this, the effect of C-O and C=O on the interfacial bonding force increased to the maximum, and the ratio between oxygen and carbon (O/C) was 0.26. In addition, the polar/surface energy ratio showed the highest value of 32.29% at 10 s, and the interfacial bonding force showed the maximum value of 32 MPa at 10 s, which is about 15% better than that of commercial carbon fiber, and was determined to be the optimal PA6-based sizing condition.

2. When the PP-based MPS sizing agent was used, both C-O and C=O as well as O=C-O, at a concentration of 0.5 wt.% and treatment time 10 s, were subjected to a dehydration–condensation reaction with -SiOH, which broke the bonds between carbon and oxygen and introduced oxygen-rich methacrylate groups (H_2C=C(CH_3)CO_2H) into the broken bonds, resulting in a significant increase in C-O and C=O and a significant increase in O/C to 0.51. Further, this study recorded a polar/surface free energy ratio of about 38% at 10 s, and the interfacial bonding force maximally increased to about 27%,

compared to commercial carbon fiber, which was judged to be the optimal PP-based sizing condition.

The interfacial shear strength characteristics reported in this study are comparable to those of commercial carbon fiber, which are typically coated with epoxy sizing agents. This outcome is attributed to the resizing of the carbon fibers in this study, achieved by selecting sizing agents that had excellent chemical coordination with each resin after surface treatment with nitric acid, an optimal condition indicated in previous studies [37]. In the future, this study plans to produce carbon composites by impregnating recycled carbon fibers and thermoplastic resins under optimal conditions to evaluate whether their mechanical properties are equivalent to those of commercial carbon composites, which can contribute to the commercialization of automotive parts built with these materials.

Author Contributions: Conceptualization, H.L. and G.K.; Methodology, D.K.; Validation, G.K.; Formal analysis, M.K.; Investigation, H.L.; Writing—original draft, H.L.; Writing—review & editing, G.K. and D.K.; Supervision, D.K. All authors have read and agreed to the published version of the manuscript.

Funding: This study was supported by the Improvement Strategy of Material & Component Technology Development Program (No. 20012924) funded by the Ministry of Trade, Industry, and Energy in Republic of Korea.

Institutional Review Board Statement: Not applicable.

Data Availability Statement: Not applicable.

Conflicts of Interest: The authors declare no conflict of interest.

References

1. Ma, K.; Chen, P.; Wang, B.; Cui, G.; Xu, X. A study of the effect of oxygen plasma treatment on the interfacial properties of carbon fiber/epoxy composites. *J. Appl. Polym. Sci.* **2010**, *118*, 1606–1614. [CrossRef]
2. Jin, F.-Y.; Park, S.-J. Preparation and characterization of carbon fiber-reinforced thermosetting composites: A review. *Carbon Lett.* **2015**, *16*, 67–77. [CrossRef]
3. Chin, K.-Y.; Shiue, A.; Wu, Y.-J.; Chang, S.-M.; Li, Y.-F.; Shen, M.-Y. Graham Leggett. Studies on recycling silane controllable recovered carbon fiber from waste CFRP. *Sustainability* **2022**, *14*, 700. [CrossRef]
4. He, J.M.; Huang, Y.D. Effect of silane-coupling agents on interfacial properties of CF/PI composites. *J. Appl. Polym. Sci.* **2007**, *106*, 2231–2237. [CrossRef]
5. Qing, W.; Zhao, R.; Zhu, J.; Wang, F. Interfacial improvement of carbon fiber reinforced epoxy composites by tuning the content of curing agent in sizing agent. *Appl. Surf. Sci.* **2020**, *504*, 144384. [CrossRef]
6. Bin, Y.; Jiang, Z.; Tang, X.-Z.; Yue, C.Y.; Yang, J. Enhanced interphase between epoxy matrix and carbon fiber with carbon nanotube-modified silane coating. *Compos. Sci. Technol.* **2014**, *99*, 131–140. [CrossRef]
7. Yao, Y.; Li, M.; Wu, Q.; Dai, Z.; Gu, Y.; Li, Y.; Zhang, Z. Comparison of sizing effect of T700 grade carbon fiber on interfacial properties of fiber/BMI and fiber/epoxy. *Appl. Surf. Sci.* **2012**, *263*, 326–333. [CrossRef]
8. Wu, Q.; Li, M.; Gu, Y.; Wang, S.; Yao, L.; Zhang, Z. Effect of sizing on interfacial adhesion of commercial high strength carbon fiber-reinforced resin composites. *Polym. Compos.* **2016**, *37*, 254–261. [CrossRef]
9. Wu, Q.; Li, M.; Gu, Y.; Wang, S.; Wang, X.; Zhang, Z. Reaction of carbon fiber sizing and its influence on the interphase region of composites. *J. Appl. Polym. Sci.* **2015**, *132*, 41917. [CrossRef]
10. Yao, S.-S.; Jin, F.-L.; Rhee, K.Y.; Hui, D.; Park, S.-J. Recent advances in carbon-fiber-reinforced thermoplastic composites: A review. *Compos. B Eng.* **2018**, *142*, 241–250. [CrossRef]
11. Kim, G.; Lee, H.; Kim, K.; Kim, D.U. Effects of heat treatment atmosphere and temperature on the properties of carbon fibers. *Polymers* **2022**, *14*, 2412. [CrossRef]
12. Song, J.-H. Tensile strength of polypropylenes carbon fiber composite for heat treatment conditions. *J. Korean Soc. Mech. Technol.* **2020**, *22*, 107–111. [CrossRef]
13. Sun, H.; Guo, G.; Memon, S.A.; Xu, W.; Zhang, Q.; Zhu, J.-H.; Xing, F. Recycling of carbon fibers from carbon fiber reinforced polymer using electrochemical method. *Compos. Part A Appl. Sci. Manuf.* **2015**, *78*, 10–17. [CrossRef]
14. Zomer, D.; Simaafrookhtch, S.; Vanclooster, K.; Dorigato, A.; Ivens, J. Forming-behavior characterization of cross-ply carbon fiber/PA6 laminates using the bias-extension test. *Compos. Part A Appl. Sci. Manuf.* **2023**, *168*, 107436. [CrossRef]
15. Yan, C.; Zhu, Y.; Liu, D.; Xu, H.; Chen, G.; Chen, M.; Cai, G. Improving interfacial adhesion and mechanical properties of carbon fiber reinforced polyamide 6 composites with environment-friendly water-based sizing agent. *Compos. Part B Eng.* **2023**, *258*, 110675. [CrossRef]

16. Ateeq, M. A review on recycling technique and remanufacturing of the carbon fiber from the carbon fiber polymer composite: Processing, challenges, and state-of-arts. *Compos. Part C Open Access* **2023**, in press. [CrossRef]
17. Jung, M.-J.; Park, M.-S.; Lee, S.; Lee, Y.-S. Effect of e-beam radiation with acid drenching on surface properties of pitch-based carbon fibers. *Appl. Chem. Eng.* **2016**, *27*, 319–324. [CrossRef]
18. Park, S.-J.; Choi, W.-K.; Kim, B.-J.; Min, B.-G.; Bae, K.-M. Effects of sizing treatment of carbon fibers on mechanical interfacial properties of nylon 6 matrix composites. *Elastom. Compos.* **2010**, *45*, 2–6.
19. Jian, L. Effect of sizing agent on interfacial properties of carbon fiber-reinforced PMMA composite. *Compos. Adv. Mater.* **2021**, *30*, 1–6. [CrossRef]
20. Fiore, V.; Orlando, V.; Sanfilippo, C.; Badagliacco, D.; Valenza, A. Effect of Silane Coupling treatment on the adhesion between polyamide and epoxy based composites reinforced with carbon fibers. *Fibers* **2020**, *8*, 48. [CrossRef]
21. Liu, W.B.; Zhang, S.; Hao, L.F.; Jiao, W.C.; Yang, F.; Li, X.F.; Wang, R.G. Properties of carbon fiber sized with poly(phthalazinone ether ketone) resin. *J. Appl. Polym. Sci.* **2013**, *128*, 3702–3709. [CrossRef]
22. Yu, S.; Oh, K.H.; Hwang, J.Y.; Hong, S.H. The effect of amino-silane coupling agents having different molecular structures on the mechanical properties of basalt fiber-reinforced polyamide 6,6 composites. *Compos. B Eng.* **2019**, *163*, 511–521. [CrossRef]
23. Li, M.; Pan, B.; Liu, H.; Zhu, L.; Fan, X.; Yue, E.; Li, M.; Qin, Y. Interfacial tailoring of basalt fiber/epoxy composites by metal–organic framework based oil containers for promoting its mechanical and tribological properties. *Polym. Compos.* **2023**, *44*, 4757–4770. [CrossRef]
24. Zhang, Y.; Zhang, Y.; Liu, Y.; Wang, X.; Yang, B. A novel surface modification of carbon fiber for high-performance thermoplastic polyurethane composites. *Appl. Surf. Sci.* **2016**, *382*, 144–154. [CrossRef]
25. Gao, B.; Du, W.; Hao, Z.; Zhou, Z.; Zou, D.; Zhang, R. Bioinspired modification via green synthesis of mussel-inspired nanoparticles on carbon fiber surface for advanced composite materials. *ACS Sustain. Chem. Eng.* **2019**, *7*, 5638–5648. [CrossRef]
26. Dong, Y.; Yu, T.; Wang, X.-J.; Zhang, G.; Lu, J.H.; Zhang, M.-L.; Long, S.-R.; Yang, J. Improved interfacial shear strength in polyphenylene sulfide/carbon fiber composites via the carboxylic polyphenylene sulfide sizing agent. *Compos. Sci. Technol.* **2020**, *190*, 108056. [CrossRef]
27. *ASTM D3822-01*; Standard Test Method for Tensile Properties of Single Textile Fibers. ASTM: West Conshohocken, PA, USA, 2001.
28. *ASTM C1557-20*; Standard Test Method for Tensile Strength and Young's Modulus of Fibers. ASTM: West Conshohocken, PA, USA, 2020.
29. *ASTM D1331-20*; Standard Test Methods for Surface and Interfacial Tension of Solutions of Paints, Solvents, Solutions of Surface-Active Agents, and Related Materials. ASTM: West Conshohocken, PA, USA, 2020.
30. Ku, S.G.; Kim, Y.S.; Kim, D.W.; Kim, K.S.; Kim, Y.C. Effect of silane coupling agent on physical properties of polypropylene (PP)/kenaf fiber (KF) felt composites. *Appl. Chem. Eng.* **2018**, *29*, 37–42.
31. Lee, M.; Kim, Y.; Ryu, H.; Baeck, S.-H.; Shim, S.E. Effects of silane coupling agent on the mechanical and thermal properties of silica/polypropylene composites. *Polymer* **2017**, *41*, 599–609. [CrossRef]
32. Qiu, B.; Li, M.; Zhang, X.; Chen, Y.; Zhou, S.; Liang, M.; Zou, H. Carboxymethyl cellulose sizing repairs carbon fiber surface defects in epoxy composites. *Mater. Chem. Phys.* **2021**, *258*, 123677. [CrossRef]
33. Sun, N.; Zhu, B.; Cai, X.; Yu, L.; Yuan, X.; Zhang, Y. Enhanced interfacial properties of carbon fiber/polyamide composites by in-situ synthesis of polyamide 6 on carbon fiber surface. *Appl. Surf. Sci.* **2022**, *599*, 153889. [CrossRef]
34. Peng, Q.; Li, Y.; He, X.; Lv, H.; Hu, P.; Shang, Y.; Wang, C.; Wang, R.; Sritharan, T.; Du, S. Interfacial enhancement of carbon fiber composites by poly(amido amine) functionalization. *Compos. Sci. Technol.* **2013**, *74*, 37–42. [CrossRef]
35. Gao, M.; Xu, Y.; Wang, X.; Sang, Y.; Wang, S. Analysis of electrochemical reduction process of graphene oxide and its electrochemical behavior. *Electroanalysis* **2016**, *28*, 1377–1382. [CrossRef]
36. Jiao, W.; Cai, Y.; Liu, W.; Yang, F.; Jiang, L.F.; Jiao, W.; Wang, R. Preparation of carbon fiber unsaturated sizing agent for enhancing interfacial strength of carbon fiber/vinyl ester resin composite. *Appl. Surf. Sci.* **2018**, *439*, 88–95. [CrossRef]
37. Han, W.; Zhang, H.-P.; Tavakoli, J.; Campbell, J.; Tang, Y. Polydopamine as sizing on carbon fiber surfaces for enhancement of epoxy laminated composites. *Compos. Part A Appl. Sci. Manuf.* **2018**, *107*, 626–632. [CrossRef]
38. Wu, Z.; Li, L.; Guo, N.; Yang, R.; Jiang, D.; Zhang, M.; Zhang, M.; Huang, Y.; Guo, Z. Effect of a vinyl ester-carbon nanotubes sizing agent on interfacial properties of carbon fibers reinforced unsaturated polyester composites. *ES Mater. Manuf.* **2019**, *6*, 38–48. [CrossRef]
39. Rame, E. The interpretation of dynamic contact angles measured by the wilhelmy plate method. *J. Coll. Interface Sci.* **1997**, *185*, 245–251. [CrossRef]
40. Sang, L.; Wang, Y.K.; Chen, G.Y.; Liang, J.C.; Wei, Z.Y. A comparative study of the crystalline structure and mechanical properties of carbon fiber/polyamide 6 composites enhanced with/without silane treatment. *RSC Adv.* **2016**, *6*, 107739–107747. [CrossRef]
41. Yang, J.; Xiao, J.; Zeng, J.; Bian, L.; Peng, C.; Yang, F. Matrix modification with silane coupling agent for carbon fiber reinforced epoxy composites. *Fibers Polym.* **2013**, *14*, 759–766. [CrossRef]

Disclaimer/Publisher's Note: The statements, opinions and data contained in all publications are solely those of the individual author(s) and contributor(s) and not of MDPI and/or the editor(s). MDPI and/or the editor(s) disclaim responsibility for any injury to people or property resulting from any ideas, methods, instructions or products referred to in the content.

Article

Short Flax Fibres and Shives as Reinforcements in Bio Composites: A Numerical and Experimental Study on the Mechanical Properties

Sofie Verstraete [1,*], Bart Buffel [1], Dharmjeet Madhav [2], Stijn Debruyne [3] and Frederik Desplentere [1]

[1] Research Group ProPoliS, Department of Materials Engineering, KU Leuven Campus Bruges, Spoorwegstraat 12, 8200 Bruges, Belgium

[2] Surface and Interface Engineered Materials, Department of Materials Engineering, KU Leuven Campus Bruges, Spoorwegstraat 12, 8200 Bruges, Belgium

[3] Research Group M-Group, Department of Mechanical Engineering, KU Leuven Campus Bruges, Spoorwegstraat 12, 8200 Bruges, Belgium

* Correspondence: sofie.verstraete@kuleuven.be

Citation: Verstraete, S.; Buffel, B.; Madhav, D.; Debruyne, S.; Desplentere, F. Short Flax Fibres and Shives as Reinforcements in Bio Composites: A Numerical and Experimental Study on the Mechanical Properties. *Polymers* 2023, 15, 2239. https://doi.org/10.3390/polym15102239

Academic Editors: Abderrahmane Ayadi, Patricia Krawczak and Chung Hae Park

Received: 10 February 2023
Revised: 4 May 2023
Accepted: 6 May 2023
Published: 9 May 2023

Copyright: © 2023 by the authors. Licensee MDPI, Basel, Switzerland. This article is an open access article distributed under the terms and conditions of the Creative Commons Attribution (CC BY) license (https://creativecommons.org/licenses/by/4.0/).

Abstract: The complete flax stem, which contains shives and technical fibres, has the potential to reduce the cost, energy consumption and environmental impacts of the composite production process if used directly as reinforcement in a polymer matrix. Earlier studies have utilised flax stem as reinforcement in non-bio-based and non-biodegradable matrices not completely exploiting the bio-sourced and biodegradable nature of flax. We investigated the potential of using flax stem as reinforcement in a polylactic acid (PLA) matrix to produce a lightweight, fully bio-based composite with improved mechanical properties. Furthermore, we developed a mathematical approach to predict the material stiffness of the full composite part produced by the injection moulding process, considering a three-phase micromechanical model, where the effects of local orientations are accounted. Injection moulded plates with a flax content of up to 20 V% were fabricated to study the effect of flax shives and full straw flax on the mechanical properties of the material. A 62% increase in longitudinal stiffness was obtained, resulting in a 10% higher specific stiffness, compared to a short glass fibre-reinforced reference composite. Moreover, the anisotropy ratio of the flax-reinforced composite was 21% lower, compared to the short glass fibre material. This lower anisotropy ratio is attributed to the presence of the flax shives. Considering the fibre orientation in the injection moulded plates predicted with Moldflow simulations, a high agreement between experimental and predicted stiffness data was obtained. The use of flax stems as polymer reinforcement provides an alternative to the use of short technical fibres that require intensive extraction and purification steps and are known to be cumbersome to feed to the compounder.

Keywords: flax fibres; injection moulding; mechanical properties; short fibre reinforced thermoplastics (SFRT); bio composites

1. Introduction

To address the end-of-life challenges of thermoset composite materials, the use of thermoplastic as matrix material is receiving more and more attention in academia and industry. Thermoplastic matrices allow faster cycle times and are applicable in large volume processes such as extrusion, compression moulding or injection moulding. The latter two processes allow the production of complex parts at relatively low production costs. However, these production processes restrict the reinforcement to short fibres. Short fibre-reinforced thermoplastics are therefore used in applications where ultimate strength or stiffness are not the main requirements [1].

As the awareness for sustainable resources is growing, recent trends show an increasing interest in the use of natural fibres as a bio-based alternative to synthetic fibres [2–6].

Flax fibres with mechanical properties that even outperform the specific stiffness of glass fibres are among the most commonly used natural fibre reinforcements [7]. The cultivation of flax fibres is characterised by a low environmental impact and relatively short sowing-to-harvesting cycle of approximately 100 days. Cultivation factors such as flax species, geographical location, nutrition, temperature, season and local climate conditions affect the mechanical properties and overall quality of natural fibres [8,9]. To cope with this inevitable variation in properties, special regulations and monitoring of cultivation are applied. This ensures the high quality of flax and reduces the variation in mechanical properties [10]. After harvesting, the flax stems are carefully and evenly spread on the soil to undergo a dew-retting process. The flax fibres and the wooden parts, also called shives, are hereby separated by a natural process of pectin degradation that binds the fibres with shives [11]. After retting, flax stem is broken, and the shives are separated from the fibres by a scutching process. Finally, flax fibres are hackled to detangle and straighten producing individual technical fibres that are suited for use in the textile industry or long-fibre reinforced composites [12,13].

Further advances towards "green composites" require the development and utilisation of alternative bio-based polymers. These polymers are synthesised out of renewable feedstock such as sugar, starch, cellulose, lignin and show an interesting combination of environmental friendliness and mechanical properties [14]. A starch-based polylactic acid (PLA) biopolymer can achieve Young's modulus of 3 GPa, tensile modulus of up to 70 MPa and impact strength of 2.5–3 kJ/m^2 [15]. The literature indicates that adding short flax fibres to a PLA matrix preserves the possibility of using the material in an injection moulding process, while the tensile modulus increases up to 5 GPa at a volume fraction of 20% [16,17].

Shives are separated from the flax stem during the production process of flax fibres and represent agricultural waste that is currently used as animal bedding, particleboard, thermal insulation material for buildings [18] or even as an alternative to the widely used wood flour plastic composites (WPC) [19]. About 70 to 85% (depending on the flax variety) of the flax stem consists of shives [20]. As France, Belgium and the Netherlands combined have a total flax cultivation area of over 124,000 ha, it is estimated that about 400,000 tonnes of flax shives was available in 2018. This amount is expected to further increase due to the growing interest in the application of natural fibres [21].

Shives were found to be responsible for 30% of the bending stiffness of the dry flax stem [20]. Nuez et al. investigated the potential of these flax shives as reinforcement in a polymetric matrix and found an improvement in the mechanical properties of a flax shive/polypropylene composite [22]. In another study by Soete et al., the fibre breakage of the flax stem during an injection moulding process was investigated [23]. They showed that the high shear forces during the compounding and injection moulding process are responsible for fibre breakage, resulting in rupture of the flax fibres and the wooden parts or flax shives that are adjacent to each other in the composite. Results of the tensile tests showed an improvement in mechanical properties when used in a PP matrix [24]. Results of the tensile tests on hackled flax fibre-reinforced PP showed an improvement in mechanical properties [24].

Since the use of flax shives as well as the whole flax stem as reinforcements in a PP matrix have shown promising results, there is a potential for high-value application of this agricultural waste product. Therefore, the present study focuses on the use of flax shives as well as the whole stem in a bio-based PLA matrix. The obtained composite is then 100% bio-based and biodegradable. By using the whole stem of the flax plant as reinforcement, energy-consuming production steps to extract the technical fibres are avoided. This leads to a reduction in the production cost of the material while maintaining the interesting mechanical properties of the composites.

The effect of the reinforcements strongly depend on the size of the fibres. In the case of continuous fibre-reinforced composites, application of nano particles in a polymetric matrix result in an increase in mechanical (tensile, flexural, fatigue, ...) properties or flame retardant improvement [25–30], or they can act to improve the mechanical properties for long fibre composites [31]. For short fibre composites, an optimal design of the fibre orientation extends the life of the composite system while lowering production cost and minimising the fibre volumetric percentage [32]. When a short fibre composite is produced through injection moulding, it is well documented that the fibre orientation is determined by the flow of the material. In this way, the typical skin–shell–core orientation is obtained [1]. The orientation of the short fibres mainly depends on the rheological properties of the molten compound, which is directly affected by the aspect ratio and volume fraction of fibre in the compound. This results in non-isotropic and non-orthotropic mechanical properties of the composite. The ratio of the minimum and maximum value of tensile modulus is defined as the degree of anisotropy in the material, which increases with increasing fibre volume fraction [33]. The literature reports a higher tensile modulus and strength along the flow direction of injection moulded parts [34,35]. The coupling of the injection moulding production process and the obtained mechanical properties is inevitable and must be considered during the design phase of a part. When used in larger assemblies for semi-structural applications, considering the local anisotropic material properties is mandatory [36].

Therefore, in the present study, the stiffness ratio along the two orthogonal directions was considered to evaluate the degree of anisotropy, and a three-phase [37] multi-layer analytical model was established to predict the mechanical stiffness along all orientations of an injection moulded plate. The approach is based on the Mori–Tanaka (MT) micromechanics model combined with general laminate theory (GL) and considers the aspect ratio, mechanical properties and process-induced orientations of the flax fibres and shives.

2. Materials and Methods

2.1. Materials

The thermoplastic matrix material considered in this study is a commercially available polylactic-acid (PLA) 4043D Ingeo biopolymer (manufactured by NatureWorks, Minneapolis, MN, USA) with a melt flow index (MFI) of 6 g/10 min (210 °C, 2.16 kg) and a density of 1.24 g/cm^3.

Flax straw (consisting of technical flax fibres and flax shives) and flax shives were provided by ABV (Algemeen Belgisch Vlasbond, Kortrijk, Belgium) and used as reinforcement to the PLA material. Short E-glass fibres are used as reference material to benchmark the performance of the flax reinforcements. The initial properties of the considered fibres are summarised in Table 1.

Table 1. Physical properties of the raw fibre materials.

Fibre Type	E-Modulus (GPa)	Tensile Strength (GPa)	Density (g/cm³)	Initial Length (mm)	Initial Diameter (µm)
Glass fibre [38]	70	3.5	2.6	10	10
Flax fibre [23]	45	0.77	1.5	10	±1500

2.2. Production Process of the Bio-Based Composite

The composites are produced using a two-step production process. First, a compounding step is used to blend the reinforcement into the PLA matrix. Subsequently, samples are produced through injection moulding. Prior to each processing step, the PLA matrix and the flax reinforcements are dried with a Moretto pressurised air dryer at 80°C for 6 h. A Leistritz ZSE18MAXX (Nürnberg, Germany) corotating twin-screw laboratory extruder was used to shear-blend the components and produce pellets with a length of 5 mm. The details of the compounding step are shown in Table 2. Compounds of PLA with up to 15 V% flax shives, up to 20 V% straw flax (flax shives + flax fibres) and up to 20 V% glass fibres were prepared. In the following sections, the volume fraction is referred to as the volume percentage and is therefore indicated with the unit %. The compounding screw elements were selected to avoid excessive shear forces to reduce fibre breakage.

Table 2. Compounding parameters.

Parameter	Value
Main feed	150 rpm
Melt pressure	61 bar
Melt temperature	195 °C
Die temperature (⌀ 3 mm)	185 °C
Screw configuration (length: 36D)	
• 2D	• Feeding PLA
• 12–16D	• Vent
• 16–20D	• Feeding fibres
• 28–32D	• Vacuum

After the second drying step of the compound (80 °C, 6 h), a batch of 15 plates of 85 mm × 85 mm × 3 mm was produced using an Arburg 320S injection moulding machine (Loßburg, Germany). The studied compounds exist of virgin PLA, PLA with 5-, 10- and 15% flax shives (FS), 5-, 10- and 20% straw flax fibres (FF) and 10- and 20% of glass fibres (GF). Tensile bar samples are milled out of the injection moulded plates at different orientations with respect to the flow direction of the melt. These tensile bars are used for mechanical testing (Figure 1a). The parameters of the injection moulding process are summarised in Table 3.

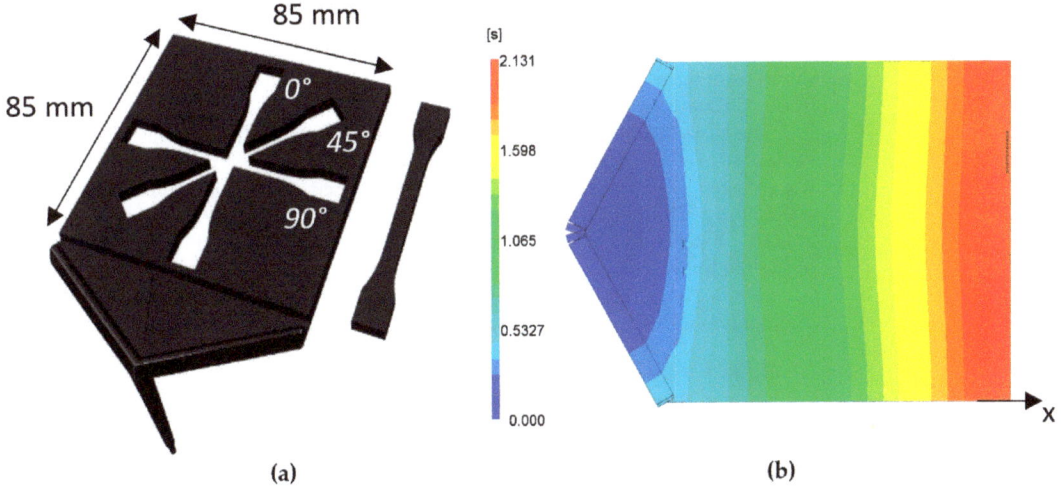

Figure 1. Plate mould geometry showing (**a**) the production of tensile bars (ISO 527-1BA) under different orientations for material characterisation and (**b**) the filling time of the plate, indicating a parallel flowing front during injection moulding process.

A fan-gate injection sprue is used to obtain a parallel flow front in the plate, as shown in Figure 1b. This parallel flow front ensures a high fibre orientation along the flow direction, which allows for the investigation of the tensile modulus of composites along different orientations with respect to the flow direction.

Table 3. Injection moulding parameters.

Parameter	Value
Melt temperature	190 °C
Injection volume	30.7 cm^3
Injection pressure	1065 bar
Injection time	5.8 s
Holding pressure	750 bar
Holding time	15 s
Cooling water temperature	25 °C
Mould temperature	25 °C
Temperature profile in cylinder	
• T1	30 °C
• T2	160 °C
• T3	170 °C
• T4	190 °C
• T5	190 °C
• T6	190 °C

2.3. Fibre Morphology

During twin screw compounding, high shear forces are exerted by the screw onto the material. These shear forces separate the technical fibres from the shives, but also inevitably cause unwanted fibre breakage and damage [24]. At each step of the production process, the morphology and dimensions of the fibres and shives were determined using an extraction process. This comprises dissolving the PLA material in chloroform for 24 h. The shives were separated from the fibres based on density, and the chloroform was removed from each component by filtration.

The straw flax structure was examined by optical laser microscopy. The fibre was embedded in an epoxy resin (EpoxyFix for embedding materialographic specimens, Struers, Kopenhagen, Denmark), and the surface was polished using a diamond paste with grain size of 1 μm. The composition was determined by weighting the two components from at least 3 batches and at different volume fractions, as described by [23]. Dimensions of the reinforcements, such as aspect ratio (AR) and volume fraction, were measured using a reflection mode optical laser microscope (VK1050, Keyence, Mechelen, Belgium) in combination with an image processing algorithm in MATLAB.

2.4. Mechanical Characterisation

ISO 527-1BA dog bone specimens were produced, using a milling process, at different orientations relative to the flow direction (0°, 30°, 45°, 60° and 90°) from the plates (PLA, PLA + 5-, 10- and 15% FS, PLA + 5-, 10-, 20% FF and PLA + 10- and 20% GF). Three specimens of each compound were tested at room temperature using an Instron 3367 two-column tensile bench equipped with a 30 kN load cell (Darmstadt, Germany), in combination with an extension meter (Instron 2630–125). A crosshead speed of 1.3 mm/min was used. Additional tests were performed in combination with a digital image correlation setup (stereo-DIC, Limess, Krefeld, Germany) to obtain values for Poisson's ratio. The CMOS cameras with a resolution of 2.3 Mpx were calibrated using an A6 (type: KL 50108) calibration grid. Results were measured with a frequency of 4 Hz and analysed in the ISTRA4D (4.6) software. A summary of the setup is given in Table 4, as described in [39].

Table 4. Experimental DIC settings and performance (Limess).

Parameter	3D Image Correlation
Sensor type	1/1.2" CMOS
Resolution	1920 × 1200 px (2.3 Mpx)
Pixel size	5.86 µm × 5.86 µm
Correlation criterion	Universal correlation evaluation
Optimisation residual	0.1349 pixel
Pre-smoothing applied to the images	None
Subset size	17 × 17 pixel
Shape function	Affine
Interpolation function	Bicubic polynomial
Smoothing method	Local regression (kernel size ACSP 05 × 05)

3. Theoretical Analysis

A numerical method is developed to determine the elastic properties of straw flax-reinforced thermoplastics. The result of this numerical method is compared to experimental data.

3.1. Three-Phase Mori–Tanaka (MT) Modelling

The Mori–Tanaka micromechanics approach is selected to determine the elastic properties. Since straw flax consists of flax fibres and shives, the composite is considered to be a three-phase system [40]. The different phases are illustrated in Figure 2. The first step considers the composite made of neat PLA and flax shives. In the second step, this composite is reinforced with flax fibres. In each step, the individual elastic properties and dimensions of both reinforcing components are taken into account.

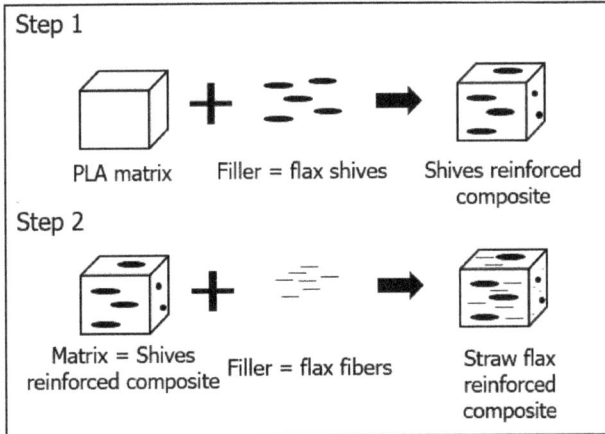

Figure 2. Setup of a Mori–Tanaka model for a three-phase system.

The stiffness tensor of the flax straw reinforced composite is calculated by the generalised Mori–Tanaka homogenisation procedure [35]. In the first step of this three-phase system, the stiffness tensor C_{comp_1} is described by a Mori–Tanaka homogenisation model presented in Equation (1). This compliance refers to the neat PLA reinforced with flax shives. The second homogenisation step, described in Equation (2), calculates the stiffness matrix C_{comp} for the full straw flax composite. In this step, the shives-reinforced composite is considered to be the matrix material and the flax fibres the reinforcement.

$$C_{comp\_1} = C_m + V_s(C_s - C_m) A^{Mori-Tanaka\_1} \qquad (1)$$

$$C_{comp} = C_{comp\_1} + V_f \left(C_f - C_{comp\_1} \right) A^{Mori-Tanaka\_2} \quad (2)$$

The subscript m represents the PLA matrix, s the flax shives and f the flax fibres. V indicates the volume fraction corresponding to the relevant composition of the fibres or shives, and $A^{Mori-tanaka}$ is the concentration strain, as defined by Mori–Tanaka [35],

$$A^{Mori-Tanaka\_x} = A \left[\left(1 - V_f \right) I + V_f A \right]^{-1} \quad (3)$$

$$A = \left[I + S \left(C_m^{-1} \right) (C_f - C_m) \right]^{-1} \quad (4)$$

where S represent the Eshelby strain tensor. This tensor is based on the assumption of an elliptic inclusion in which the width and height of the inclusion are assumed to be equal ($a_2 = a_3$) [35]. The aspect ratio of the inclusion is calculated as $\rho = \frac{a_1}{a_2}$, where the values for a_1 and a_2 are experimentally determined values for shives and technical fibres.

3.2. Orientation

Subsequently, the effect of the fibre orientation on the elastic properties is evaluated. During injection moulding, the short fibres are immersed in a highly viscous polymer melt. The final orientation distribution of the fibres is determined by the rheological behaviour of the compound and by the injection moulding process conditions. The orientation of the fibres is described by the probability distribution function $\Psi(p)$. In this function, the orientation of the fibres is associated with unit vector p. The components of p are related to the angles θ and Φ, as shown in Figure 3 [40].

$$p_1 = sin\theta \, cos\Phi \quad (5)$$

$$p_2 = sin\theta \, sin\Phi \quad (6)$$

$$p_3 = cos\theta \quad (7)$$

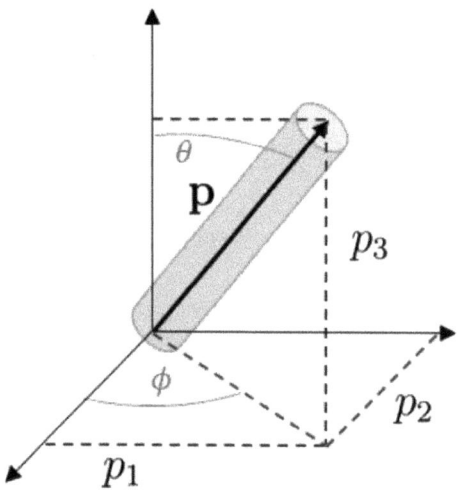

Figure 3. Coordinate system of unit vector p.

A set of second-order orientation tensors is defined by forming dyadic products of the vector p and integrating them with the distribution function over all possible directions, as given in Equation (8). Each element of the orientation tensor is described as

$a_{ij} = \oint p_i p_j \Psi(p) dp$ [40]. The subscripts i and j represent values between 1 and 3, referring to the orientations in Equations (5)–(7).

$$P = \begin{bmatrix} \langle p_1 p_1 \rangle & \langle p_1 p_2 \rangle & \langle p_1 p_3 \rangle \\ \langle p_2 p_1 \rangle & \langle p_2 p_2 \rangle & \langle p_1 p_3 \rangle \\ \langle p_3 p_1 \rangle & \langle p_3 p_2 \rangle & \langle p_1 p_3 \rangle \end{bmatrix} \approx \begin{bmatrix} a_{11} & a_{12} & a_{13} \\ a_{21} & a_{22} & a_{23} \\ a_{31} & a_{32} & a_{33} \end{bmatrix} \tag{8}$$

The fourth-order stiffness tensors T_{ijkl} according to the fibre orientation tensor P are then obtained for a transversely isotropic material assumption using Equation (9).

$$\begin{aligned} T_{ijkl}(p) = & B_1 \left(a_{ijkl} \right) + B_2 \left(a_{ij}\delta_{kl} + a_{kl}\delta_{ij} \right) \\ & + B_3 \left(a_{ik}\delta_{jl} + a_{il}\delta_{jk} + a_{jl}\delta_{ik} + a_{jk}\delta_{il} \right) \\ & + B_4 \left(\delta_{ij}\delta_{kl} \right) + B_5 \left(\delta_{ik}\delta_{jl} + \delta_{il}\delta_{jk} \right) \end{aligned} \tag{9}$$

The B-factors are 5 scalar constants related to the independent components of the transversely isotropic elasticity tensor. These are defined as $B_1 = C_{11} + C_{22} - 2C_{12} - 4C_{66}$, $B_2 = C_{12} - C_{23}$, $B_3 = C_{66} + \frac{1}{2}(C_{23} - C_{22})$, $B_4 = C_{23}$ and $B_5 = \frac{1}{2}(C_{22} - C_{23})$ [40].

The fourth-order orientation tensor a_{ijkl} in Equation (9) is calculated using Equations (10)–(12). The hybrid closure approximation is used to obtain the fourth-order orientation tensor from the second-order orientation tensor, where $f = A a_{ij} a_{ji} - B$. For orientation in a planar state, A is equal to 2, and B is equal to 1 [40].

$$\begin{aligned} a_{ijkl}^{LIN} = & \tfrac{1}{7} \left(a_{ij}\delta_{kl} + a_{ik}\delta_{jl} + a_{il}\delta_{jk} + a_{kl}\delta_{ij} + a_{jl}\delta_{ik} + a_{jk}\delta_{il} \right) \\ & - \tfrac{1}{35} \left(a_{ij}\delta_{kl} + a_{ik}\delta_{jl} + a_{il}\delta_{jk} \right) \end{aligned} \tag{10}$$

$$a_{ijkl}^{QUA} = a_{ij} a_{kl} \tag{11}$$

$$a_{ijkl}^{HYB} = (1 - f) a_{ijkl}^{LIN} + f \cdot a_{ijkl}^{QUA} \tag{12}$$

3.3. Variability of the Fibre Orientation through Thickness

Due to the injection moulding process, there is no uniform fibre orientation through the thickness of the plate [41–44]. To account for these variations, the injection moulded samples are considered as a multilayer material. In each layer, the orientation distribution is assumed to comply with transverse isotropic material properties. These values are obtained with Autodesk Moldflow simulations using a Midplane mesh consisting of 12 layers (determined by a sensitivity study) [45]. The revised Folgar–Tucker fibre orientation model was selected, introducing standard calculated values for the coefficient of interaction (C$_I$) and the thickness moment of interaction coefficient (Dz). This approach gives high accuracy of the predicted fibre orientation in a concentrated suspension, using hybrid closure [46].

The fibre orientation tensor component a_{11} represents the fibre orientation along the flow direction in the plate. In this study, the x-direction is parallel to the flow direction. The numerical results for a_{11} through the thickness of the plates (PLA reinforced with 10- and 20% straw flax fibres) are shown in Figure 4. In this figure, the through-thickness position in the 3 mm thick plates is normalised in the range (−1,1). The a_{11} values are reported for different positions ($X_1 \to X_5$) on the plate and visualised in Figure 4. These positions are chosen at specific locations in the plate, covering both the longitudinal and transverse orientations during mechanical testing. Position 1 is at the centre of the plate, positions 2 and 3 are along the orientation of the flow, position 4 is along the perpendicular orientation (y-direction), and position 5 is at an angle of 45° with respect to the x-axis. Results indicate a constant orientation distribution through the thickness over the chosen positions on the plate. Therefore, the fibre orientation tensor (FOT) at the location in the centre of the plate (X_1) is considered to be the representative distribution over the entire plate. Additionally, the fibre volume fraction affects the final fibre orientation distribution in the plate by changing the viscosity of the SFRT in the molten state [1]. For this reason, a representative FOT distribution is defined per volume fraction of a composite.

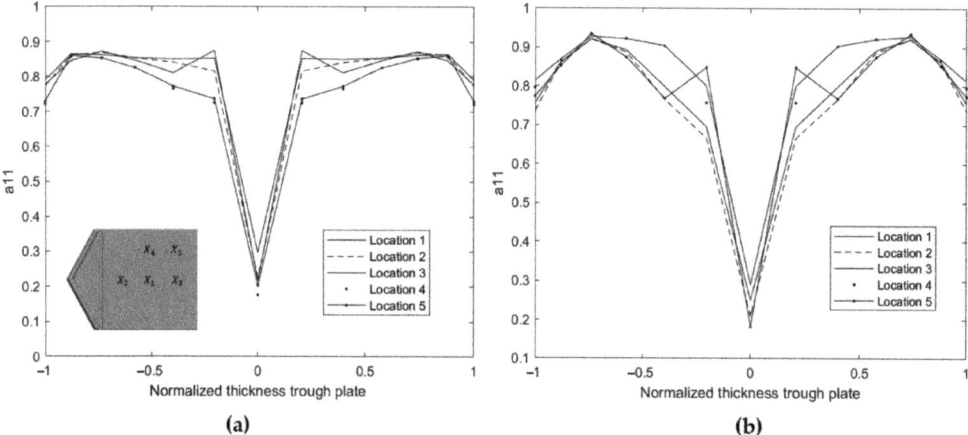

Figure 4. Orientation distribution for a11 through the (normalized) thickness of the plate at different locations, obtained via Moldflow simulations for (**a**) PLA10FF and (**b**) PLA20FF.

Next, the classical laminate theory (CLT) [47–49] is applied to obtain the material behaviour of the full SFRT, considering the 12 individual layers of the midplane mesh. The reduced stiffness matrix $[K]$ is constructed in Equation (16), using Equations (13)–(15). Where k is the number of layers, and z is the height of the bottom of the layer, for which the coordinate system is taken at the centre of the composite. According to the plain stress assumption, a reduced 3×3 matrix $[Q]$ is obtained using the fourth-order stiffness tensor $[T]$ described in Equation (9). Subscript i refers to each individual layer.

$$[A] = \sum_{i=1}^{k} [Q]_i (z_i - z_{i-1}) \tag{13}$$

$$[B] = \frac{1}{2} \sum_{i=1}^{k} [Q]_i \left(z_i^2 - z_{i-1}^2 \right) \tag{14}$$

$$[D] = \frac{1}{3} \sum_{i=1}^{k} [Q]_i \left(z_i^3 - z_{i-1}^3 \right) \tag{15}$$

$$[K] = \begin{bmatrix} A & B \\ B & D \end{bmatrix} \tag{16}$$

The 6×6 matrix $[Q]$ of the laminate of 12 layers in Equation (16) is obtained according to Equations (13)–(15). Note that injection moulding allows the presence of a limited amount of out-of-plane oriented fibres. The present model does not allow taking different fibre orientations into consideration. Since the Moldflow simulations indicate values for a_{33} to be smaller than 0.05, this assumption is reasonable.

To predict the stiffness of the composite along these different orientations, only the relevant part of the stiffness matrix $[K]$ is considered [50,51]. The reduced stiffness matrix $[Q']$ is rotated for angles Φ in the (1,2) plane. Experimental data for load cases at different angles with respect to the flow direction are compared to the calculated elastic properties under these angles in Equation (17).

$$[Q\_angle] = [T_1][Q'][T_2]^{-1} \tag{17}$$

where:

$$[T_1] = \begin{bmatrix} cos^2\Phi & sin^2\Phi & 2\cos\Phi\sin\Phi \\ sin^2\Phi & cos^2\Phi & -2\cos\Phi\sin\Phi \\ -\cos\Phi\sin\Phi & \cos\Phi\sin\Phi & cos^2\Phi - sin^2\Phi \end{bmatrix} \qquad (18)$$

$$[T_2]^{-1} = \begin{bmatrix} cos^2\Phi & sin^2\Phi & -\cos\Phi\sin\Phi \\ sin^2\Phi & cos^2\Phi & \cos\Phi\sin\Phi \\ 2\cos\Phi\sin\Phi & -2\cos\Phi\sin\Phi & cos^2\Phi - sin^2\Phi \end{bmatrix} \qquad (19)$$

4. Results

4.1. Measurement of Fibre Composition

Figure 5 presents the interior structure of a flax stem before processing. The elemental fibres are located in the outer layer of the flax stem, while the flax shives are located in the core of the stem. This observation is in line with the literature [23].

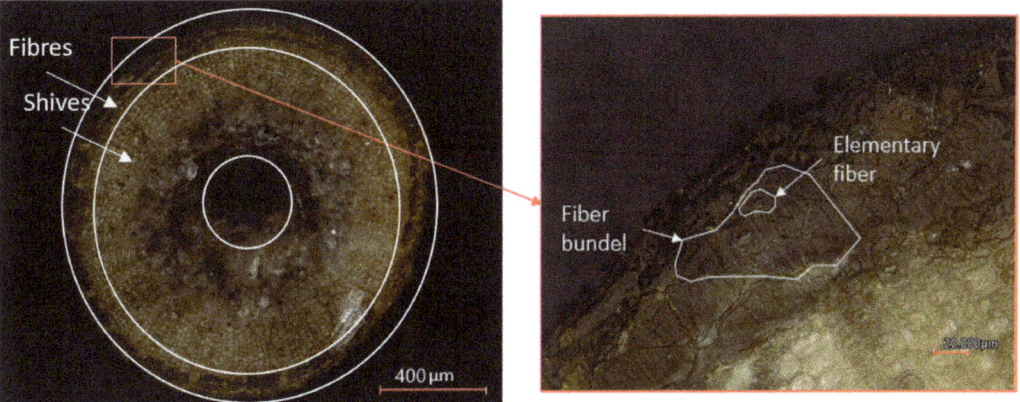

Figure 5. Microscopic image of the straw flax fibre before processing using the Keyence laser microscopy.

Fibre breakage and splitting of shives and fibres occur due to the high shear forces during processing [24]. This results in flax shives and elementary fibres separately present in the composite. The microscopic image in Figure 6 shows the cross-sectional area of the composite after injection moulding. The impregnation of the PLA matrix within the hollow structure of the flax shives is visible.

The composition of the flax straw used in this work was determined after dissolving the PLA matrix of the injection moulded plates. The results of these measurements are summarised in Table 5 and are consistent with the data in the literature. The shives content is in the range of 70 to 85 V%, depending on the type of flax [20,52].

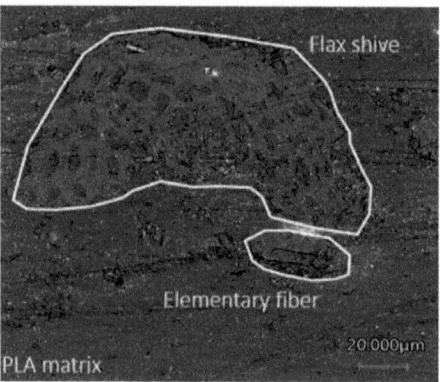

Figure 6. Microscopic image of the reinforcements in the PLA matrix after the production process.

Table 5. Composition of the str aw flax fibre.

Component	Weight Fraction	Volume Fraction
Technical fibres	25 wt%	22 V%
Flax shives	75 wt%	78 V%

4.2. Measurement of the Fibre Morphology

The aspect ratio distribution for shives and fibres is shown in Figure 7. Gravimetric separation of shives and fibres (1.5 g/cm^3 and 1.237 g/cm^3, respectively [23]) allows the evaluation of their aspect ratio separately. Both number and volume-based distributions of the aspect ratio can be used to analyse the fibre breakage during processing. However, earlier studies indicate that the distribution by volume of the fibres is preferred, as this parameter is most representative of the volume fraction of the reinforcements and, therefore, of the mechanical reinforcement properties [22,24,38]. Therefore, volume-based distribution is reported in this study.

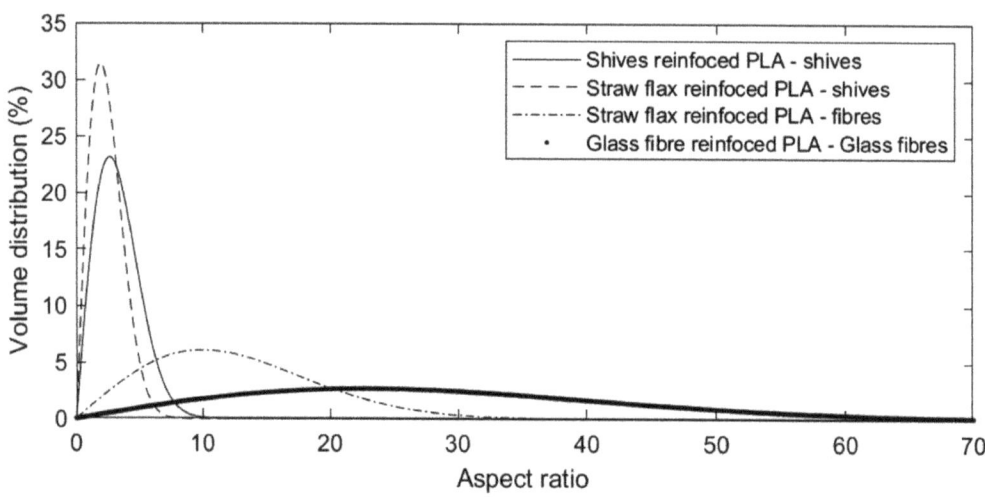

Figure 7. Volume distribution of the aspect ratio of flax shives and fibres after processing the different composite compounds.

The results summarised in Table 6 indicate a strong reduction in the dimensions of the reinforcements during processing. Average aspect ratios of 2.3 and 9.5 were measured for the flax shives and fibres, respectively. The fibre aspect ratio shows a much wider distribution, compared to the shives (Figure 7). The main difference in size between the two components is shown by the larger average values for the length and width of the flax shives. These act rather as larger particles in the flow.

Furthermore, an average fibre diameter of about 200 µm is measured for the flax fibres. It is known from the literature that the average diameter of an elementary flax fibre is between 40 and 80 µm [50]. This difference between the measured diameter and the literature values suggests that the elementary fibres are not completely separated into individual technical fibres, and that fibre bundles are also present in the compound. This reduces the apparent aspect ratio of the flax fibres, leading to an underestimate of the mechanical stiffness of the composite. To determine the effect of flax fibres on mechanical performance, the average diameter of the elementary fibres is considered. These results also indicate that separation of the fibres from the shives occurs during the compounding and injection moulding step. When technical flax fibres are used as reinforcement, an extra scutching step prior to the processing would be required.

Table 6. Properties of fibre morphology by volume averaging.

Component		Length (µm)	Diameter (µm)	Aspect Ratio (-)
Flax shive-reinforced composite				
Flax shives	Mean. Value St. dev.	1147 ± 109	521 ± 37	2.3
Straw flax-reinforced composite				
Technical fibres	Mean. Value St. dev.	567 ± 215	60 ± 17	9.5
Flax shives	Mean. Value St. dev.	992 ± 212	496 ± 99	2.1
Glass fibre composite				
E-glass	Mean St. dev.	228 ± 140	10 ± 0.1	22.6

4.3. Tensile Properties SFRT Composite

The elastic properties of the SFRT composites obtained after tensile testing are summarised in Table 7. They mainly depend on the type of reinforcement, volume fraction and orientation. Adding 15% of flax shives to the neat PLA material results in a 31% increase of stiffness in the flow direction and a degree of anisotropy of 11%, defined by the relative difference in stiffness in the first and second orientation, written as $\frac{E_{11}-E_{22}}{E_{11}} \cdot 100\%$. Lower pressure values were observed during the melt processing of the compounds for straw flax-reinforced materials. This allowed the production of straw flax-reinforced composites with a volume fraction up to 20%. This is attributed to a smaller increase in viscosity caused by the straw flax, compared to the flax shives. Adding 20% of straw flax of the PLA results in a 58% increase in stiffness and a degree of anisotropy of 21%.

Table 7. Mechanical properties of SFRT (FS: Flax Shives, FF: Straw Flax Fibres).

Material	E11 Modulus (MPa)	E22 Modulus (MPa)	V12 (-)	Degree of Anisotropy (%)
PLA	3877 (±193)	-	0.35	-
PLA + 5% FS	4186 (±96)	4077 (±239)	0.34	2.6
PLA + 10% FS	4737 (±100)	4337 (±128)	0.34	8.8
PLA + 15% FS	5070 (±307)	4596 (±515)	0.33	11.1
PLA + 5% FF	4284 (±79)	4137 (±135)	0.34	3.5
PLA + 10% FF	5091 (±158)	4563 (±133)	0.33	10.4
PLA + 20% FF	6129 (±58)	4828 (±123)	0.32	21.2

These results show the promising effect of using the whole stem as reinforcement, thus avoiding the extra scutching step beforehand. A variation of less than 5% in the mechanical properties was obtained. This low variation is attributed to well-controlled processing parameters and using fibres and shives from the same batch. The values for the Poisson's ratios were obtained by taking the ratio of the average strain in the first and second direction over the surface of the tensile bar, as shown in Figure 8.

The mechanical properties of glass fibre-reinforced PLA are summarised in Table 8. The stress strain curves are visualized in Figure 9. The increase in tensile stiffness of these composites is slightly higher, compared to the straw flax-reinforced PLA. However, if the density of the reinforcement is considered (2.6 g/cm^3 [38] and 1.3 g/cm^3 [23], respectively), the flax-reinforced composite shows a higher specific stiffness than the glass fibre-reinforced composite. A higher aspect ratio of the glass fibres was measured using the same technique as for the straw flax-reinforced composites. This higher aspect ratio is mainly caused by the small diameter of the glass fibres (10 μm), compared to the diameter of the flax fibres. This explains the higher degree of anisotropy. Furthermore, as the flax shives possess an aspect ratio of only 2.1, they rather act as a filler that increases the stiffness in both directions at a lower degree of anisotropy, compared to the fibre-filled composites.

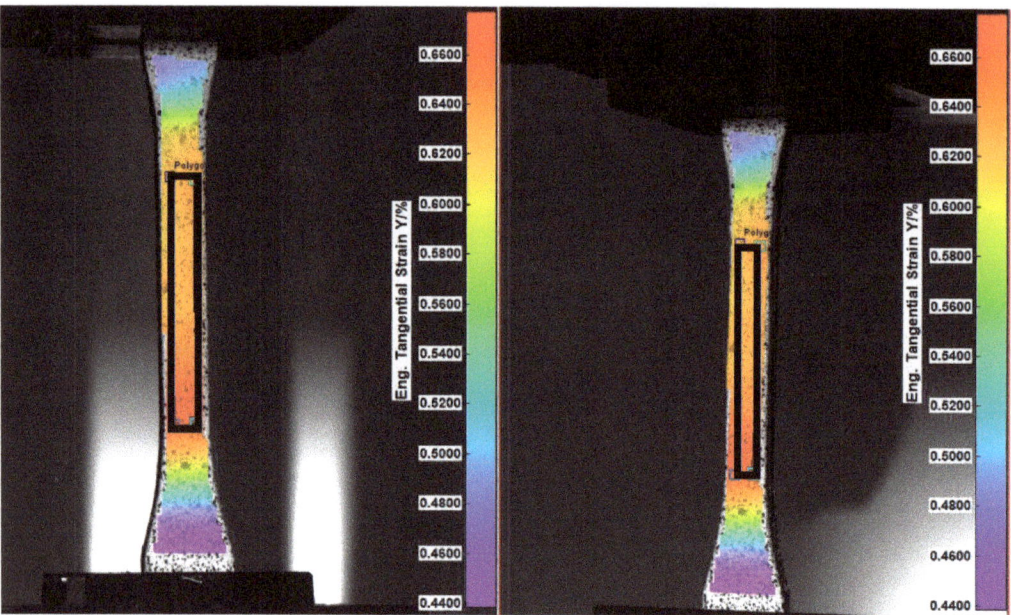

Figure 8. Stereo-vision DIC measurement on tensile bar with defined region for averaging of strain to determine the Poisson's ratio.

Table 8. Specific stiffness (FF: straw Flax Fibres, GF: Glass Fibres).

Material	Young's Modulus E_{11} (MPa)	Specific Young's Modulus E_{11}/ρ	Stiffness Ratio (%)
PLA + 10% FF	5091 (±158)	4085	10.4
PLA + 20% FF	6129 (±58)	4895	21.2
PLA + 10% GF	5211 (±103)	3787	27.0
PLA + 20% GF	6475 (±240)	4282	36.1

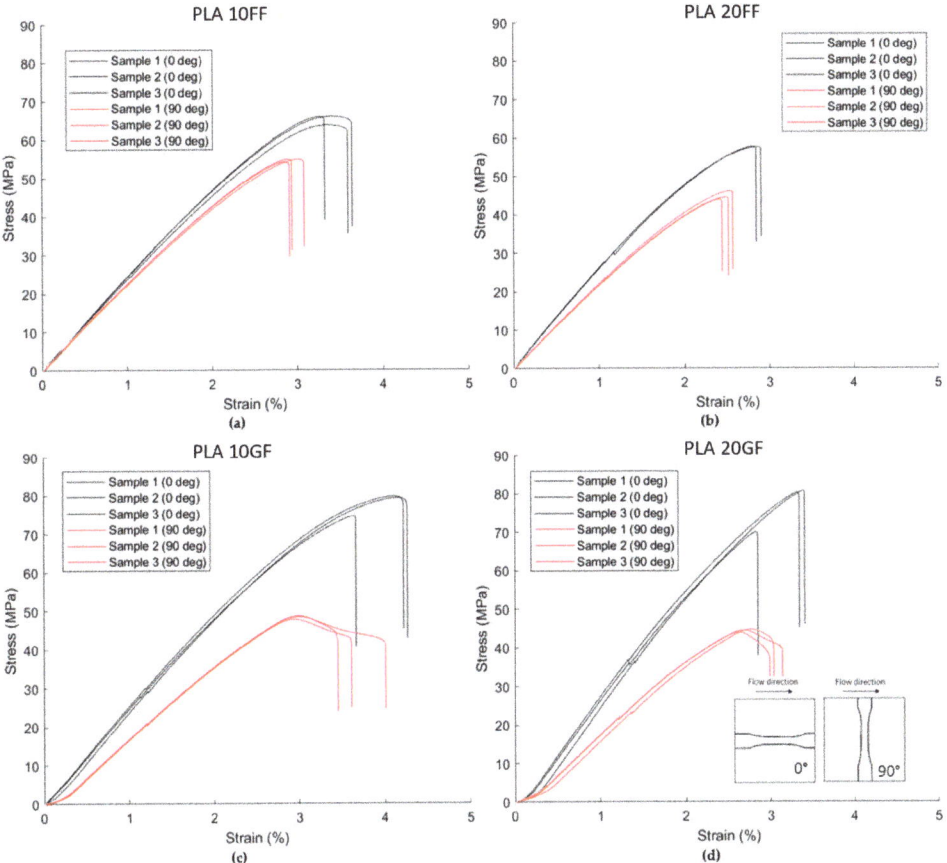

Figure 9. Tensile stress–strain curves in longitudinal and transversal direction according to ISO 527-1BA for (**a**) PLA10FF, (**b**) PLA20FF, (**c**) PLA10GF and (**d**) PLA20GF.

4.4. Stiffness Estimation of Flax Shives

The literature on the effect of flax shives on a PP matrix indicates that a micromechanics-based model is well suited to determine the elastic properties of the composite [23]. No experimental data on the mechanical properties of flax shives are currently available in the literature. To obtain the stiffness of the flax shives used within this study, a micromechanics-based reverse engineering approach is applied. Due to the low aspect ratio of the flax shives, the effect of the varying orientations trough thickness is negligible. Therefore, a generalised two-phase Mori–Tanaka approach for unidirectional (UD) is used. The error between the predicted tensile modulus C using the generalised Mori–Tanaka homogenisation approach and experimental data for the tensile modulus in both longitudinal and transversal directions at different volume fractions was minimised. The minimisation of the cost function in Equation (20) is obtained by a nonlinear optimisation function in MATLAB. By optimising the cost function, a stiffness of 22.2 GPa was found for the shives. Figure 10 indicates the correlation between the experimental values and the predictive model optimised according to these data.

$$Cost(C_{shives}) = \min_{i=0\% \rightarrow 15\%shives} \left(C_{mori-tanaka}(i) - C_{experimental}(i) \right)^2 \qquad (20)$$

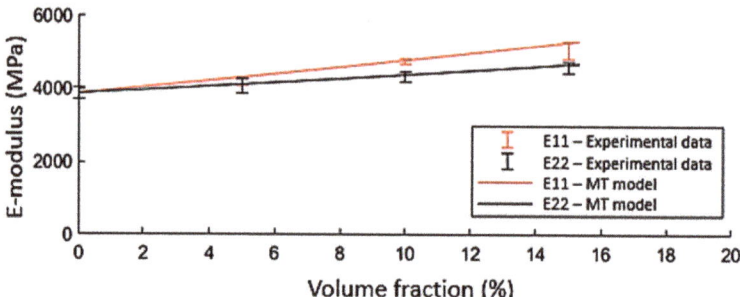

Figure 10. Comparison between the generalised Mori–Tanaka model and the experimental data for flax shives reinforced composites in both the longitudinal and transverse orientation. The presented lines for the Mori–Tanaka model are obtained after optimisation of the cost function.

4.5. Fibre Orientation

Considering the straw flax reinforced composites, the mechanical properties are initially modelled by assuming a highly aligned SFRT using the proposed generalised Mori–Tanaka approach for a three-phase system. The stiffness of the flax fibre bundles used in this work is reported to be 43.5 GPa in the literature [24]. The other necessary input values were obtained in the experimental part of this study. When the composite is assumed to be a highly aligned SFRT composite material (UD assumption), with no variations in fibre orientation through the thickness, the obtained stiffness data overestimate the experimental data for the longitudinal direction and underestimate the data in the transverse direction (Figure 11). This indicates the expected importance of the through-thickness variation in fibre orientation, caused by the injection moulding process. The use of the multilayer approach (multi-layer setup) leads to a better correspondence of the model to the experimental data. The overestimated stiffness predictions are reduced by lower orientations across the plate thickness.

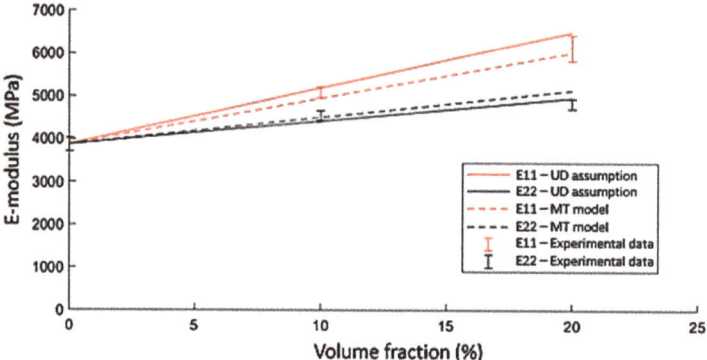

Figure 11. Results of the analytical modelling approach using a UD-based assumption and by considering it to be a laminate consisting of 12 layers with different fibre orientations.

The varying fibre orientation through the thickness causes non-negligible deviations in tensile modulus with respect to a unidirectional (UD) orientation of the fibres [46]. The experimental results of tensile stiffness and strength at 0, 30, 45, 60, and 90° are summarised in Table 9. Since the analytical model allows the consideration of any fibre orientation, a comparison of the modelling approach with the experimental data is made. A fibre rotation matrix for the cases between 0° and 90° was used to obtain stiffness data along different directions. The results are shown in Figure 12.

Table 9. Mechanical properties of SFRT under different orientations (FF: straw Flax Fibres).

Orientation	PLA + 10% FF		PLA + 20% FF	
Angle on Flowing Front	E (MPa)	σ (MPa)	E (MPa)	σ (MPa)
0°	5091 (±158)	65.3 (±1.2)	6129 (±58)	58.0 (±0.0)
30°	5051 (±73)	59.3 (±2.8)	5544 (±130)	49.6 (±2.3)
45°	4897 (±243)	51.3 (±0.9)	5263 (±247)	48.3 (±2.3)
60°	4675 (±36)	54.3 (±0.4)	5037 (±23)	44.3 (±2.3)
90°	4546 (±113)	54.4 (±0.4)	4829 (±123)	44.8 (±1.0)

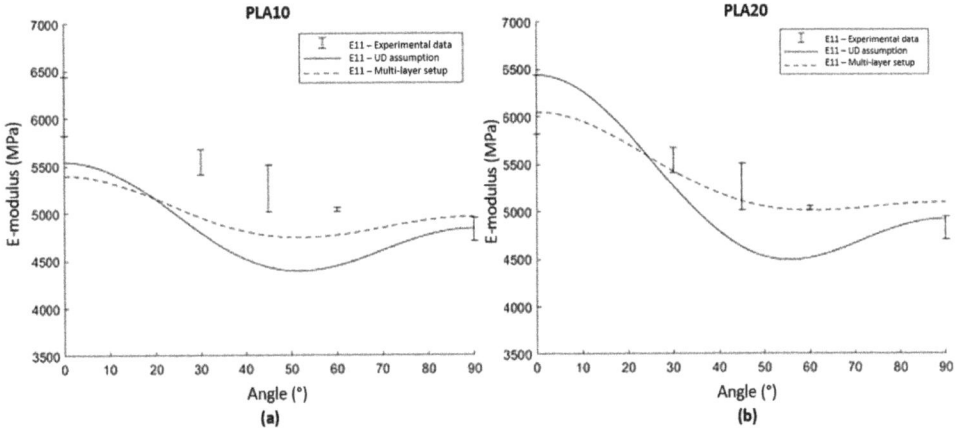

Figure 12. Comparison of the UD modelling approach, the proposed multi-layer approach and experimental data for (a) PLA reinforced with 10% straw flax and (b) PLA reinforced with 20% straw flax.

Experimental data are compared with the two modelling approaches. When the UD orientation of the SFRT is assumed, there is a rapid decrease in stiffness as the angle deviates from 0°. When the multilayer approach is used, the stiffness remains at a higher value over a wider range of angles. In this approach, the effect of local fibre orientations through the thickness increases the stiffness in different orientations and thus results in a reduction of this drop in stiffness for off-axis orientations. This highly corresponds to the experimental data, which indicates this effect by showing a slow decrease in stiffness over varying angles with the flow front. Results of this study clearly indicate the importance of taking local orientations into account for short fibre reinforced injection moulded composites, as otherwise the anisotropy over the full composite will be overestimated. The deviation between the UD assumption and the experimental data was shown to be even more pronounced considering the stiffness under various angles with respect to the flow orientation.

5. Discussion and Conclusions

Considering the entire flax stem as reinforcement for thermoplastic composites requires less production steps and reduces the cost of time and energy. The potential of flax shives, available as agricultural waste, and straw flax (consisting of fibres and shives) as reinforcing materials in a thermoplastic matrix is investigated in this study. The use of these bio-based reinforcements to produce lightweight and high-performance composites was found to be promising in previous studies that focus on non-renewable feedstock-based matrices such as PP/MAPP [21–23]. This work has investigated the applicability of this type of reinforcement in a polylactic acid (PLA) matrix to obtain a fully bio-based material. Moreover, the elastic material properties using a multilayer micromechanical approach were predicted.

Results show that the tensile stiffness of the material increases with increasing volume fraction of the straw flax. These results included an increased degree of anisotropy due to fibre orientation caused by the injection moulding process. For straw flax-reinforced PLA, an increase in tensile strength was observed up to a fibre volume fraction of 10%, after which there was a decrease. Compared to chopped glass fibre-reinforced PLA, the low density of natural fibres leads to a higher specific composite stiffness in the proposed bio-based composite. In addition, the presence of the flax shives in the compound also leads to a lower degree of anisotropy, as they behave more as fillers to increase the stiffness of the matrix.

Furthermore, an analytical modelling approach based on the Mori–Tanaka micromechanical model was set up and validated. By minimising the cost function of flax shives reinforced PLA for different volume fractions, the stiffness of flax shives was estimated via a reverse engineering approach. Assuming that the material is a three-phase system, the dimensions and mechanical properties of both the flax shives and the flax fibres present in the flax stem were considered separately. Injection moulding causes a typical through-thickness variation in fibre orientation, which is taken into account with a multilayer approach, with each layer considered to have its own fibre orientation. This approach resulted in a high correlation of the predicted stiffness with the experimental data in the longitudinal and transverse directions. Assuming complete UD orientation gives an overestimation of the mechanical properties indicating the importance of taking the actual fibre orientation into account, even if the deviation from UD seems rather small. The UD assumption overestimates the drop in stiffness at increasing angle values. The multilayer approach presented in this work was found to be better suited to predict the stiffness of the injection moulded SFRT composites along different orientations.

Author Contributions: Conceptualization, S.V.; Methodology, S.V.; B.B., S.D. and F.D.; Investigation, S.V.; Writing original draft, S.V.; Writing—review & editing, S.V., B.B., D.M. and S.D.; Funding acquisition, B.B., S.D. and F.D.; Supervision, F.D. All authors have read and agreed to the published version of the manuscript.

Funding: This research was funded by INTERREG France-Wallonia-Vlaanderen via the ATHENS project (1.1.351).

Institutional Review Board Statement: Not applicable.

Data Availability Statement: The data presented in this study are available on request from the corresponding author.

Conflicts of Interest: The authors declare no conflict of interest.

References

1. Monfared, V. Problems in short-fiber composites and analysis of chopped fiber-reinfoced materials. In *New Materials in Civil Engineering*; Elsevier: Amsterdam, The Netherlands, 2020.
2. Yildizhan, S.; Calik, A.; Serin, H. Bio-composite materials: A short review of trends, mechanical and chemical properties, and applications. *Eur. Mech. Sci.* **2018**, *2*, 83–91. [CrossRef]
3. Bharath, K.N.; Basavarajappa, S. Applications of biocomposite materials based on natural fibers from renewable resources: A review. *Sci. Eng. Compos. Mater.* **2016**, *23*, 123–133. [CrossRef]
4. Vijayan, R.; Krishnamoorthy, A. Review on Natural Fiber Reinfoced Composites. *Mater. Today Proc.* **2019**, *16*, 897–906. [CrossRef]
5. Akampumuza, O.; Wambua, P.M.; Ahmed, A.; Li, W.; Qin, X.-H. Review of the applications of biocomposites in the automotive industry. *Polym. Compos.* **2016**, *38*, 2553–2569. [CrossRef]
6. Witayakran, S.; Smitthipong, W.; Wangpradid, R.; Chollakup, R.; Clouston, P. Natural Fiber Composites: Review of Recent Automotive Trends. In *Reference Module in Materials Science and Materials Engineering*; Elsevier: Amsterdam, The Netherlands, 2017; pp. 166–174. [CrossRef]
7. Rahman, Z. Mechanical and damping performances of flax fibre composites–A review. *Compos. Part C Open Access* **2020**, *4*, 100081. [CrossRef]
8. Komuraiah, A.; Kumar, N.S.; Prasad, B.D. Chemical composition of Natural fiers and its influence on their mechanical properties. *Mech. Compos. Mater.* **2014**, *50*, 359–376. [CrossRef]
9. Dittenber, D.B.; GangaRao, H.V. Critical review of recent publications on use of natural composites in infrastructure. *Compos. Part A Appl. Sci. Manuf.* **2012**, *43*, 1419–1429. [CrossRef]

10. Baley, C.; Gomina, M.; Breard, J.; Bourmaud, A.; Davies, P. Variability of mechanical properties of flax fibres for composite reinfocements. A review. *Ind. Crop. Prod.* **2020**, *145*, 111984. [CrossRef]
11. Djemiel, C.; Grec, S.; Hawkins, S. Characterization of bacterial and fungal community dynamics by high*-throughput sequencing (HTS) metabarcoding during flax daw-retting. *Front. Microbiol.* **2017**, *8*, 2052. [CrossRef]
12. Yan, L.; Chouw, N.; Jayaraman, K. Flax fibre and its composites–A review. *Compos. Part B* **2014**, *56*, 296–317. [CrossRef]
13. Barillari, F.; Chini, F. Biopolymers—Sustainability for the Automotive Value-added Chain. *ATZ Worldw.* **2020**, *122*, 36–39. [CrossRef]
14. Ilyas, R.A.; Zuhri, M.Y.M.; Aisyah, H.A.; Asyraf, M.R.M.; Hassan, S.A.; Zainudin, E.S.; Sapuan, S.M.; Sharma, S.; Bangar, S.P.; Jumaidin, R.; et al. Natural fiber-reinforced polylactic acid, polylactic acid blends and their composites for advanced applications. *Polymers* **2022**, *14*, 202. [CrossRef] [PubMed]
15. Aliotta, L.; Gigante, V.; Coltelli, M.B.; Cinelli, P.; Lazzeri, A. Evaluation of mechanical and interfacial properties of bio-composites based on poly(lactic acid) with natural cellulose Fibers. *Int. J. Mol. Sci.* **2019**, *20*, 960. [CrossRef] [PubMed]
16. Aliotta, L.; Gigante, V.; Coltelli, M.-B.; Cinelli, P.; Lazzeri, A.; Seggiani, M. Thermo-Mechanical properties of PLA/Short Flax Fiber Biocomposites. *Appl. Sci.* **2019**, *9*, 3797. [CrossRef]
17. Foulk, J.; Akin, D.E.; Dodd, R.; Ulven, C. Production of flax fibers for biocomposites. In *Cellulose Fibers: Bio- and Nano-Polymer Composites*; Springer: Berlin/Heidelberg, Germany, 2011.
18. Le Duigou, A.; Deux, J.M.; Davies, P.; Bayley, C. PLLA/flax mat/balsa bio-sandwich envirnomental impact and simplified life cycle analysis. *Appl. Compos. Mater.* **2012**, *19*, 363–378. [CrossRef]
19. Goudenhooft, C.; Bourmaud, A.; Baley, C. Varietal section of flax over time: Evolution of plant architecture related to influence on the mechanical properties of fibers. *Ind. Crops Prod.* **2017**, *97*, 56–64. [CrossRef]
20. Evons, P. Production of fiberboards from shives collected after continuous fiber mechanical extraction from oleaginous flax. *J. Nat. Fibres* **2018**, *16*, 453–469. [CrossRef]
21. Nuez, L.; Beaugrand, J.; Shah, D.U.; Mayer-Laigle, C.; Bourmaud, A.; D'arras, P.; Baley, C. The potential of flax shives as reinforcements for injection moulded polypropylene composites. *Ind. Crop. Prod.* **2020**, *148*, 112324. [CrossRef]
22. Nuez, L.; Magueresse, A.; Lu, P.; Day, A.; Boursat, T.; D'Arras, P.; Perré, P.; Bourmaud, A.; Baley, C. Flax xylem as composite material reinforcement: Microstructure and mechanical properties. *Compos. Part A* **2021**, *149*, 106550. [CrossRef]
23. Soete, K.; Desplentere, F.; Lomov, S.V.; Vandepitte, D. Variability of flax fibre morphology and mechanical properties in injection moulded short straw flax fibre-reinfoced PP composites. *J. Compos. Mater.* **2016**, *51*. [CrossRef]
24. Tanguy, M.; Bourmaud, A.; Beaugrand, J.; Gaudry, T.; Baley, C. Polypropylene reinforcement with flax or jute fibre; Influence of microstructure and constituents properties on the performance of composite. *Compos. Part B Eng.* **2018**, *139*, 64–74. [CrossRef]
25. Zaghloul, M.M.Y.; Mohamed, Y.S.; El-Gamal, H. Fatigue and tensile behaviors of fiber-reinforced thermosetting composites embedded with nanoparticles. *J. Compos. Mater.* **2018**, *53*, 709–718. [CrossRef]
26. Zaghloul, M.M.Y.; Zaghloul, M.Y.M.; Zaghloul, M.M.Y. Experimental and modeling analysis of mechanical-electrical behaviors of polypropylene composites filled with graphite and MWCNT fillers. *Polym. Test.* **2017**, *63*, 467–474. [CrossRef]
27. Zaghloul, M.M.Y. Influence of flame retardant magnesium hydroxide on the mechanical properties of high density polyethylene composites. *J. Reinf. Plast. Compos.* **2017**, *36*, 1802–1816. [CrossRef]
28. Yousry, M.; Zaghloul, M.; Mahmoud, M. Developments in polyester composite materials–An in-depth review on natural fibres and nano fillers. *Compos. Struct.* **2021**, *278*, 114698, ISSN 0263-8223. [CrossRef]
29. Fuseini, M.; Zaghloul, M.M.Y. Investigation of Electrophoretic Deposition of PANI Nano fibers as a Manufacturing Technology for corrosion protection. *Prog. Org. Coat.* **2022**, *171*, 107015. [CrossRef]
30. Moustafa, Z. Mechanical properties of linear low-density polyethylene fire-retarded with melamine polyphosphate. *J. Appl. Polym. Sci.* **2018**, *135*, 46770. [CrossRef]
31. Zaghloul, M.; Zaghloul, Y.; Zaghloul, Y. Physical analysis and statistical investigation of tensile and fatigue behaviors of glass fiber-reinforced polyester via novel fibers arrangement. *J. Compos. Mater.* **2023**, *57*, 147–166. [CrossRef]
32. Ansari, F.; Granda, L.A.; Joffe, R.; Berglund, L.A.; Vilaseca, F. Experimental evaluation of anisotropy in injection molded polypropylene/wood fiber biocomposites. *Compos. Part A Appl. Sci. Manuf.* **2017**, *96*, 147–154. [CrossRef]
33. Holmström, P.H.; Hopperstad, O.S.; Clausen, A.H. Anisotropic tensile behaviour of short glass-fibre reinforced polyamide-6. *Compos. Part C Open Access* **2020**, *2*, 100019. [CrossRef]
34. Garofalo, E.; Russo, G.M.; Di Maio, L.; Incarnato, L. Modelling of mechanical behavior of polyamide nanocomposite fibers using a three-phase Halpin-Tsai model. *e-Polymers* **2009**. [CrossRef]
35. Tucker, C.L.; Liang, E. Stiffness predictions for unidirectional short-fiber composites: Review and evaluation. *Compos. Sci. Technol.* **1999**, *59*, 655–671. [CrossRef]
36. Verstraete, S.; Desplentere, F.; Debruyne, S. Evaluating the influence of short fiber reinforced thermoplastic composites produced by injection molding on the stress distribution in an adhesively bonded joint using a multi-scale numerical modeling approach. In Proceedings of the 2nd International Conference on Industrial Applications of Adhesives 2022, Carvoeiro, Portugal, 3–4 March 2022; Proceedings in Engineering Mechanics. Springer: Cham, Switzerland; pp. 101–114. [CrossRef]
37. Jones, E.M.C.; Iadicola, M.A. *A Good Practices Guide for Digital Image Correlation*; International Digital Image Correlation Society, 2018. [CrossRef]

38. Dai, X.Q. 10–Fibers, Biomechanical Engineering of Textiles and Cloting. Woodhead Publishing Ltd: Cambridge, UK; Abington Hall: Abington, UK, 2006.
39. Ashrafi, B. Theoretical and Experimental Investigations on the Elastic Properties of Carbon Nanotube-Reinfoced Polmer Thin Films. 2008. Available online: https://www.researchgate.net/publication/30003294_Theoretical_and_experimental_investigations_of_the_elastic_properties_of_carbon_nanotube-reinforced_polymer_thin_films (accessed on 25 March 2023).
40. Avanti, S.; Tucker, C. The use of Tensors to describe and predict fiber orientation in short fiber composites. *J. Rheol.* **1987**, *31*, 751–784.
41. Kugler, S.K.; Kech, A.; Cruz, C.; Osswald, T. Fiber Orientation Predictions—A Review of Existing Models. *J. Compos. Sci.* **2020**, *4*, 69. [CrossRef]
42. Ogah, A.O.; Afiukwa, J.N. Characterization and comparison of mechanical behavior of agro fier-filled high-density polyethylene bio-composites. *J. Reinf. Plast. Compos.* **2014**, *33*, 37–46. [CrossRef]
43. Réquilé, S.; Goudenhooft, C.; Bourmaud, A.; Le Duigou, A.; Baley, C. Exploring the link between flexural behavior of hemp and flax stems and fibre stiffness. *Ind. Crops. Prod.* **2018**, *113*, 179–186. [CrossRef]
44. Bernasconi, A.; Cosmi, F.; Dreossi, D. Local anisotropy analysis of injection moulded fibre reinforced polymer composites. *Compos. Sci. Technol.* **2008**, *68*, 2574–2581.
45. Caton-Rose, P.; Hine, P.; Costa, F.; Jin, X.; Wang, J.; Parveen, B. Measurement and predictions of short glass fibre orientation in injection moulding composites. In Proceedings of the ECCM15–15th conference on composite materials, Venice, Italy, 24–28 June 2012.
46. Vinson, J.R.; Sierakowski, R.L. Anisotropic Elasticity and Composite Laminate Theory. In *The Behavior of Structures Composed of Composite Materials*; Solid Mechanics and Its Applications; Springer: Berlin/Heidelberg, Germany, 2008.
47. Aboudi, J.; Arnold, S.M.; Bednarcyk, B.A. *Micromechanics of Composite Materials, a Generalized Multiscale Analysis Approach*; Elsevier: Amsterdam, The Netherlands, 2013.
48. Mallick, P.K. *Fiber-Reinfoced Composites, Materials, Manufacturing and Design*; CRC Press: Boca Raton, FL, USA, 2008.
49. Tsai, S.W.; Hahn, H.T. *Introduction to Composite Materials*; Routledge: Boca Raton, FL, USA, 1980.
50. Lu, N.; Swan, R.H., Jr.; Ferguson, I. Composition, structure, and mechanical properties of hemp fiber reinforced composite with recycled high-density polyethylene matrix. *J. Compos. Mater.* **2012**, *46*, 1915–1924. [CrossRef]
51. Aufrere, C. Plenary Lecture II: Current Advances, Needs and Future Challenges in High-Volume Automotive Composite Structures [FAURECIA] | Rooms: Sevilla (1,2 and 3), ECCM Conference 16. 2014. Available online: http://www.escm.eu.org/eccm16/cgif00f.html?idexp=HN6FE&main=progetaglance (accessed on 25 March 2023).
52. Di Giuseppe, E. Reliability evaluation of automated analysis, 2D scanner, and micro-tomography methods for measuring fiber dimensions in polymer-lignocellulosic fier composites. *Compos. Part A–Appl. Sci. Manuf.* **2019**, *90*, 320–329. [CrossRef]

Disclaimer/Publisher's Note: The statements, opinions and data contained in all publications are solely those of the individual author(s) and contributor(s) and not of MDPI and/or the editor(s). MDPI and/or the editor(s) disclaim responsibility for any injury to people or property resulting from any ideas, methods, instructions or products referred to in the content.

Article

Describing and Modeling Rough Composites Surfaces by Using Topological Data Analysis and Fractional Brownian Motion

Antoine Runacher [1,2], Mohammad-Javad Kazemzadeh-Parsi [3], Daniele Di Lorenzo [1,4], Victor Champaney [1], Nicolas Hascoet [1], Amine Ammar [3] and Francisco Chinesta [1,4,*]

[1] PIMM Laboratory and ESI Group Chair, Arts et Metiers Institute of Technology, 151 Boulevard de Hopital, 75013 Paris, France
[2] IPC, 2 Rue Pierre et Marie Curie, 01100 Bellignat, France
[3] LAMPA Laboratory and ESI Group Chair, Arts et Metiers Institute of Technology, 2 bd du Ronceray, 49035 Angers, France
[4] ESI Group, 3 Rue Saarinen, 94150 Rungis, France
* Correspondence: francisco.chinesta@ensam.eu

Abstract: Many composite manufacturing processes employ the consolidation of pre-impregnated preforms. However, in order to obtain adequate performance of the formed part, intimate contact and molecular diffusion across the different composites' preform layers must be ensured. The latter takes place as soon as the intimate contact occurs and the temperature remains high enough during the molecular reptation characteristic time. The former, in turn, depends on the applied compression force, the temperature and the composite rheology, which, during the processing, induce the flow of asperities, promoting the intimate contact. Thus, the initial roughness and its evolution during the process, become critical factors in the composite consolidation. Processing optimization and control are needed for an adequate model, enabling it to infer the consolidation degree from the material and process features. The parameters associated with the process are easily identifiable and measurable (e.g., temperature, compression force, process time, ...). The ones concerning the materials are also accessible; however, describing the surface roughness remains an issue. Usual statistical descriptors are too poor and, moreover, they are too far from the involved physics. The present paper focuses on the use of advanced descriptors out-performing usual statistical descriptors, in particular those based on the use of homology persistence (at the heart of the so-called topological data analysis—TDA), and their connection with fractional Brownian surfaces. The latter constitutes a performance surface generator able to represent the surface evolution all along the consolidation process, as the present paper emphasizes.

Keywords: topological data analysis—TDA; composite consolidation; rough surfaces; fractional Brownian surfaces

1. Introduction

Many manufacturing processes are based on the consolidation of composites' pre-impregnated preforms that are available in the form of sheets or tapes. In both cases, the consolidation requires putting a new sheet (or tape) in contact with the one already laid, which now constitutes the so-called substrate [1]. If the temperature at the sheet–substrate contact level is high enough, the polymer viscosity becomes low enough to ensure the asperities flow under the externally applied pressure (from a press or from a roller in the case of automated tape laying as sketched in Figure 1). Thus, the initial asperities flow and spread, inducing the roughness smoothing and the increase in the intimate contact. As soon as the intimate contact occurs, at a temperature ensuring molecular mobility and during a time period long enough to ensure molecular reptation across the interface, consolidation is attained, and ideally bulk properties are recovered at the sheet interfaces.

Analysis and modeling of the roughness evolution are critical points for evaluating the quality of parts [2], to predict the remaining roughness that will result in a residual porosity in the composited laminate [3], and to evaluate tribological [4] or mechanical properties [5]. Other processes, such as machining, can induce noticeable roughness in the parts' surfaces [6].

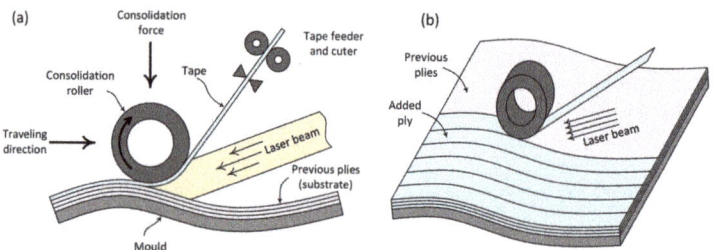

Figure 1. The automated tape placement (ATP) composite manufacturing: (**a**) physical phenomena; (**b**) processing.

The design, optimization and control of the process involves description, modeling and simulation issues [7], revisited below.

1.1. Rough Surface Representation

Different approaches have been proposed to represent a rough surface in view of consolidation analyses. The simplest consists of representing the asperities as rectangular elements of a given height and width [8,9]. To consider more realistic representations, fractal surfaces have also been employed, where several characteristic dimensions are hierarchically present [10–15]. Finally, the most realistic option consists of acquiring the real surface itself, on which physics-based models can be applied. The use of wavelet representations is a valuable route, because of the fact that they favor a multi-resolution description, and that a simple description consisting of rectangles is easily manipulable when, for example, the Haar wavelet is considered [7].

All these descriptions enable surface representation for solving realistic physical models on them; however, as discussed later, they do not allow to easily perform surface comparisons or classifications. The latter, in general, require the use of adequate metrics, different to the usually considered Euclidean metrics. The use of dynamic time wrapping [16] could help in some cases, but it remains too macroscopic when different description scales co-exist (fine and coarse surface features), as is the case when considering micro and macro roughness.

When looking for more intrinsic descriptors, extracting roughness features that could serve to compare and classify surfaces, statistical descriptors [17–21] represent a first natural choice. However, in many cases such descriptors are quite far from the physics involved in the consolidation process, and, even if they are adequate for describing roughness from a geometrical point of view, they seem limited in representing the physics-based surface properties as well as the time-evolution during processing.

Indeed, there exist many advanced and richer descriptors of microstructures, such as the ones reported in a study by Torquato [22]. Richer descriptions should retain much more information but at the same time be more intrinsic than when one proceeds on the real surfaces, as previously mentioned. Thus, real surfaces could be expressed in alternative spaces by using well-experienced transformations, such as Fourier, discrete cosine transform, wavelets, . . ., or by learning the appropriate transformation such as *Code2Vector* [23] performs, for instance.

Recently, alternative transformations emphasizing topological persistence, which are at the heart of the so-called TDA—topological data analysis (revisited later), are emerging and proving their value [24–29]. The so-called persistence diagram $\mathcal{PD}$ and its associated persistence image $\mathcal{PI}$ describe the topology of data in a very convenient way, such that surfaces exhibiting the same topology result in a similar $\mathcal{PI}$. Persistence images, being defined in a vector space, can be easily compared by using adequate metrics (e.g., Euclidean, Wasserstein, ...) or manipulated by using standard machine learning technologies such as convolutional neural networks—CNN [30].

In [31] we proved that TDA can be used for classifying surfaces according to their topological content, out-performing previous studies making use of statistical descriptors [17], and proved it is possible to use that information to infer process-induced properties. By analyzing usual statistical roughness descriptors and TDA-based descriptors, respectively, in [17,31], the superiority of the latter was observed.

Another possible route consists of extracting the main modes representing these images by performing linear (principal component analysis—PCA) or nonlinear dimensionality reduction based on manifold learning [32], as successfully accomplished in [33], or by employing auto-encoders [34].

The main issue related to the use of statistical descriptors and transformation procedures is the possibility of generating synthetic surfaces according to these descriptors.

A valuable surface generator consists of exploiting the link between the random nature of roughness and Brownian motion. Generating time series, curves or surfaces from Brownian motion has been widely considered. However, Brownian motion seems, in some cases, too limited, in particular when processing occurs and the surface is strongly modified. In those cases, fractional Brownian motion seems a valuable route. Fractional motion is closely related to fractional diffusion, which, in turn, makes use of fractional calculus. Anomalous diffusion, also called non-Brownian diffusion or fractional diffusion, occurs when the mean square displacement scale with a non-integer power of time, this depending on the exponent value, defines surdiffusion (long-jumps) or subdiffusion (long-rest) [35].

Generating random surfaces by using fractional Brownian motion becomes a valuable route, and, even more interesting, this motion (as well as the resulting surfaces) is characterized from a single scalar, the so-called Hurst index. Fractional Brownian surfaces represent a timely research topic widely considered [36–40].

Thus, one could expect that modeling the Hurst index from some features, enables constructing models and makes it possible to reconstruct surfaces compatible with the features, a key route for material and process optimization and for processing control. This point represents the main goal of the present paper.

1.2. Original Contribution and Paper Outline

The present paper represents a step forward in the description of machine learning-based process modeling and process-induced properties.

Its main original contributions concern: (i) proving that TDA is able to discriminate fractional Brownian surfaces, the latter characterized by the Hurst index, in a very efficient manner, as soon as a regression model is developed for predicting the Hurst index from TDA-based persistence images; (ii) evaluating the effect of the composite compression on the persistence image, proving the ability of TDA to discriminate original and compressed surfaces; (iii) evidencing that the surface compression increases the Hurst index; and (iv) emphasizing the ability of fractional Brownian motion to generate synthetic surfaces with controlled topology, enabling further statistical analyses.

The paper is organized as follows. After the present introduction, the next section revisits the main methodologies employed in the present research work, in particular topological data analysis and fractional Brownian surface generation. Then, Section 3 addresses numerical analyses proving the ability of TDA to infer the Hurst index of fractional Brownian surfaces, as well as the evolution of topology and Hurst index during surface compression.

2. Methods

2.1. Topological Data Analysis

When TDA is applied on a time series, it transforms the time series into an image defined in a vector space, the so-called persistence image, that could be considered as a topological descriptor, enabling topological metrics to serve in comparing curves, or even to be employed as inputs in learned models.

For that purpose, and as explained in [31,41], when using a level-set-based filtration, neighboring local minima/local maxima are paired, leading to the so-called min–max pairs. Each min–max constitutes a point of the persistence diagram. The persistence diagram consists of a two-dimensional representation, where, on its horizontal axis, the minimum value of each data-pair is reported, while the associated maxima are displayed on the vertical axis. Because the maximum is always higher than the associated minimum, the axes of the persistence diagram are usually labeled as birth and death axes, respectively.

For the random profile depicted in Figure 2, $(11,14)$, $(7,9)$ and $(9,10)$ represent the three neighboring local minima/local maxima.

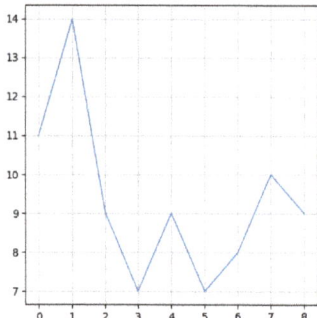

Figure 2. Random profile.

Because of the nature of maximum and minimum (or birth and death) the points in the persistence diagram are located in the upper bisector defining the space $y \geq x$, with x and y referring to birth and death (horizontal and vertical), respectively. The persistence diagram related to the profile depicted in Figure 2 is shown in Figure 3.

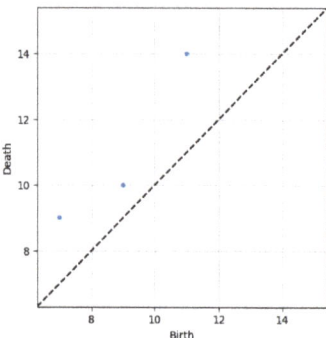

Figure 3. Typical persistence diagram $\mathcal{PD}$.

The higher the distance of a point to the diagonal $x = y$, the more persistent the topology of the considered point (maximum to minimum difference). Points located close to the diagonal contain a small persistent topology, which in many cases can be associated to noise (or finer physical scales).

Another important point concerns the invariance properties that the topology provides. If we consider a point on the persistence diagram depicted in Figure 3, for example the $(11, 14)$-pair, this point is associated with local-minimum/local-maximum neighboring data points. All the pairs with a minimum value equal to 11, and the neighboring local maximum equal to 14, are associated with the referred $(11, 14)$-pair, all of them represented by the same data-point in the persistence diagram, independently of their location on the time axis, or the time delay between the minimum and maximum constituting each pair.

Thus, two identical time series, but one of them translated in the time axis with respect to the second one, are represented by the same persistence diagram. Two time series, one of them consisting of a dilation of the other, have the same persistence diagram. These invariance properties are very important when the topology is more important than the location at which these topological events take place.

When addressing heat and momentum transfer on rough interfaces, TDA seems a very promising descriptor that extracts geometrical features closely related to the physics operating on it.

When topologies are close but not identical, their associated persistence diagrams consist of two set of points. To evaluate the distance between them, appropriate metrics must be employed for comparing the distance between both point clouds. Thus, the Wasserstein distance consists of a simple and efficient choice for measuring the distance between two persistence diagrams.

Another derived representation consists of the so-called persistence image. Prior to its introduction, an intermediate transformation is needed. Instead of representing birth–death, an equivalent representation consists of representing birth–lifetime, the latter defined as the difference between the death and the birth. The main advantage of using the lifetime diagram instead of the usual persistence diagram, is that in the former points fill the whole 2D space instead of only the upper bisector ($y \geq x$). The lifetime diagram $\mathcal{LT}$ associated with the persistence diagram $\mathcal{PD}$ shown in Figure 3 is given in Figure 4.

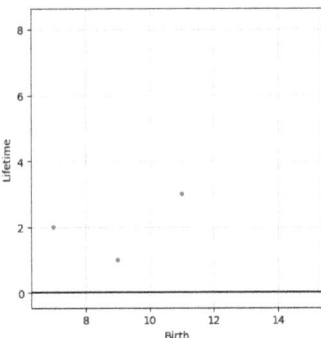

Figure 4. Lifetime diagram $\mathcal{LT}$.

Now, the last step consists of smoothing the point representation involved in the lifetime diagram, to transform a point into a 2D function. For that purpose, and as described in [31], points in the lifetime diagram are replaced by a Gaussian function centered at each point of the lifetime diagram $\mathcal{LT}$.

Next, the so-called persistent image $\mathcal{PI}$, defined in a vector space is constructed. For that purpose we consider a continuous piece-wise differentiable non-negative weight function $w(x, y)$ and a bivariate normal distribution $g_{x,y}(u, v)$ centered at each point (x, y), from which a function $\rho(u, v)$ can be defined at each point (u, v) in the lifetime domain [31]

$$\rho(u, v) = \sum_{(x,y) \in \mathcal{LT}} w(x, y) \, g_{(x,y)}(u, v). \tag{1}$$

Now, the domain is partitioned in a series of non-overlapping subdomains covering it, the so-called pixels P_i, $i = 1, 2, \cdots$, where function $\rho(u, v)$ is averaged to define the persistence image $\mathcal{PI}$, according to

$$\mathcal{PI}(P_i) = \iint_{P_i} \rho(u, v) \, du \, dv. \tag{2}$$

A typical persistence image related to a profile larger than the one shown in Figure 2 is illustrated in Figure 5. Now, this being a quite standard image, one is tempted to use it as input in learning procedures, in particular those making use of convolutional neural networks—CNN [33], even if many other options exist. Other possibilities consist of applying dimensionality reduction, either linear (e.g., PCA) or nonlinear (e.g., auto-encoders), to facilitate the neural network training, that now makes use of smaller input data, or of applying the weights of PCA modes when using the PCA or the data mapped into the latent reduced space when considering auto-encoders.

Figure 5. Typical persistence image.

2.2. From Brownian Diffusion to Anomalous Diffusion

This section revisits the derivation of Brownian diffusion and its connection with Brownian motion, viewed as a random walk. Then, in the case of anomalous diffusion, fractional Brownian motion will lead to the so-called fractional Brownian trajectories and surfaces.

2.2.1. Brownian Diffusion

Following Einstein's works, the particle displacement Δ in the unbounded one-dimensional axis x is represented by a random variable whose probability density reads $\phi(\Delta)$.

The particle's balance considers both, the local time evolution

$$\rho(x, t + \tau) = \rho(x, t) + \frac{\partial \rho(x, t)}{\partial t} \tau + \Theta(\tau^2), \tag{3}$$

and the transport in space

$$\rho(x, t + \tau) = \int_{\mathbb{R}} \rho(x + \Delta, t) \phi(\Delta) d\Delta. \tag{4}$$

Developing $\rho(x + \Delta, t)$

$$\rho(x + \Delta, t) = \rho(x, t) + \frac{\partial \rho(x, t)}{\partial x} \Delta + \frac{1}{2} \frac{\partial^2 \rho(x, t)}{\partial x^2} \Delta^2 + \Theta(\Delta^3) \tag{5}$$

and taking into account the normality $\int_{\mathbb{R}} \phi(\Delta)d\Delta = 1$ and symmetry $\int_{\mathbb{R}} \Delta\, \phi(\Delta)d\Delta = 0$ conditions, the final result is

$$\rho(x,t) + \frac{\partial \rho(x,t)}{\partial t}\tau = \rho(x,t) + \frac{1}{2}\frac{\partial^2 \rho(x,t)}{\partial x^2} \int_{\mathbb{R}} \Delta^2 \phi(\Delta)d\Delta, \qquad (6)$$

or

$$\frac{\partial \rho(x,t)}{\partial t}\tau = D\frac{\partial^2 \rho(x,t)}{\partial x^2}, \qquad (7)$$

with

$$D = \frac{1}{2\tau} \int_{\mathbb{R}} \Delta^2 \phi(\Delta)d\Delta. \qquad (8)$$

The integration of the diffusion equation from the localized (at $x = 0$) initial condition $\rho(x, t = 0) = \delta_0(x)$, leads to the Gaussian distribution

$$\rho(x,t) = \frac{1}{\sqrt{4\pi Dt}} e^{-\frac{x^2}{4Dt}}, \qquad (9)$$

whose main square displacement scales linearly with the time

$$\langle x^2 \rangle = 2Dt. \qquad (10)$$

2.2.2. Brownian Motion

The diffusion equation can be also derived from a random walk. For illustrating the derivation, we consider a grid on the x-axis, with step Δx being the time step Δt. At each time step, each particle is assumed to jump to one of its two neighboring sites with the same probability.

Thus, the balance at the j-site (grid node) reads

$$W_j(t + \Delta t) = \frac{1}{2}W_{j+1}(t) + \frac{1}{2}W_{j-1}(t), \qquad (11)$$

where $W_j(t)$ is the probability of having the particle at site j at time t.

By developing $W_i(t + \Delta t)$, $W_{j+1}(t)$ and $W_{j-1}(t)$ according to

$$\begin{cases} W_i(t + \Delta t) = W_j(t) + \left.\frac{\partial W_j(t)}{\partial t}\right|_t \Delta t + \Theta(\Delta t^2) \\ W_{j+1}(t) = W_j(t) + \left.\frac{\partial W(t)}{\partial x}\right|_j \Delta x + \frac{1}{2}\left.\frac{\partial^2 W(t)}{\partial x^2}\right|_j \Delta x^2 + \Theta(\Delta x^3) \\ W_{j-1}(t) = W_j(t) - \left.\frac{\partial W(t)}{\partial x}\right|_j \Delta x + \frac{1}{2}\left.\frac{\partial^2 W(t)}{\partial x^2}\right|_j \Delta x^2 - \Theta(\Delta x^3) \end{cases}, \qquad (12)$$

the balance results in the usual diffusion equation

$$\frac{\partial W}{\partial t} = D\frac{\partial^2 W}{\partial x^2}, \qquad (13)$$

with D defined in the limit of infinitesimal Δx and Δt by

$$D = \frac{\Delta x^2}{2\Delta t}. \qquad (14)$$

2.3. Fractional Brownian Motion

The equivalence between Brownian diffusion and Brownian motion motivates the modeling of anomalous diffusion (which results in the fractional diffusion equation [35]) by using fractional Brownian motion.

In fractional Brownian motion, also known as fractal Brownian motion, the increments are not independent, in contrast to Brownian motion.

For that purpose, we consider the continuous Gaussian process $B_H(t)$ (the index $\bullet_H$ refers to the Hurst index), with $B_H(t=0) = 0$, having a null expectation at each time, and defined by the covariance

$$\mathbb{E}(B_H(r), B_H(s)) = \frac{1}{2}\Big(|r|^{2H} + |s|^{2H} - |r-s|^{2H}\Big), \tag{15}$$

with $H = 0.5$ in the case of Brownian motion.

For generating a random trajectory we consider a set of time instants $t_1, t_2, \ldots, t_n$ and construct the matrix $\mathbf{R}$ (symmetric positive–definite), whose component R_{ij} given by

$$R_{ij} = \frac{1}{2}\Big(|t_i|^{2H} + |t_j|^{2H} - |t_i - t_j|^{2H}\Big). \tag{16}$$

The singular value decomposition—SVD of $\mathbf{R}$ writes $\mathbf{R} = \mathbf{VDV}^T$. Thus, we can define the matrix $\mathbf{S} = \mathbf{VD}^{1/2}\mathbf{V}^T$.

Now, we generate n values following a standard Gaussian distribution (zero mean and unit variance), grouped in vector $\mathbf{G}$. Finally, this results in the associated fractional values in vector $\mathbf{F}_H$:

$$\mathbf{F}_H = \mathbf{SG}. \tag{17}$$

3. Numerical Tests

3.1. Topological Data Analysis

Different pre-impregnated composite sheets, delivered by the same provider and prepared with the same materials and under the same process conditions, were analyzed, to check the existence of topological variability on their respective surfaces. A Bruker optical profilometer was used to extract the surface roughness profiles along different directions on the sheet surfaces, in particular along the direction of fibers, along the perpendicular direction to those fibers, and along long fibers that constitute the pre-impregnated composite sheets reinforcement.

Roughness profiles were measured at 17 different locations on both surfaces of each composite sheet. The locations of those measured zones and their dimensions are illustrated in Figure 6, where the fibers' orientation is also indicated.

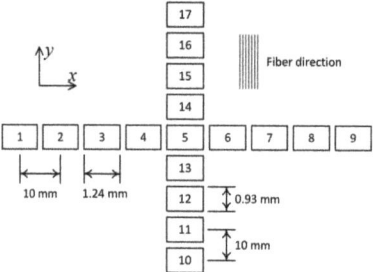

Figure 6. Measurement locations on the composite sheets' surfaces.

The persistence images associated with the different profiles measured along a fiber's direction and along its perpendicular direction, were obtained. A supervised random forest classifier involving 400 trees was trained to discriminate the profiles related to each of the three available composite sheets. A total of 80% of the 1020 available profiles was considered in the training, whereas the remaining 20% served to test the trained classifier performances. Figure 7 shows the associated confusion matrix related to the test set. This figure proves that the three sheet surfaces exhibited similar roughness from a topological viewpoint, and consequently the trained classifier failed to discriminate to which sheet each profile in the test set belongs.

Different classifiers were tested, all them exhibiting similar performances. To conclude on the reasons of the poor classifier performance, different persistence images extracted from the three sheets were analyzed and it was noticed that the differences between the images coming from the same sheets were similar to the differences between the images of different sheets. Thus, it can be concluded that the poor classification capabilities are due to the nature of the surfaces, instead of being a consequence of the classifiers' performance. In these analyses, the Wasserstein distance was considered for quantifying the differences between the topological description of the surfaces.

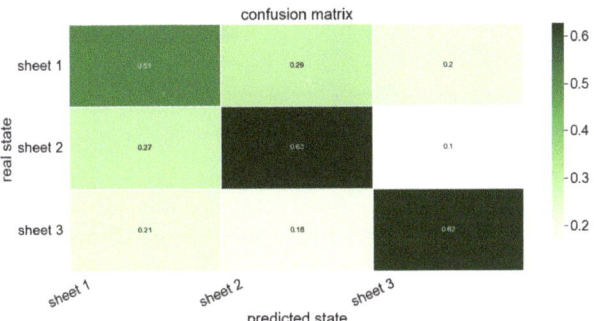

Figure 7. Confusion matrix related to the profiles of the three composite sheets.

However, during composite processing, the sheet surfaces will experience a compression affecting the roughness. Even if, as commented, no difference is expected between the different sheets, a difference is expected concerning the roughness topology before and after compression. To prove that discrimination ability, a sample of each sheet was heated to 200 degrees Celsius and compressed by using a pneumatic press, during 5 min, for ensuring the flow of surface asperities.

Figure 8 compares two characteristic surface topographies before and after compression, from which a noticeably higher smoothness can be seen on the compressed surface. Then, different rough profiles were measured by again using the optical profilometer in both directions (with respect to the fiber orientation).

As the roughness is significantly higher in the perpendicular direction to the fiber orientation, our analyses mainly focus on profiles along that perpendicular direction.

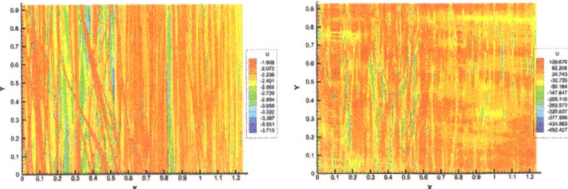

Figure 8. Surface topography before (**left**) and after (**right**) compression.

Figure 9 shows the confusion matrix associated with a random forest classifier (involving 400 trees), trained from 1016 profiles and tested on 254 profiles that were excluded from the training set. It can be concluded that TDA perfectly discriminates original and compressed surfaces.

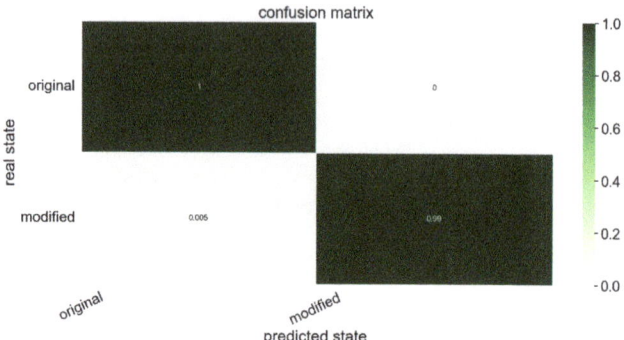

Figure 9. Confusion matrix related to the profiles of the three composite sheets.

To better visualize the roughness-topology change induced by the compression, Figure 10 shows three persistence images, one extracted from each of the three composite sheets along the perpendicular direction to the long fibers, where no major differences are noticed between them.

These figures also reveal an almost mono-disperse topological content, which explains the localized pattern present in the persistence images.

On the other side, Figure 11 compares the roughness topologies before and after compression, where a significant difference can clearly be noticed; the topology is more poly-disperse after compression. On the original surface the asperities' height seems quite mono-disperse, with the associated localization in the persistence image. When compressing the surface, the highest asperities start to be compressed and the asperities' height starts exhibiting a larger spectrum, with the associated effect on the persistence image.

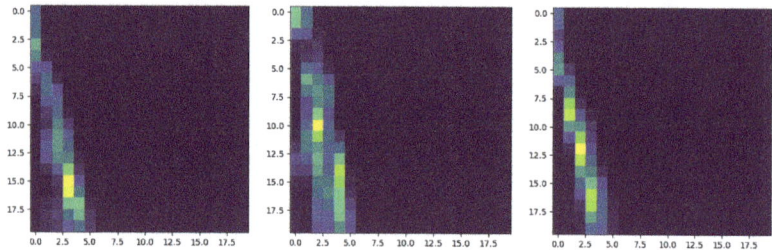

Figure 10. Persistence images related to three roughness profiles, each extracted from one of the three composite sheets.

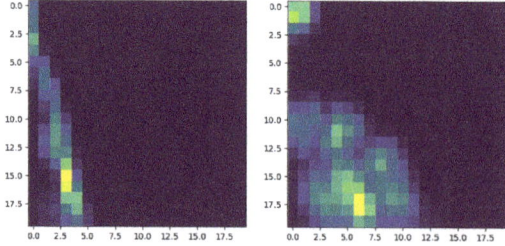

Figure 11. Characteristic persistence images before (**left**) and after (**right**) compression.

3.2. Fractional Brownian Surfaces

This section aims at generating different roughness profiles by using fractional Brownian motion, characterized by different Hurst indexes. Then, a classifier is trained with the associated TDA surfaces, and the classifier's performance is tested on the surfaces composing the test set.

Figure 12 shows three surfaces generated from a fractional Brownian motion with $H = 0.25$, $H = 0.5$ and $H = 0.75$. No major differences are noticeable at first glance; however, a topological footprint is expected to exist, enabling an efficient classification.

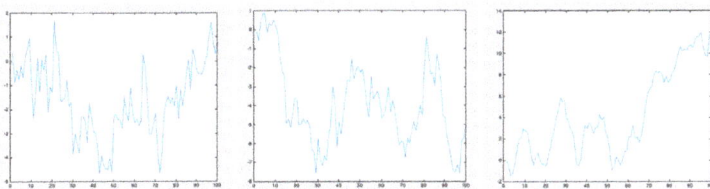

Figure 12. Profiles generated by a fractional Brownian motion characterized by $H = 0.25$ (**left**), $H = 0.5$ (**center**) and $H = 0.75$ (**right**).

To prove the Brownian behavior of profiles generated with $H = 0.5$, and the anomalous diffusion for $H < 0.5$ and $H > 0.5$, we generated 1000 trajectories and evaluated their mean square displacement evolution. Figure 13 superposes 30 different profiles generated with $H = 0.25$, $H = 0.5$ and $H = 0.75$. Figure 14 depicts the evolution of the mean square displacement. As expected, it can be noticed that for $H = 0.5$ (Brownian motion) the mean square displacement evolves linearly.

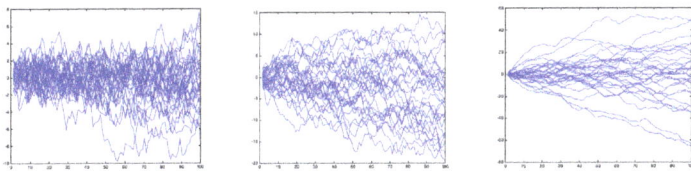

Figure 13. Population of 30 profiles generated by a fractional Brownian motion characterized by $H = 0.25$ (**left**), $H = 0.5$ (**center**) and $H = 0.75$ (**right**).

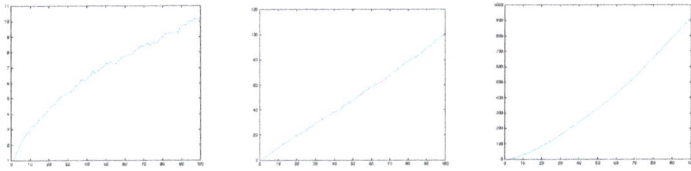

Figure 14. Mean square displacement evolution for the population of profiles generated by a fractional Brownian motion characterized by $H = 0.25$ (**left**), $H = 0.5$ (**center**) and $H = 0.75$ (**right**).

Now, the persistence images associated with each profile are obtained, and a part of them (2400 images) used for training a random forest classifier (involving 400 trees), which is then applied for classifying 600 persistence images composing the test set.

The confusion matrix associated with the classifier, again a random forest involving 400 trees, trained from 8000 images, applied to the test set (composed of 2000 images) is shown in Figure 15 and proves the ability of the TDA to capture the different data topologies associated with the different Hurst indexes.

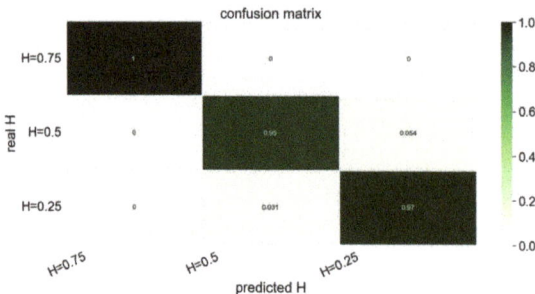

Figure 15. Confusion matrix related to the trained classifier applied to the three fractional Brownian surfaces characterized by $H = 0.25$, $H = 0.5$ and $H = 0.75$.

In order to prove the proposed model's robustness, the procedure was repeated, but now considered 10 different Hurst indexes. Again, the confusion matrix associated with the trained classifier, shown in Figure 16, proves the TDA's robustness for representing fractional Brownian surfaces.

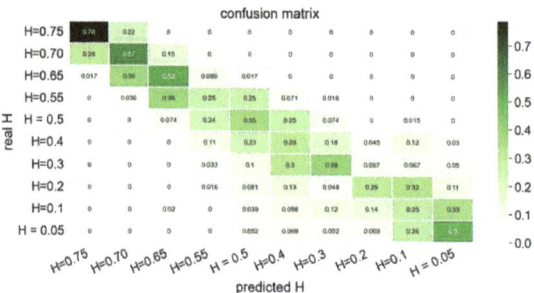

Figure 16. Confusion matrix related to the trained classifier applied to ten fractional Brownian surfaces.

3.3. Hurst Index Evolution during the Surface Compression

To better understand the effect of surface compression on the Hurst index, we considered many Brownian surfaces generated by considering $H = 0.5$. Then, we removed all the positive peaks, that is, the positive part of all the profiles was removed and replaced by a null value, for emulating the flatness induced by the compression. The asperities' flattening is compulsory for consolidation to occur, because of the fact that molecular reptation needs the intimate contact that flatness facilitates. Then, a regressor (random forest involving 400 trees) operating on the persistence images of 4800 profiles, with Hurst indexes ranging in the interval $(0.05, 0.75)$, and tested on the 1200 images composing the test set, was applied to the persistence images associated with flattened profiles. It was observed that the inferred Hurst index ranged in the interval $(0.55, 0.6)$, proving that, as expected, higher correlations are induced in the data representing the flattened profiles.

Now, the same rationale was applied to the real composites' profiles, for comparing the Hurst index distribution before and after compression. For that purpose 250 non-compressed and other 250 compressed profiles, randomly chosen from the 1270 available, were analyzed, and their Hurst indexes inferred by using the trained regressor just introduced. The histograms related to the inferred Hurst indexes are compared in Figure 17, evidencing that the compression induces a noticeable increase in the Hurst index distribution.

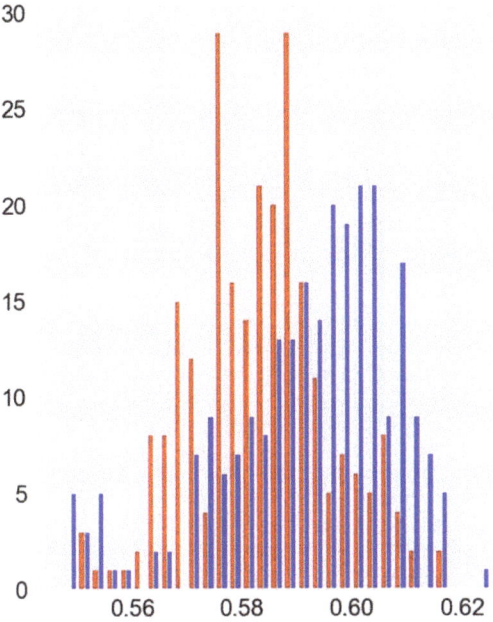

Figure 17. Inferred Hurst index histograms for 250 non-compressed (red) and 250 compressed (blue) profiles.

4. Conclusions

The present paper first proved the ability of topological data analysis to quantify the roughness evolution during composite processing involving compression, and at the same time characterize the roughness of pre-impregnated composite sheets.

Then, the link between TDA and fractional Brownian surfaces was evidenced. TDA enables classifying those surfaces according to their associated Hurst indexes.

Finally, compressed and non-compressed surfaces were characterized using a Hurst index evolution, which could represent a valuable descriptor in processing optimization and control.

Author Contributions: Conceptualization, A.R., F.C., A.A.; methodology, A.R., D.D.L., V.C.; experiments and data analysis, M.-J.K.-P.; software, N.H., A.R. All authors have read and agreed to the published version of the manuscript.

Funding: This project has received funding from the European Union Horizon 2020 research and innovation programme under grant agreement No 101007022.

Institutional Review Board Statement: Not applicable.

Informed Consent Statement: Not applicable.

Data Availability Statement: Data is available upon request.

Acknowledgments: The authors acknowledge the contribution and support of: (i) the CREATE-ID ESI-ENSAM Chair; (ii) the European Union Horizon 2020 research and innovation programme under the Marie Skłodowska-Curie grant agreement No. 956401 (XS-Meta); and (iii) the H2020 DOMMINIO project.

Conflicts of Interest: The authors declare no conflict of interest.

References

1. Chinesta, F.; Leygue, A.; Bognet, B.; Ghnatios, C.; Poulhaon, F.; Bordeu, F.; Barasinski, A.; Poitou, A.; Chatel, S.; Maison-Le-Poec, S. First steps towards an advanced simulation of composites manufacturing by automated tape placement. *Int. J. Mater. Form.* **2014**, *7*, 81–92. [CrossRef]
2. Boon, Y.D.; Joshi, S.C.; Bhudolia, S.K. Filament Winding and Automated Fiber Placement with In Situ Consolidation for Fiber Reinforced Thermoplastic Polymer Composites. *Polymers* **2021**, *13*, 1951. [CrossRef] [PubMed]
3. Song, Q.; Liu, W.; Chen, J.; Zhao, D.; Yi, C.; Liu, R.; Geng, Y.; Yang, Y.; Zheng, Y.; Yuan, Y. Research on Void Dynamics during In Situ Consolidation of CF/High-Performance Thermoplastic Composite. *Polymers* **2022**, *14*, 1401. [CrossRef]
4. Pierik, E.R.; Grouve, W.J.B.; Wijskamp, S.; Akkerman, R. Prediction of the peak and steady-state ply-ply friction response for UDC/PAEK tapes. *Compos. Part A* **2022**, *163*, 107185. [CrossRef]
5. Li, C.; Fei, J.; Zhang, T.; Zhao, S.; Qi, L. Relationship between surface characteristics and properties of fiber-reinforced resin-based composites. *Compos. Part B Eng.* **2023**, *249*, 110422. [CrossRef]
6. Rajasekaran, T.; Palanikumar, K.; Latha, B. Investigation and analysis of surface roughness in machining carbon fiber reinforced polymer composites using artificial intelligence techniques. *Carbon Lett.* **2022**, *32*, 615–627. [CrossRef]
7. Leon, A.; Argerich, C.; Barasinski, A.; Soccard, E.; Chinesta, F. Effects of material and process parameters on in-situ consolidation. *Int. J. Mater. Form.* **2019**, *12*, 491–503. [CrossRef]
8. Lee, W.I.; Springer, G.S. A model of the manufacturing process of thermoplastic matrix composites. *J. Compos. Mater.* **1987**, *21*, 1057–1082.
9. Levy, A.; Heider, D.; Tierney, J.; Gillespie, J. Inter-layer thermal contact resistance evolution with the degree of intimate contact in the processing of thermoplastic composite laminates. *J. Compos. Mater.* **2014**, *48*, 491–503. [CrossRef]
10. Borodich, F.; Mosolov, A. Fractal roughness in contact problems. *J. Appl. Math. Mech.* **1992**, *56*, 786–795. [CrossRef]
11. Ganti, S.; Bhushan, B. Generalized fractal analysis and its applications to engineering surfaces. *Wear* **1995**, *180*, 17–34. [CrossRef]
12. Leon, A.; Barasinski, A.; Chinesta, F. Microstructural analysis of pre-impregnated tapes consolidation. *Int. J. Mater. Form.* **2017**, *10*, 369–378. [CrossRef]
13. Majumdar, A.; Tien, C. Fractal Characterization and simulation of rough surfaces. *Wear* **1990**, *136*, 313–327. [CrossRef]
14. Mandelbrot, B.; Passoja, D.; Paullay, A. Fractal character of fracture surfaces of metals. *Nature* **1984**, *308*, 721–722. [CrossRef]
15. Yang, F.; Pitchumani, R. A fractal Cantor set based description of interlaminar contact evolution during thermoplastic composites processing. *J. Mater. Sci.* **2001**, *36*, 4661–4671. [CrossRef]
16. Senin, P. *Dynamic Time Warping Algorithm Review*; Technical Report; University of Hawaii at Manoa: Honolulu, HI, USA, 2008.
17. Argerich, C.; Ibáñez, R.; León, A.; Abisset-Chavanne, E.; Chinesta, F. Tape surface characterization and classification in automated tape placement processability: Modeling and numerical analysis. *AIMS Mater. Sci.* **2018**, *5*, 870–888. [CrossRef]
18. Longuet-Higgins, M. Statistical properties of an isotropic random surface. *Ser. A-Math. Phys. Sci.* **1957**, *250*, 157–174.
19. Longuet-Higgins, M. The Statistical Analysis of a Random, moving surface. *Ser. A-Math. Phys. Sci.* **1957**, *249*, 321–387.
20. Sayles, R.; Thomas, T. The spatial representation of surface roughness by means of the structure function: A practical alternative to correlation. *Wear* **1977**, *42*, 263–276. [CrossRef]
21. Yaglom, A. Volume I—Basic Results. In *Correlation Theory of Stationary and Related Random Function*; Springer: New York, NY, USA, 1987.
22. Torquato, S. Statistical Description of Microstructures. *Annu. Rev. Mater. Res.* **2002**, *32*, 77–111. [CrossRef]
23. Argerich, C.; Ibanez, R.; Leon, A.; Barasinski, A.; Chinesta, F. Code2vect: An efficient heterogenous data classifier and nonlinear regression technique. *C. R. Mécanique* **2019**, *347*, 754–761. [CrossRef]
24. Carlsson, G.; Zomorodian, A.; Colling, A.; Guibas, L. Persistence Barcodes for Shapes. In Proceedings of the 2004 Eurographics/ACM SIGGRAPH Symposium on Geometry Processing, Nice, France, 8–10 July 2004.
25. Carlsson, G. Topology and Data. *Bull. Am. Math. Soc.* **2009**, *46*, 255–308. [CrossRef]
26. Chazal, F.; Michel, B. An introduction to Topological Data Analysis: Fundamental and practical aspects for data scientists. *arXiv* **2017**, arXiv:1710.04019.
27. Oudot, S.Y. *Persistence Theory: From Quiver Representation to Data Analysis*; Mathematical Surveys and Monographs; American Mathematical Society: Providence, RI, USA, 2010; Volume 209.
28. Rabadan, R.; Blumberg, A. *Topological Data Analysis For Genomics And Evolution*; Cambridge University Press: Cambridge, UK, 2020.
29. Saul, N.; Tralie, C. Scikit-TDA: Topological Data Analysis for Python. 2019. Available online: https://github.com/scikit-tda/scikit-tda (accessed on 12 March 2023).
30. Venkatesan, R.; Li, B. *Convolutional Neural Networks in Visual Computing: A Concise Guide*; CRC Press: Boca Raton, FL, USA, 2017.
31. Frahi, T.; Yun, M.; Argerich, C.; Falco, A.; Chinesta, F. Tape Surfaces Characterization with Persistence Images. *AIMS Mater. Sci.* **2020**, *7*, 364–380. [CrossRef]
32. Lee, J.A.; Verleysen, M. *Nonlinear Dimensionality Reduction*; Springer: New York, NY, USA, 2007.
33. Yun, M.; Argerich, C.; Cueto, E.; Duval, J.L.; Chinesta, F. Nonlinear regression operating on microstructures described from Topological Data Analysis for the real-time prediction of effective properties. *Materials* **2020**, *13*, 2335. [CrossRef] [PubMed]
34. Hinton, G.E.; Zemel, R.S. Autoencoders, minimum description length and Helmholtz free energy. In *Advances in Neural Information Processing Systems 6 (NISP 1993)*; Morgan-Kaufmann: Burlington, MA, USA, 1993; pp. 3–10.

35. Chinesta, F.; Abisset, E. *A Journey Around the Different Scales Involved in the Description of Matter and Complex Systems*; SpringerBrief: Cham, Germany, 2017.
36. Bardet, J.M.; Surgailis, D. Measuring the roughness of random paths by increment ratios. *Bernoulli* **2011**, *17*, 749–780. [CrossRef]
37. Gelbaum, Z.; Titus, M. Simulation of Fractional Brownian Surfaces via Spectral Synthesis on Manifolds. *arXiv* **2013**, arXiv:1303.6377v1.
38. Kroese, D.P.; Botev, Z.I. Spatial Process Generation. *arXiv* **2013**, arXiv:1308.0399v1.
39. Rabiei, H.; Coulon, O.; Lefevre, J.; Richard, F. Surface regularity via the estimation of fractional Brownian motion index. *IEEE Trans. Image Process.* **2020**, *30*, 1453–1460. [CrossRef]
40. Stein, M.L. Fast and Exact Simulation of Fractional Brownian Surfaces. *J. Comput. Graph. Stat.* **2002**, *11*, 587–599. [CrossRef]
41. Frahi, T.; Chinesta, F.; Falco, A.; Badias, A.; Cueto, E.; Choi, H.Y.; Han, M.; Duval, J.L. Empowering Advanced Driver-Assistance Systems from Topological Data Analysis. *Mathematics* **2021**, *9*, 634. [CrossRef]

Disclaimer/Publisher's Note: The statements, opinions and data contained in all publications are solely those of the individual author(s) and contributor(s) and not of MDPI and/or the editor(s). MDPI and/or the editor(s) disclaim responsibility for any injury to people or property resulting from any ideas, methods, instructions or products referred to in the content.

Article

Assessing Intra-Bundle Impregnation in Partially Impregnated Glass Fiber-Reinforced Polypropylene Composites Using a 2D Extended-Field and Multimodal Imaging Approach

Sujith Sidlipura, Abderrahmane Ayadi * and Mylène Lagardère Deléglise

IMT Nord Europe, Institut Mines-Télécom, Univ. Lille, Centre for Materials and Processes, 59000 Lille, France
* Correspondence: abderrahmane.ayadi@imt-nord-europe.fr

Abstract: This study evaluates multimodal imaging for characterizing microstructures in partially impregnated thermoplastic matrix composites made of woven glass fiber and polypropylene. The research quantifies the impregnation degree of fiber bundles within composite plates manufactured through a simplified compression resin transfer molding process. For comparison, a reference plate was produced using compression molding of film stacks. An original surface polishing procedure was introduced to minimize surface defects while polishing partially impregnated samples. Extended-field 2D imaging techniques, including polarized light, fluorescence, and scanning electron microscopies, were used to generate images of the same microstructure at fiber-scale resolutions throughout the plate. Post-processing workflows at the macro-scale involved stitching, rigid registration, and pixel classification of FM and SEM images. Meso-scale workflows focused on 0°-oriented fiber bundles extracted from extended-field images to conduct quantitative analyses of glass fiber and porosity area fractions. A one-way ANOVA analysis confirmed the reliability of the statistical data within the 95% confidence interval. Porosity quantification based on the conducted multimodal approach indicated the sensitivity of the impregnation degree according to the layer distance from the pool of melted polypropylene in the context of simplified-CRTM. The findings underscore the potential of multimodal imaging for quantitative analysis in composite material production.

Keywords: compression molding; polymer matrix composites; thermoplastic resin; microstructural analysis; porosity; polarized light microscopy; fluorescence microscopy; scanning electron microscopy; multimodality

Citation: Sidlipura, S.; Ayadi, A.; Lagardère Deléglise, M. Assessing Intra-Bundle Impregnation in Partially Impregnated Glass Fiber-Reinforced Polypropylene Composites Using a 2D Extended-Field and Multimodal Imaging Approach. *Polymers* **2024**, *16*, 2171. https://doi.org/10.3390/polym16152171

Academic Editor: Vincenzo Fiore

Received: 25 June 2024
Revised: 25 July 2024
Accepted: 26 July 2024
Published: 30 July 2024

Copyright: © 2024 by the authors. Licensee MDPI, Basel, Switzerland. This article is an open access article distributed under the terms and conditions of the Creative Commons Attribution (CC BY) license (https://creativecommons.org/licenses/by/4.0/).

1. Introduction

Compression resin transfer molding (CRTM) is a process within liquid resin transfer molding to manufacture polymer matrix composites and is suitable for thermoplastic matrices, such as polypropylene (PP). For thermoplastic matrices, processing includes thermal regulation to bring the polymeric resin to its liquid state, which is then forced to flow through a fabric of reinforcing fibers, filling spaces between individual fibers at a micro-scale and between fiber bundles at a meso-scale. The impregnation vector during CRTM can be controlled by applying compressive mechanical pressure or a controlled displacement along the direction of the preform's thickness (macro-scale). This pressure reduces unfilled spaces between the reinforcing filaments starting from the micro-scale and alters the preform's permeability; at the same time, it promotes resin flow by creating streamlined pathways between and within bundles of reinforcing fibers. This antagonistic mechanism, where mechanical compression reduces permeability but facilitates filling of empty spaces, requires an in-depth understanding to effectively saturate the initially dry preform and reduce porosity (i.e., unfilled spaces and trapped air bubbles). For detailed insights into how physical forces interact with fluid dynamics to optimize resin distribution and quality in CRTM, refer to [1]. A significant challenge in CRTM processes is monitoring the flow of resin and the degree of impregnation, particularly in the thickness direction of a dry preform, known as the through thickness flow. This through thickness flow scenario is complex due to factors such as gravity, pressure gradients, permeability, temperature gradients, and capillary effects within the fabric's microstructure. Simplifying assumptions can be considered where the through thickness flow can be simplified to the case where a fully saturated liquid zone (i.e., a pool of polymeric resin in the liquid state) transitions to unfilled, non-saturated fabric layers. In addition, when the saturated liquid zone is located below the preform, the combination of high-viscosity thermoplastic resins (compared to thermosets) and gravity allows for the neglect of capillary effects at the micro-scale of individual fibers.

Isothermal compaction can help to limit the presence of thermal gradients. The scenario involving pressure gradients and permeability variations is introduced by the authors in another study [2] and is referred to as simplified-CRTM. This displacement-controlled thermo-compression method (i) simulates CRTM's impregnation phase, (ii) bypasses the injection stage of the molten polymeric resin, (iii) assumes isothermal conditions, and (iv) allows the resin to flow transversally through the preform's thickness. Based on general considerations from the scientific literature about CRTM processes, to capture the anisotropic and non-linear behavior of resin flow, advanced computational fluid dynamics and experimental monitoring of the microstructure are needed to better understand these dynamics, ensuring the even saturation of the resin throughout the fibrous preform and mitigating potential residual porosity. Due to the complexity of real-time monitoring of fluid dynamics at multiple scales, post-manufacturing characterization of the microstructure can help to assess the porosity and degree of impregnation as a first step toward more complex analyses, passing through intermediate state impregnation levels. A literature survey focused on microstructure characterization of polymer matrix composites using 2D imaging techniques is provided in Table 1. This survey highlights studies on polymer matrix composite materials, including carbon fiber-reinforced polymers (CFRPs) and glass fiber-reinforced polymers (GFRPs), utilizing 2D imaging techniques such as optical microscopy (OM) and scanning electron microscopy (SEM). The analysis indicates 2D imaging techniques are mainly destructive, involving surface preparation and sample cutting to conduct subsurface observations and provide information about the microstructure in the form of multidimensional spatial arrays known as "images". Authors frequently associate both destructive and non-destructive imaging methods to analyze microstructure-related aspects, including residual porosity, morphological aspects of fiber bundles or porosity, and impregnation. For porosity characterization, Purslow [3] utilized OM and SEM to assess porosity in CFRPs at the micro-scale, focusing on quantification and distribution, while Liu et al. [4] characterized porosity shape, size, and location at the meso- and micro-scales, linking these observations to processing parameters. Abdelal et al. [5] compared methods for characterizing porosity in glass fiber-reinforced composites (GFRCs), primarily using OM and μCT (micro-computed tomography). Gagani et al. [6] used OM to determine porosity location and quantification in GFRPs, and Ekoi et al. [7] investigated fatigue-induced damage in 3D-printed CFRPs using SEM. Zou et al. [8] examined residual porosity in aramid fiber-reinforced polymers using SEM and μCT. Regarding bundle morphology, Kabachi et al. [9] performed image-based characterization at the macro- and meso-scales using OM, while Breister et al. [10] studied bundle interactions in vinyl ester polymer using OM and SEM. Liu et al. [11] investigated high fiber volume fraction thermoplastics using SEM. For impregnation and resin flow, Ishida et al. [12] used OM during the compression process of thermoplastic composites, capturing images at different holding times to track impregnation and flow front progression at the meso- and micro-scales. Little et al. [13] analyzed oven-cured CFRP samples using density measurement, OM, environmental SEM, and μCT, while Eliasson et al. [14] used OM with neural network-based segmentation for void characterization in CFRP laminates. This non-extensive literature review indicates that various 2D imaging techniques are effective for both the qualitative and quantitative inspection of polymer matrix composites. However, several gaps remain unaddressed or insufficiently documented: (i) Studies lack comparative analyses of the same region of interest (ROI) under different imaging techniques. Indeed, multimodal imaging approaches, commonly used in the biological and medical fields, combine information from different images obtained through multiple techniques to create a richer and more accurate synthetic 2D image, overcoming the limitations of individual techniques [15–21]. The use of multimodal imaging in the context of polymer matrix composites is poorly documented (refer to Appendix S1 for a brief literature review exploring keyword co-occurrence relationships [22], and graphical illustrations created using the open access software VOSviewer, version 1.6.20 [23]). (ii) The analysis of partially impregnated composite samples using 2D destructive imaging techniques requires extensive surface preparation, such as polishing, yet the quality of these prepared surfaces is poorly documented. (iii) The observation of the entire thickness of composite samples is often needed, but there is a lack of consideration regarding the trade-off between the field of view, the smallest detectable detail, and the achievable resolution. (iv) Abdelal et al. [5] compared methods for characterizing porosity in GFRCs, employing ultrasound, burn-off tests, serial sectioning with OM, and μCT. They used a fluorescent dye mixed with epoxy as a resin mount and with talc powder on polished surfaces to highlight voids, although fluorescence microscopy (FM) was not used due to the lack of a UV source.

Table 1. Overview of microscopy-based imaging techniques for characterizing polymer matrix composites focused on references [3–14]. For each reference, the (X) mark indicates the considered imaging technique, length scale, and focus of the analysis. Shaded cells indicate the non-considered aspects.

Ref.	Techniques Used				Scale			Analysis Focus		
	OM	SEM	FM	Other	Micro	Meso	Macro	Porosity	Bundles	Impregnation
[3]	X	X			X			X		
[4]	X				X	X		X		
[5]	X			X				X		
[6]	X					X	X	X		
[7]		X		X	X	X	X	X		X
[8]		X		X				X		
[9]	X					X	X		X	
[10]	X	X			X	X	X		X	
[11]		X						X		
[12]	X	X			X	X				X
[13]	X	X		X						
[14]	X							X		

The current research explores the potential of multimodal 2D imaging techniques for qualitative and quantitative microstructure analyses of glass fiber-reinforced polypropylene matrix composites manufactured using isothermal compression molding to generate partially impregnated composite plates. A first manufacturing configuration based on film stacking is used to generate a reference composite plate formed of alternating thermoplastic films and woven unidirectional glass fiber plies. A second configuration based on simplified-CRTM is then used to manufacture two other plates by varying the compaction ratio during displacement-controlled impregnation. The aim is to generate composite plates with different impregnation levels to examine how the polypropylene (PP) matrix saturates a preform of six layers of woven glass fiber that are stacked as $[0/90]_3$, aiming to understand resin distribution and localize porosity. The study addresses the potential of multimodal imaging (i.e., using different microscopy techniques) for analyzing the same composite microstructure for two main objectives: (i) quantifying porosity (i.e., unfilled spaces within the polymeric matrix) and (ii) assessing the degree of impregnation (i.e., saturation degree of intra-bundle spaces post-manufacturing) at a meso-scale of cross-sections of glass fiber bundles. The targeted techniques include optical microscopy using polarized light (PLM) for surface quality inspections and multimodality based on fluorescence microscopy (FM) and scanning electron microscopy (SEM) for quantitative analyses. The integration of these techniques provides a comprehensive assessment of the composites' internal structure, contributing to the optimization of other manufacturing processes such as CRTM, including through thickness flow-dominated impregnation scenarios.

2. Materials and Methods

2.1. Materials

The matrix material employed was a commercial-grade polypropylene (PP) thermoplastic, identified as PPC13442 by Total® (France), featuring a density of 0.905 g/cm^3, a melting point of 165 °C, and a melt flow index of 100 g per 10 min. This PP was provided in form of pellets and was processed into thin films through thermocompression, yielding films with a mean thickness of 0.57 mm ($\pm$0.03 mm). For reinforcement, an experimental unidirectional (UD) woven glass fiber (GF) provided by Chomarat® (France), referenced as JB111, were utilized. This GF exhibited a density of 2.55 g/cm^3 and an areal density of 1054 g/m^2. Rectangular UD plies measuring 375 × 375 mm^2 were manually sectioned from larger length-scale rolled sheets. Rectangular composite plates with the dimensions of 375 × 375 mm^2 were fabricated, each incorporating six of UD woven glass fiber plies arranged in a [0/90]$_3$ sequence and seven thin sheets of PP matrix.

2.2. Methods

2.2.1. Fabrication of Composite Plates

The goal of this section is to manufacture glass fiber-reinforced polymer composites (GFRPs) as partially impregnated plates, each with a different level of PP matrix saturation. As a reminder, the aim of the current research study is to evaluate multimodal imaging for assessing porosity and not optimizing the manufacturing of composite plates. In this context, three composite plates with varying impregnation levels were fabricated using isothermal compaction molding using an industrial-scale press (Pinette PEI®, France) capable of 120 tons of controlled force and precise top mold displacement to within $\pm$0.1 mm (Figure 1a). During manufacturing, the press executed programmable cycles for temperature, displacement, and force, ensuring controlled manufacturing conditions. The glass fiber-reinforced polypropylene composite plates were produced using the same mold schematically illustrated in Figure 1b and which is equipped with a venting part allowing the evacuation of excessive polymer. The targeted theoretical heights of the composite plates were modified while maintaining the exact same constituents (i.e., seven PP films and six GF plies), as shown in Figure 1c,d, and can be verified from the column indicating the initial masses before manufacturing in Table 2. A first plate was manufactured using a film stacking configuration (Figure 1c), alternating glass fiber (GF) plies and PP films. A fixed compression force of 21 kN was applied to the top part of the mold (Figure 1b), and the mold's cavity-height evolution during the process was recorded (Figure 1e). At this imposed level of compressive force, the preform was barely compacted (compared to the other plates within this same study), and impregnation was primarily due to the fixed compression force. Film stacking was selected to limit the compression level and maintain the fiber bundles undisturbed, based on force-driven control of the compression process. As the polymer melted, the mold gap was expected to progressively decrease due to the force-controlled process. The final thickness of the obtained plate was considered as the reference height for the other composite plates and for assessment of the compaction ratio (Cr) according to Equation (1):

$$Cr = \frac{h_{ref} - h_{final}}{h_{ref}} \qquad (1)$$

where h_{ref} is height of the plate manufactured using configuration 1 and h_{final} is height of consecutive plates manufactured using configuration 2.

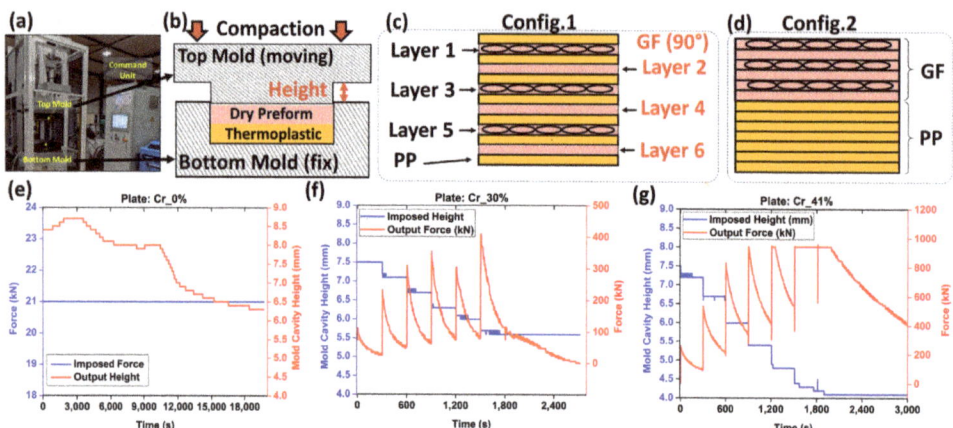

Figure 1. (**a**) Manufacturing platform: Pinette PEI press. (**b**) Schematic representation of the mold used. (**c**) Film stacking configuration 1 used for compression molding. (**d**) Film stacking configuration 2 used for simplified-CRTM. Recorded force-mold cavity heights for plates: (**e**) compaction ratio (Cr), Cr_0%, (**f**) Cr_30%, and (**g**) Cr_41%.

Table 2. Control parameters before and after the manufacturing of the composite plates. V_f refers to the measured final fiber volume fractions after manufacturing, while V_f^* refers to the targeted fiber volume fractions set before manufacturing the composite plates.

Plate	Lay-Up	Config.	Manufacturing Weight (g)		V_f^* (%)	Metrological Control		Cr (%)	V_f (%)	Microscopy Control		Cr (%)	V_f (%)	Burn-Off Test V_f (%)	
						Thickness (mm)				Thickness (mm)					
			Initial	Final		Avg	StDev			Avg	StDev			Avg	StDev
Cr_0%		Film Stacking	1397	1350	42.2	6.1	0.08	0	41.6	6.2	0.01	0	40.0	38.6	0.1
Cr_30%	[0/90]₃	Simplified-CRTM	1408	1081	45.2	4.2	0.13	30.7	60.1	4.4	0.01	30.2	57.4	58.4	0.9
Cr_41%		Simplified-CRTM	1392	1037	63.2	3.5	0.08	41.9	71.7	3.7	0.09	40.9	67.8	64.9	0.4

The other two plates were manufactured using isothermal compression molding similar to simplified-CRTM by positioning all solid-state PP sheets below the stack of GF plies (Figure 1d). Impregnation levels were controlled through a displacement-controlled process, targeting the compaction ratio of the preform as in a previous study by the authors [24]. The heating configuration was similar to film stacking, with isothermal compactions conducted at 215 (±2) °C. As the PP melted, a one-sided fluid pool formed below the preform. Due to the high viscosity of PP (compared to thermoset resins), capillary-driven flow into the preform gaps was restricted, making mechanical pressure during compression the primary driver for the impregnation of the GF bundles. Variations in the impregnation quality were influenced by the balance between the preform's mechanical deformability and reduced permeability under compression, potentially resulting in residual porosity, including dry zones and trapped air bubbles. Some porosity may also result from PP shrinkage post-cooling [25]. No distinction is made between the causes of residual porosity in the context of the current study. This procedure is denoted by simplified-CRTM through the following sections. The simplified-CRTM procedure involved a ten-segment program, which included temperature-regulated heating, displacement-controlled compression, and cooling steps. Initially, two consecutive heating segments were carried out, with conductive heating using calorific oil to stabilize the mold temperature at 100 °C. This was followed by electric heating to achieve 215 (±2) °C. Once the target temperature was reached, six displacement-controlled segments, each lasting 300 s, were automatically activated to conduct staged compaction, progressively impregnating the reinforcement and achieving the targeted final thickness. The final two segments focused on cooling the plates while maintaining the final mold height, reducing the temperature to around 40 °C before opening the mold and extracting the plate. This displacement-controlled manufacturing configuration ensured the control of the final thicknesses of the plates, and thus, control over compaction ratios, which help in varying the impregnation levels. The input (i.e., displacement) and output (i.e., force) of the manufacturing process for the manufactured plates using simplified-CRTM are detailed in Figure 1f,g.

The final thickness control of the manufactured plates was achieved through metrological inspections, providing global average thicknesses from at least five locations on each manufactured plate. Additionally, optical microscopy inspections, including at least ten control points, were conducted on one localized sample extracted from the center of each composite plate (Figure 2a). Based on these microscopy inspections, compaction ratios of 0%, 30%, and 41% were obtained using Equation (1), as detailed in Table 2. A limited difference between the global metrological and local microscopy inspections was observed, confirming a relative uniformity in the thickness of the manufactured composite plates. Since the imaging techniques in the following sections will focus on these same samples used for optical microscopy-based height inspection, the corresponding compaction ratios were used to establish a simplified nomenclature for the composite plates: Cr_0%, Cr_30%, and Cr_41%, as denoted in Table 2. This nomenclature will be consistently used to reference the plates in subsequent sections. The final fiber volume fractions (V_f) of all three composite plates post-manufacturing, compared to the target volume fractions (V_f^*) set before manufacturing, are reported in Table 2. Additional burn-off tests, according to ASTM D 2584, were conducted similarly to those reported in another study by the authors of [24] to quantify the V_f within the manufactured thermoplastic composite plates. As provided in Table 2, the obtained fiber volume fractions were 38.6%, 58.4%, and 64.9%, which were within same range of microscopy- and metrology-based evaluations. For each of the manufactured plates, a certain amount of polypropylene (PP) escaped the mold's cavity, with weight losses of 3.4%, 23.2%, and 25.5% for the Cr_0%, Cr_30%, and Cr_41% plates, respectively. The significant losses observed in plates manufactured using the simplified-CRTM configuration can be explained by the unconstrained boundaries of the preform within the mold's cavity, allowing in-plane polymer squeeze flows at the edges. As this study focuses on the microstructure inspection of partially impregnated composite plates using multimodal imaging, and not on correlating microstructure with manufacturing parameters, the limited constraints applied to the edges of the preforms during isothermal compression molding were considered beyond the scope of the current study.

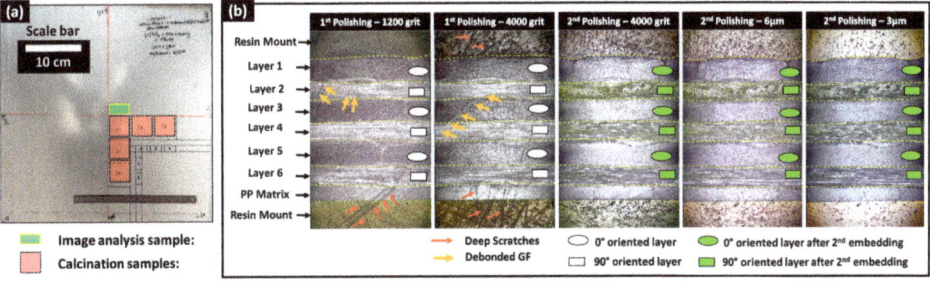

Figure 2. (**a**) Locations and geometrical sizes of extracted microscopy and burn-off samples. (**b**) Evolution of surface state during polishing operations: after the first (white symbols) and second (green symbols) applications of fluorescent dye-enriched resin mount.

2.2.2. Mechanical Polishing of Partially Impregnated Composite Samples

After manufacturing the composite plates, one sample of 10×20 mm^2 from the central zone of each composite plate, as illustrated in Figure 2a, was cut using a water-lubricated diamond saw and dried for 12 h at 40 °C to eliminate residual water. The surface preparation protocol for these dried samples relied on resin embedding and mechanical polishing to expose the subsurface within the microstructure. The mechanical polishing plane was considered perpendicular to the 0°-oriented bundles (in layers 1, 3, and 5) and tangential to the 90°-oriented bundles (in layers 2, 4, and 6). Mechanical polishing is mainly used for well-impregnated glass fiber-reinforced polypropylene composites to achieve a glossy mirror surface with controlled surface roughness [26,27]. However, in the presence of residual porosity and unsaturated regions within the preform where glass fibers are not fully impregnated by the polymeric resin, mechanical damage to unstable fibers limits surface quality control. Indeed, during polishing operations, glass fibers are subjected to quasi-static compressive forces (from the sample holder) and dynamic shear forces (from the rotating polishing discs), causing the breakage of single fibers and debonding of poorly impregnated fibers (Figure 2). To the best of the authors' knowledge, no study has explicitly addressed the challenge of polishing partially impregnated thermoplastic composites. To address the specific challenge of polishing samples extracted from partially impregnated composite materials, this section introduces a four-step surface preparation protocol, which is illustrated in Figure 3. The aims are: (i) minimizing glass fiber breakage, particularly within fiber bundles perpendicular to the polishing surface (i.e., oriented at 0°), and (ii) helping the detection of process-induced residual porosity during image-based inspection after surface preparation while minimizing surface damage within these bundles. Surface control during the surface preparation protocol was principally qualitative, and quantitative inspections such as surface roughness control were considered beyond the scope of the current study. The first step involved a non-classic resin embedding operation of the cut and dried samples, by applying a fluorescent dye-enriched resin mount. EpoFix (Struers®, Denmark) resin and EpoDye (Struers®, Denmark) fluorescent agent were mixed respecting a ratio of 5 g dye per 1000 mL resin before adding the hardener with respect of a volume mixing ratio of 15 to 2. The mixture was degassed in a vacuum chamber for 20 min and then immediately poured into a circular mold of approximately 30 mm of inner diameter and 20 mm depth containing a carefully oriented and clamped composite sample. The mixture was then left to cure at room temperature for 12 h to ensure full polymerization. The second step involved height-controlled serial polishing using abrasive discs of 500, 1200, and 4000 grits and an automated polishing machine (TegraMin 20, Struers®, Denmark). This equipment allows the control of parameters including the compression force applied on the sample, rotational speeds of the sample and the disc, and the duration of each polishing step.

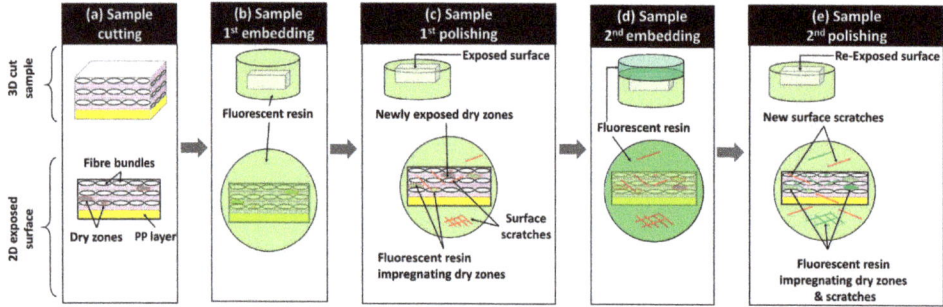

Figure 3. Synthetic overview of the proposed four-step surface preparation protocol.

The inspections visually checked the 0°-oriented bundles where the limited presence of surface scratches and relatively good circularity of single fibers are signs of acceptable results at 4000 grit. In the case of the 90°-oriented bundles, as the polishing exposes new surfaces over time, their quality was not considered for surface control during this second step of the protocol, where a significant amount of broken and debonded single fibers are observed. These broken fibers in vicinity of 90°-oriented bundles caused the deterioration of the surface quality, notably deep scratches within polymer-rich zones and in 0°-oriented bundles. An illustrative example of surface state progression during polishing and following grit changes is provided in Figure 2b. The 500 grit, which is considered the roughest, was considered to remove about 0.3 mm/min from the composite sample to be away from any potential structural damage caused during the sample cutting operation. Polishing parameters for each grit level were selected through an extensive trial-and-error procedure combined with polarized light microscopy inspections of the surface using a microscope equipped with a digital camera.

After finishing the polishing at 4000 grit and qualitatively inspecting the quality of the 0°-oriented bundles, a third step of the surface preparation procedure was applied. It consisted of a second embedding of the exposed surface using the fluorescent dye-enriched epoxy mixture in the same proportions as in the first step. Precautions were taken to add an around 5 mm thick layer without changing the outer boundaries (i.e., bottom surface and cylindrical side wall) of the existing resin mount. The purposes of these steps are to (i) mechanically constrain all exposed and poorly impregnated glass fibers and (ii) seal new exposed porosity (i.e., including dry zones, trapped air bubble cavities) by the conducted sequential polishing steps and potential surface damage, such as scratches and localized fiber breakage. The fourth step of the surface preparation procedure consists of repolishing, assisted with the metrological control of the heights at a precision level of ± 0.005 mm, to re-expose the same surface reached by the end of the second step using 500-, 1200-, and 4000-grit discs. This step removes the excess resin from the second embedding (i.e., the third step) while preserving the same microstructure exposed at 4000 grit by the end of the first polishing. The control of this operation is based on comparisons with the reference height of the resin mount after the first polishing step at 4000 grit. Additionally, microscopy inspections were simultaneously conducted on 0°-oriented bundles while ensuring that the fluorescence-enriched resin maintains transversely oriented single fibers within 90° bundles, limiting their debonding. For the glass fibers within the 0° fiber bundles, the interstitial spaces from dry zones or localized surface micro-cracks caused by the polishing from step 2 were expected to be infiltrated by the fluorescence-enriched resin, thus limiting further damage to fiber circularity and breakage. The polishing operation at 4000 grit was stopped once the target surface was fully exposed with a distance control of ± 0.005 mm. Two finishing steps, using diamond particles of 6 μm and 3 μm, were then applied to refine surface quality and enhance the circularity of fibers, particularly in the 0°-oriented bundles (Figure 2b). It is worth noting that the sequence of polishing parameters (force, rotation speeds, and durations) used in this study cannot be considered unique or optimal, but it is considered reliable enough to guarantee similarity of surface states with minimal fiber breakage within the 0°-oriented bundles and minimal fiber debonding from the 90°-oriented bundles when applied to the considered partially impregnated samples. Additionally, all steps of the polishing procedure included ultrasonic cleaning in ethanol for the removal of material residues and surface cleaning. The main qualitative and quantitative focus will be on layers where bundles are oriented at 0°. For the layers where fiber bundles are oriented at 90°, the fluorescence-enriched resin is expected to saturate all unfilled spaces, enhancing the global surface quality. However, this improvement comes with the drawbacks of (i) hiding the core of these bundles for PLM and SEM observations and (ii) limiting the extraction of quantitative information using FM due to the local saturation with the fluorescence-enriched mounting epoxy resin, which is expected to cause localized brightness effects.

2.2.3. Microstructure Characterization Using 2D Multimodal Imaging Techniques

The characterization of microstructural features in partially impregnated composite materials requires detailed information about the localization of the material constituents (i.e., GF and PP) and structural integrity correlated with manufacturing defects such as porosities, including trapped air bubbles or unsaturated zones termed dry zones. To this end, the current study focused on laboratory-scale 2D imaging equipment based on surface inspection using a comprehensive suite of imaging modalities. A numerical microscope (Axio Zoom V16, Zeiss®, Germany) equipped with a white light source, a fluorescence light illuminator (HXP 200 C), and a 20-megapixel microscope camera (Invenio20EIII, DeltaPix®, Denmark) was used. This setup is suitable for collecting RGB-encoded images from both polarized light microscopy (PLM) and fluorescence microscopy (FM) modes. This microscope, with an automated displacement-controlled stage and an X80 magnification lens, allowed for extended field imaging across all six layers of fiber bundle cross-sections throughout the sample's thickness (Z-direction), covering between five and six complete bundle cross-sections within each layer. Precautions were taken to localize the bundle cross-sections at least by excluding one bundle from each of the cut edges (by the diamond saw) in the through thickness direction of the prepared composite sample. A scanning electron microscope (JCM6000, Jeol®, Japan) operating in Backscattered Electron Detector—Composition mode (BED-C mode) was used at X200 magnification to collect extended-field images (Appendix S2, in Supplementary Materials). SEM samples required additional gold-based metallization. The increased magnification for SEM observations was deliberate to maximize details at the scale of single fibers. The lack of an automated stage in the SEM device necessitated operator-assisted manual translations for capturing overlapping images. The collected images followed a snake-like column-based trajectory, allowing later assembly (i.e., extended-field image reconstruction) using a numerical stitching operation to generate a complete overview of the targeted microstructure. All image acquisitions from the three microscopy techniques (PLM, FM, and SEM) were carried out on composite samples polished up to 3 μm according to the preparation protocol (refer to Section 2.2.2). The integrative use of multimodal imaging techniques provides a robust framework for characterizing the microstructure of partially impregnated composite samples beyond the classical use of optical microscopy (OM) in Figure 4a. Indeed, PLM was used to control the state of the polished surfaces, augmenting contrast in anisotropic materials. PLM can reveal subtle differences on the observed surface, where birefringence indicates the orientation and integrity of fibers and the polished surface. This imaging technique excels in identifying areas with structural anomalies, such as surface scratches or rough zones, which are critical for assessing the quality of the polishing process (Figure 4b). FM is expected to enhance porosity visualization in composites by using a resin mount enriched with fluorescent dye. This resin infiltrates open porosities and/or dry zones. UV light excites the dye's UV-sensitive molecules, which emit light at varying wavelengths based on the dye concentration (Figure 4c). The intensity of fluorescent-rich zones depends on the EpoDye concentration, corresponding to the depth of the damage, mainly due to process-induced porosities or, secondarily, to surface preparation-induced damage. SEM in the BED-C mode was used for the detailed compositional analysis of the polished surface of the composite samples. This technique involves directing an electron beam onto the sample surface, where heavier elements (such as glass fiber) cause more electrons to scatter back. The BED-C mode captures these backscattered electrons to generate high-resolution images that provide contrast based on atomic number variations. This method is especially effective for mapping the distribution of different phases (i.e., resin, PP matrix, and glass fiber) within the samples (Figure 4d). SEM in the BED-C mode delivers precise compositional mapping, essential for analyzing glass fibers.

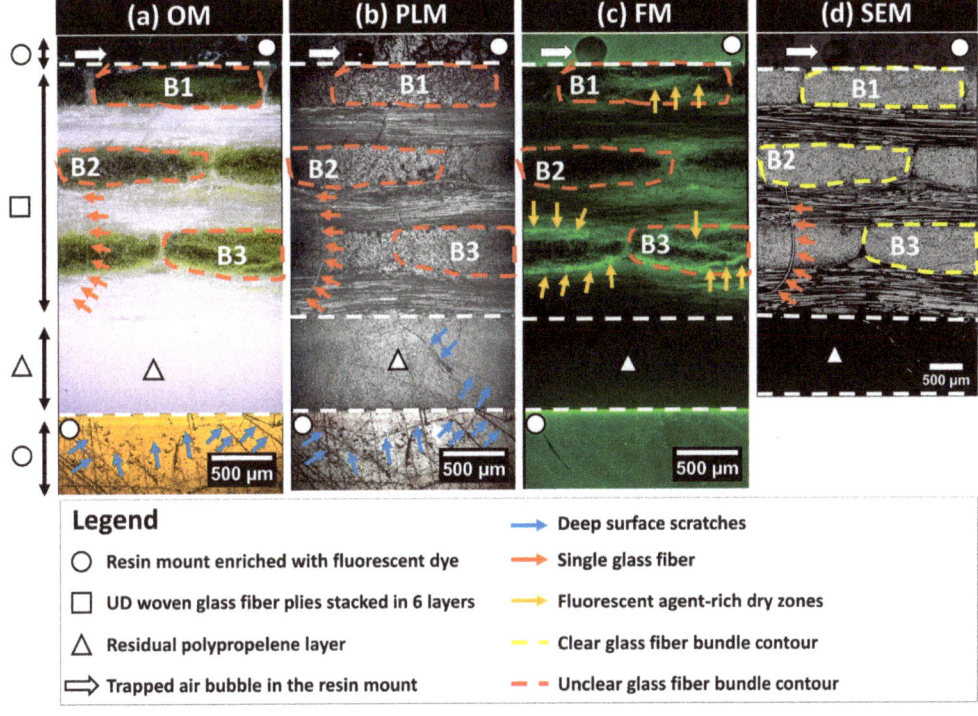

Figure 4. Images obtained using optical microscopy (OM), polarized light microscopy (PLM), fluorescence microscopy (FM), and scanning electron microscopy (SEM) of the same surface polished to 4000 grit following the application of the first fluorescent dye-enriched resin mount.

This multimodal approach allows for a comprehensive analysis of the same microstructure, with each technique providing unique insights that contribute to a detailed assessment about the material. The data are encoded in a 2D spatial array (i.e., numerical images), where each pixel represents specific physical properties. Based on a comparative qualitative inspection of the same zone of interest within the exposed surface of the microstructure using the OM, PLM, FM, and SEM techniques as illustrated in Figure 4, surface defects such as scratches are predominantly discernible with PLM. For instance, surface scratches are clearly detectable in the residual polypropylene layer marked by the triangle symbol in Figure 4. SEM images are more reliable for glass fiber localization and the clear identification of the fiber bundle contours, as seen in the representative bundles B1, B2, and B3 illustrated in Figure 4. SEM significantly contributes to detecting single fibers, where small red arrows in Figure 4d point to a single fiber that was torn out due to polishing from a 90°-oriented bundle. The same single fiber is barely detectable using OM but can also be detected using PLM. Since the single fiber is not impregnated with the fluorescent resin, it is undetectable using FM. FM clearly highlights the zones of localization of the fluorescent agent, as seen from bundle B3 in Figure 4, where high brightness branches indicate a significant presence of the fluorescence-enriched epoxy, indirectly indicating a poor impregnation of the fiber bundle core by PP. Bundle B2 in Figure 4 appears to have a high level of impregnation, as the presence of bright green color in Figure 4c is faint, with numerous black spots indicating exposed glass fiber surfaces. This observation is supported by corresponding OM and SEM observations.

3. Post-Processing Multimodal Images

The subsequent sections detail a workflow for post-processing multimodal images, with a primary emphasis on SEM and FM extended-field images, progressing to bundles of glass fibers. Figure 5 depicts this workflow at the macro-scale of the full thickness of the microscopy-inspected samples. It is worth noting that no filtering, histogram normalization, or brightness corrections were applied to the raw images (Table S2, in Appendix S2 of Supplementary Materials) to preserve the authenticity of the captured data. The process begins by stitching single-tiled images from each imaging techniques to form extended-field images (cf. Section 3.1). Given that extended-field images based on PLM and FM have an X80 magnification, SEM images are resized to match this magnification by reducing their original X200 magnification, typically using interpolation methods to maintain image quality. All full-scale and extended-field images from the same composite sample are then registered to ensure their accurate alignment with the reconstructed extended-field SEM image as the reference image (cf. Section 3.2). These registered images are then compiled into a stack and subsequently cropped. The cropped FM and SEM images are subjected to pixel classification using Random Forest classifiers, a machine learning technique that classifies pixels based on various features extracted from the images (cf. Section 3.3). The processed images are stored in a five-layer stack.

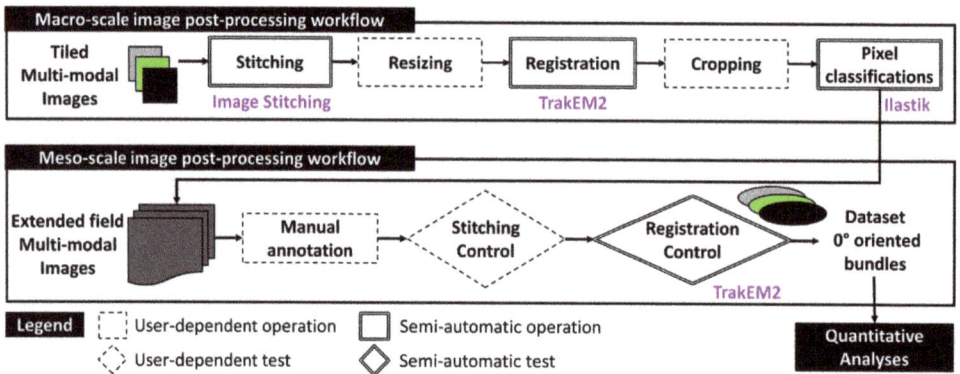

Figure 5. Schematic representation of the post-processing workflows at the macro- and meso-scales for generating bundle datasets from input tiled multimodal images.

At the meso-scale, this stack for each composite sample is manually annotated to identify and segregate fiber bundles oriented at 0° into distinct five-layer stacks. Manual annotation follows strict criteria to ensure consistency, with each stack meticulously inspected for stitching imperfections and undergoing local adjustments to ensure precise alignment. The final bundle images are compiled into a dataset containing about 15 individual 0°-oriented bundles from each composite sample. This quantity is expected to create a robust dataset for both qualitative and quantitative analyses. Additional details on the operations within this two-scale workflow, including the specific algorithms and software used for stitching, registration, and pixel classification, will be elaborated upon in the following sections.

3.1. Macro-Scale Stitching: Reconstruction of Extended-Field and Full-Scale Images

Stitching operations enable the merging of elementary images to visualize larger areas of interest while preserving the pixel resolution of the input elementary (i.e., tiled images). This method is essential for conducting principally qualitative analyses at the macro- and meso-scales (i.e., physical scale of the plate thickness and the respective individual glass fiber bundles) and then achieving the precise pixel classification of pixels and segmentation of microstructures principally at the micro-scale of the single glass fibers. An illustrative example of the considered image stitching method is provided in Figure 6, which depicts a stitched grid of (3 × 3) SEM local images, including the acquisition path of these images. During the acquisition, each image included an overlapping area ranging between 10 and 25%, covering identical microstructural details, such as single fibers, to aid in creating the extended-field image. This stitching method utilizes the open-source Image Stitching of ImageJ/Fiji plugin [28,29], which relies on the Fourier Shift theorem to compute in-plane (X,Y) translations between a grid of 2D images. The corresponding computations are based on a predetermined path for collecting these images and leverages cross-correlation measurements to determine the optimal overlap, thereby reconstructing extended-field images from numerous tiled input images. The Image Stitching plugin requires prior inputs, including the number of columns and rows, the expected overlap range between tiles, and the tolerated minimum and maximum displacement thresholds. The plugin can automatically compute the optimal overlaps, and then merges the images based on different methods, including the linear blending of gray levels within the common intersection areas between adjacent images. It is this same procedure that was applied for generating the full-field stitched images presented in Figure 7 based on the corresponding elementary tiled images provided in Table S2, in Appendix S2 of Supplementary Materials.

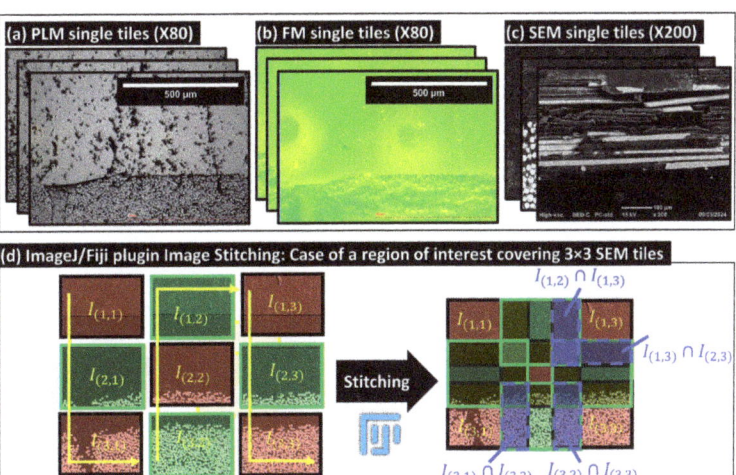

Figure 6. Illustrative examples of collected tiled images: (**a**) polarized light microscopy (PLM), (**b**) fluorescence microscopy (FM), and (**c**) scanning electron microscopy (SEM). (**d**) Example of the stitching operation for a grid of 3 × 3 SEM local images, demonstrating the creation of an extended-field image from individual tiles.

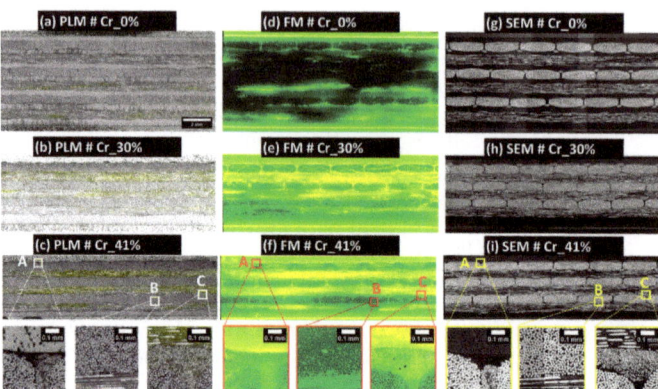

Figure 7. Overview of the final extended-field images of the samples generated using an image processing workflow based on the open-source software ImageJ/Fiji. The images underwent stitching, resizing, registration, and cropping operations. All extended-field images in (**a**–**i**) conform to the scale bar indicated in (**a**). Localized details illustrating full-scale are highlighted in zones A, B, and C. Images (**a**–**c**) were obtained using PLM, images (**d**–**f**) using FM, and images (**g**–**i**) using SEM.

3.2. Macro-Scale: Resizing and Registration Operations of Extended-Field Images

Utilizing the same Zeiss microscope for both PLM and FM simplified the registration process due to the similarity in acquiring the grids of tile images (Table S2, in Appendix S2 of Supplementary Materials), thereby facilitating the alignment of their correspondent extended-field images. SEM images were resized for congruence with the sample's plate thickness, as determined by the PLM reference image. The resized SEM images, based on bicubic interpolation using ImageJ/Fiji functionality, offered enhanced clarity and contrast for visualizing glass fibers and served as reference images for the rigid registration of the corresponding PLM and FM images. This registration involved translations in the XY plane and rotations about the out-of-plane axis. The registration process aligned control points, primarily individual glass fibers and the outer shapes of fiber bundles. Rigid registration operations utilized the TrakEM2 plugin from ImageJ/Fiji, which supports both the manual and semi-automatic alignment of image stacks through rigid transformations. These transformations can be stored as ".xml" files, and the aligned images can be exported post-registration to the same input image-encoding formats. Additional details about the TrakEM2 plugin are available in [30]. Post-registration, the images were grouped into stacks of images (i.e., a multilayer image) and then cropped to eliminate extraneous zero-value pixels at the edges and to focus on a region of interest at the macro-scale containing between five and six complete fiber bundles in the top layer of the composite plate (referred to as layer 1 and containing 0°-oriented bundles). Figure 7 provides a visual summary of the resized and registered images for the three analyzed samples, while Table S2 (in Appendix S2 of Supplementary Materials) outlines the dimensions and pixel sizes of the images post-stitching, resizing, registration, and cropping.

The obtained extended field PLM images, as illustrated in Figure 7a–c, facilitate the qualitative inspection of the polished material surfaces. These images reveal the presence of numerous scratches, particularly within polypropylene-rich areas and fluorescence-enriched epoxy resin layers used for embedding the composite sample. Based on PLM images and SEM scans of the 3 μm polished surface according to the four-step surface preparation procedure (see Section 2.2.2), the presence of single fibers pulled out from the 90°-oriented bundles and covering the adjacent 0°-oriented layers of fiber bundles is absent. The absence of such debonded single fibers on the final surface post-polishing qualitatively indicates the improved impregnation of dry zones within the exposed composite layer by polishing, thus underscoring the efficacy of the developed four-step polishing technique.

Additionally, a localized examination of the fiber bundles showed an acceptable level of circularity for individual fibers within the 0°-oriented bundles (check zones A, B, and C in Figure 7).

3.3. Macro-Scale: Random Forest-Based Pixel Classification

The pixel classification methodology was applied using the open-source software Ilastik (version 1.3.3) [31,32], developed by Sommer et al. [33], which incorporates integrated machine learning approaches designed for users with limited machine learning experience. This method employed Random Forest classifiers for analyzing high-dimensional image data [34]. Ilastik accommodates both mono-channel (8-bit) and multichannel (RGB) inputs. Initially, users select up to nine data features, including color/intensity, edge detection, and texture, which Ilastik automatically computes at various scales using Gaussian smoothing variance. Users then graphically select and label a few pixels using a virtual brush tool, assigning specific labels to each pixel class. These annotated pixels and their associated features are used by the Random Forest classifier to generate a decision tree matching the training labels, allowing for real-time feedback and iterative improvements to enhance the classifier accuracy. Ilastik outputs label predictions and accuracy levels, assisting in the evaluation of classifier performance. To optimize classifier convergence, random pixels from each extended field image were selected to adjust for variations in contrast and brightness, as detailed in Figure 7. Computations were executed on a Windows workstation with an Intel Xeon E5-1650 v2 CPU, 64 GB RAM, and an Nvidia Quadro K2000 GPU, completing the classification process within approximately 30 min per extended-field image. In the case of SEM extended-field images, five specific pixel labels were utilized: center of single glass fibers, edge of glass fibers, zones of rich PP such as residual PP layers, fluorescence-enriched epoxy outside the sample, and unfilled spaces at the ground level of 8-bit grayscale images (see Figure 7). Given that the study's primary focus is quantitative analysis, pixel classification in SEM images primarily targeted pixels at the center of glass fibers to facilitate separation within densely packed 0°-oriented fiber bundles, enabling the calculation of local distances to neighboring fibers and the generation of distance maps between the centers of single glass fibers. Figure 8 shows input SEM images and segmented glass fibers, noting local contrast variations due to the sensitivity of the retro-diffused electron probe during image acquisition. Despite these variations, the high gray level values of glass fibers in 8-bit SEM images enabled accurate pixel segmentation (see Figure 8, zones A, B, and C). These contrast variations were considered non-critical for the Random Forest classifier-based machine learning pixel classification. Post-segmentation, SEM images were used to delineate the physical limits of fiber bundles and systematize the nomenclature from left to right for complete bundles and from top to bottom between the layers of UD plies principally oriented perpendicularly to the observation plane (Figure 8a–c). The similarity in nomenclature designation (L_iM_j) of bundles does not imply any bundle order correspondence between the composite samples; it is used solely for simplicity.

In RGB-encoded FM extended-field images, pixel classification utilizes a five-label scheme through Ilastik with a Random Forest classifier, focusing on the intensity of the green color to indicate fluorescent agent concentrations. This methodology categorizes fluorescence into five levels: 0 (no agent), 1 (low saturation), 2 (medium-low saturation), 3 (medium-high saturation), and 4 (maximum brightness indicating pixel saturation). Level 0, depicting the absence of fluorescence, is prominently seen in Figure 9a, highlighting the sample Cr_0% known for high-quality impregnation and minimal dry zones. Level 4, often at the interface between the robing resin and the composite samples as shown in Figure 9a–c, represents maximum brightness and potential optical interference effects. Level 1 indicates partial staining on PP layers, likely due to abrasive polishing, signifying thin residual layers of the embedding matrix or slightly stained PP. Level 2, associated with medium-low saturation, typically appears in resin zones outside the main composite structure. Level 3, characterized by a yellow-dominated hue representing medium-high saturation, is visible in the 90°-oriented layers in both Figure 9b,c, marking dry zones or debonded and broken single glass fibers, as identified by the scratching marks on the PP layers depicted in Figure 7a. Figure 9 provides a comprehensive view of these fluorescence levels across three composite samples, with segmented images in Figure 9d–f illustrating variations in impregnation levels and resin distribution. This classification framework aids in the precise segmentation and analysis of the composite microstructure. Due to limited visibility in 90°-oriented fiber bundles caused by excessive fluorescent resin residues from polishing, qualitative inspections will focus on 0°-oriented bundles.

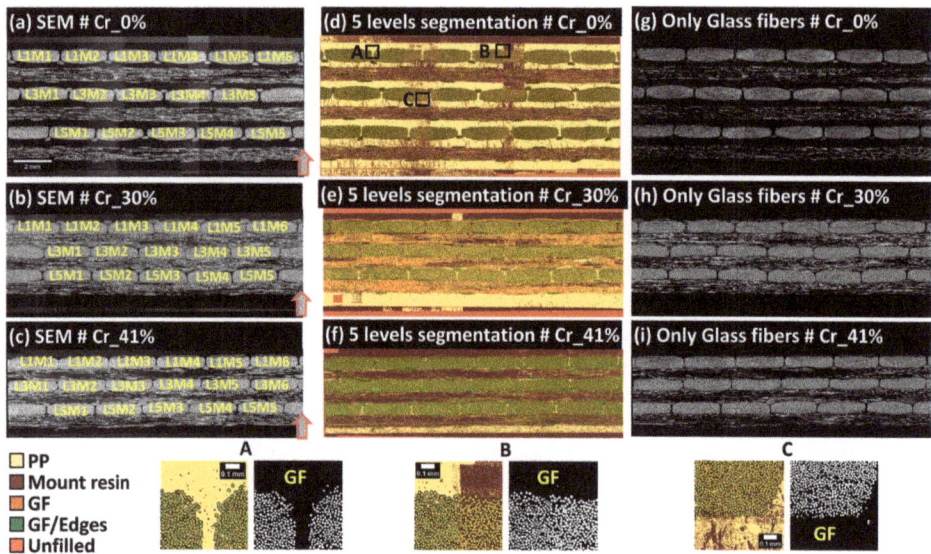

Figure 8. Final extended field SEM images of samples Cr_0%, Cr_30%, and Cr_41% are shown in (**a**–**c**), respectively, alongside five-label pixel classification based on Illastik (version 1.3.3) in (**d**–**f**) and segmented glass fiber pixels in (**g**–**i**). Bundle nomenclatures are based on SEM images, with glass fibers extracted using a five-label Random Forest pixel classification. Illustrative classifications of pixels in zones A, B, and C are provided with corresponding GF segmentations. Arrows denote the orientation of the plate's thickness within the manufacturing mold. The similarity in nomenclature designation (LiMj) of bundles does not consider any bundle order correspondence between the plates. All extended-field images in (**a**–**i**) conform to the scale bar indicated in (**a**).

In Figure 9d, the Cr_0% sample shows predominant levels 0 and 1, indicating limited fluorescent agent presence and suggesting well-exposed single glass fibers with minimal polishing damage and high saturation by PP from the film-stacking manufacturing method. In contrast, the Cr_30% sample in Figure 9b presents high impregnation quality in 0°-oriented layer 5, adjacent to the bottom PP layer, likely due to PP fluidity during manufacturing. Progressing from bottom to top through layers 5 to 3 and then to 1, mixed levels of 1, 2, and 3 indicate partial impregnation. As the compaction ratio increases from Figure 9e–f, the impregnation quality appears to improve, although level 1 bundles near the top exhibit higher impregnation at the edges, suggesting limited through thickness polymer flow. This analysis potentially supports a justification of the through thickness flow in a simplified-CRTM process, though further analysis is required to correlate macro-scale compaction ratios and impregnation scenarios in the samples.

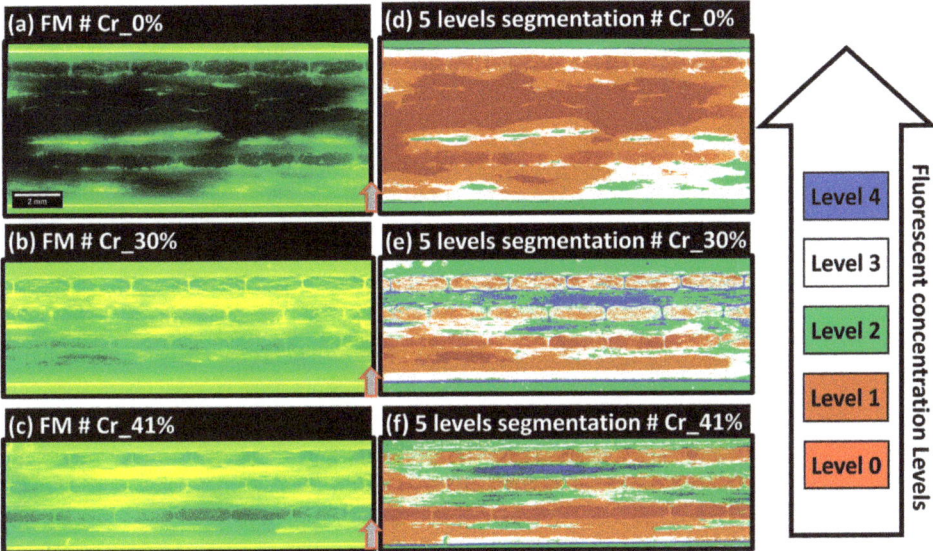

Figure 9. Final extended-field FM images of samples Cr_0%, Cr_30%, and Cr_41% are shown in (**a**–**c**) respectively, accompanied by segmented concentration level domains in (**d**–**f**) respectively. Arrows indicate the orientation of the plate's thickness within the manufacturing mold. All extended-field images in (**a**–**f**) conform to the scale bar indicated in (**a**).

3.4. Meso-Scale: Workflow Applied to 0°-Oriented Bundles

The main focus of the current section based on meso-scale inspections is focused on 0°-oriented fiber bundles to conduct the quantification of glass fiber content, porosity, and consequently, polypropylene as a third possible component. In the specific context of conducting meso-scale quantitative analyses of the segmented images obtained from FM and SEM, all image processing operations relied on finalized extended-field image and their respective segmented images illustrated in Figures 7–9. All post-processing procedures conducted at the meso-scale of fiber bundles were based on open source Fiji/ImageJ software, open source Ilastik (version 1.3.3), and MATLAB R2022a software (Mathworks Inc., MA, USA).

3.5. Meso-Scale: Inspection of Stitched Images

First, the manual annotation and extraction of fiber bundle contours were conducted based on local inspection at the meso-scale of the extended field output images following the macro-scale image post-processing workflow, including stitching, resizing, rigid registrations, cropping, and pixel classification operations. The main challenges were related to: (i) controlling the output of automatic stitching operations of the large grids of collected tiled images (see Table S2, in Appendix S2 of Supplementary Materials) and (ii) checking the local precision of the rigid registration operations conducted at the macro-scale. Indeed, the considered macro-scale control procedure was based on the extended field images from the three microscopy techniques (PLM, FM, and SEM), as illustrated in Figure 7. First, the three images of the same microstructure generated from the three microscopy techniques (PLM, FM, and SEM) and the segmented images of SEM and FM were all merged into a five-layer single stack (i.e., pile of 2D images), all encoded into the RGB format. Then, manual annotations were conducted using a polygon-based manual contouring of fiber bundles using ImageJ/Fiji to extract all individual bundles from the images based on the visual delimitation of each single bundle limit based on SEM images.

The visual inspection of each bundle was then conducted to check for the presence of critical stitching imprecisions. Subsequently, rigid registration was applied to the five-layer stack (of SEM, PLM, FM, SEM segmented, and FM segmented images) to further correct any potential imprecision at the scale of the single fibers. The main critical imprecision consisted of misalignment at the interface of overlapping individual tiles, causing a shadow effect or the local discontinuity of the microstructure. The detection of such imperfections is considered an exclusion criterion for a few fiber bundles, which will not be included in the following quantitative analyses. In this study, only the bundles L1M6 and L5M5 extracted from the sample Cr_0% were excluded, and the corresponding imperfections are illustrated in Figure 10. This problem of limited precision in stitching is expected when using large datasets of individual tiles. The rest of the bundles are considered within the required precision range to conduct the quantitative analyses. In the context of FM and PLM images, the image acquisition included the automatic coordinate-based collection of individual tiles, and it was noted that stitching misalignments were absent with respect to the considered stitching approach using the Image Stitching plugin of ImageJ/Fiji. A second challenge observed after macro-scale stitching is the presence of local changes in grayscale brightness levels between the tiles, particularly in SEM images, as seen in the macro-scale extended-field images in Figure 7. This effect is due to long-duration acquisitions ranging from 8 to 16 h (according the total grid for each sample indicated in Table S2, in Appendix S2 of Supplementary Materials), which are associated with inevitable drift of the incident electron beam and performance stability of the electron detector of the used SEM equipment. As indicated earlier, no histogram normalization or filtering operations were applied to the raw tiled images or the reconstructed extended-field images. To check the performance of Random Forest classifiers in overcoming such effects, a representative bundle, L5M3, from the Cr_0% sample was considered. Detailed visuals of the segmentation procedure are provided in Figure 11. Two regions of interest, ROI 1 and ROI 2 (Figure 11a), were extracted from the raw SEM image of the bundle, and the corresponding histograms of grayscale levels were generated (Figure 11b). The obtained histograms show that equipment drift over time (as the right and left sides of this particular bundle are from two different tiled images) causes a shift in histogram peaks from left to right, with the peak around grayscale level 250 corresponding to glass fiber pixels and the peak around grayscale level 50 corresponding to polymer-rich zones. This shift is considered insignificant during pixel classification, as shown by the classified pixels from ROI 1 and ROI 2 (Figure 11c), and mainly affects the distinction between PP and the epoxy-based mount resin. This intensity-related local variability was deemed negligible in this study. However, as a precaution, quantitative analyses will evaluate the glass fiber content from each bundle to estimate the uncertainty caused by such brightness change-related image artifacts.

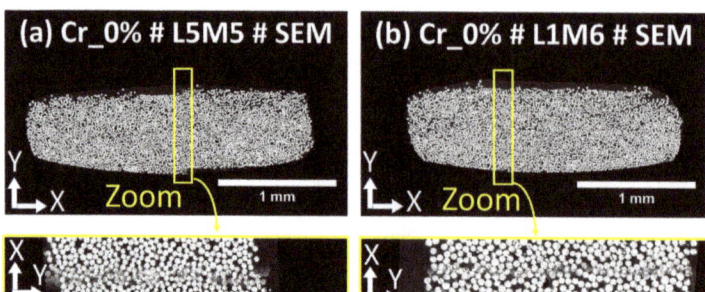

Figure 10. Examples of misalignment detected locally after the macro-scale automatic stitching of the single tiles. (**a**) corresponds to bundle L5M5 from the Cr_0% sample, and (**b**) corresponds to bundle L1M6 from the Cr_0% sample. Rectangles indicate zones of limited stitching precision, marked by shadows from non-overlapping single glass fibers due to the linear blending of misaligned adjacent image tiles.

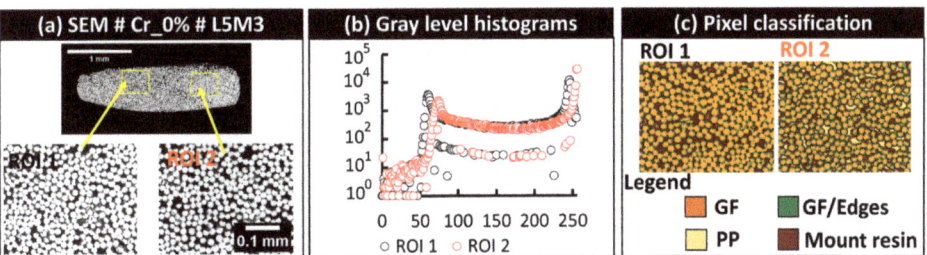

Figure 11. (a) Effect of local brightness in SEM images shown through two regions of interest, (b) the corresponding grayscale histograms, and (c) the corresponding Random Forest pixel classification output.

After conducting all the inspection operations, a clear visual overview of all extracted individual bundles oriented at 0° from SEM and FM extended-field images is presented in Figure S3a–c in Appendix S3 of Supplementary Materials. The PLM images were mainly collected to check the surface state following the polishing procedure and will not be used during the meso-scale quantitative analyses, and the corresponding single bundles are not illustrated.

3.6. Meso-Scale: Quantitative Analysis Workflow of 0°-Oriented Fiber Bundles

The conducted quantitative analyses were based on the extracted bundle SEM, FM, and corresponding segmented images. First, pixels corresponding to the central pixels of single glass fibers were extracted and converted to binary masks, as illustrated in Figure 12c. In a second step, the SEM images of fiber bundles were subjected to a two-step procedure for the object classification of the extracted single fibers using the open-source Ilastik software (Version 1.3.3), which allows the attribution of a single identifier to each fiber (Figure 12d). The objective of this operation was to check the number of single glass fibers within a bundle and to extract the center of each fiber. Based on interpolation between the coordinates of the fiber centers, an initial bundle contour was identified. This contour overestimated the bundle area, particularly in bundles with non-packed fibers, such as those from the top layer of the composite sample Cr_41%, where the limited impregnation of layer 1 (see Figure 7) was discussed in the section on the qualitative inspection of the corresponding macro-scale-segmented FM image. To overcome this limitation, normalized distances between the center of each single glass fiber and its nine adjacent neighboring fibers were evaluated using MATLAB, and the corresponding maps of normalized distances were provided, as illustrated in Figure 12e and in Figure S3a–c in Appendix S3 of the Supplementary Materials. Maps of normalized distances were first considered without any filtering to define the largest bundle contour passing by the center of all circumferential single fibers. In a second step, a maximum normalized distance threshold of 0.4 was judged appropriate by the authors to objectively define a narrower contour for all fiber bundles, excluding parts of single fibers with normalized distances higher than 0.4, while conserving the shape of the packed bundles. An illustration of the two generated contours, one large and one narrow, is shown in Figure 12f and in Figure S3a–c in Appendix S3 of the Supplementary Materials. These two contours were then superimposed on the segmented FM image of the bundle to extract the pixels within each contour representing levels 2, 3, and 4 of fluorescent concentration. These levels were considered indicators of limited impregnation, based on the macro-scale qualitative analysis of the segmented images. Next, the pixels corresponding to the segmented glass fibers in Figure 12c were subtracted from the masks representing levels 2, 3, and 4 of the FM images to define the zones outside the single glass fibers, representing either porosity or areas of limited impregnation. A second subtraction of the total bundle area, defined by each of the previously identified contours in Figure 12f, removed the glass fiber pixels and the limited impregnation pixels, leaving the remaining pixels to represent the PP-impregnated areas. This meso-scale workflow resulted in two sets of quantitative values for each bundle (one from the large contour and one from the narrow contour). The same process was repeated and generalized to all bundles identified as suitable for quantitative analysis. In this context, all distance maps and the two contours are provided in Figure S3a–c in Appendix S3 of the Supplementary Materials.

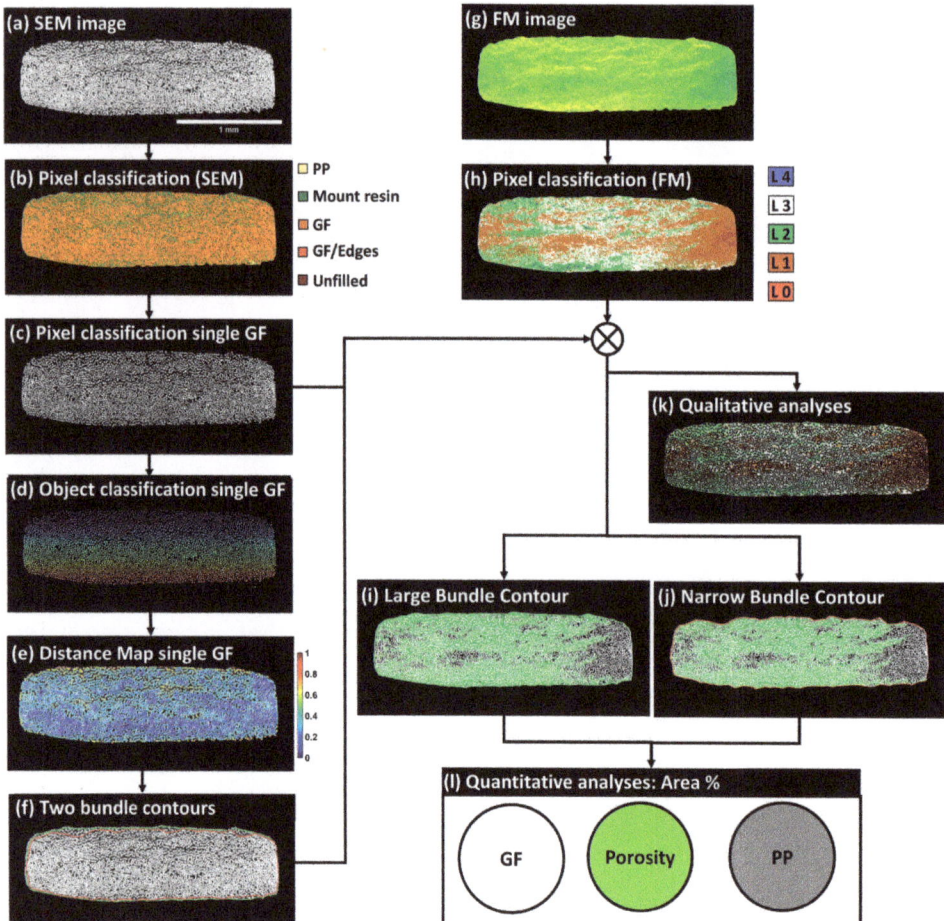

Figure 12. Suggested imaging workflow at the meso-scale of fiber bundles for the quantification of area fractions of GF, PP, and porosity. (**a**) SEM image of bundle L1M1 from sample CR_41%, (**b**) Pixel classification of the SEM image, (**c**) Segmented GF pixels, (**d**) Object classification of a single GF, (**e**) Distance map of a single GF, (**f**) Two bundle contours: the contour in green represents a large bundle, and the contour in red represents a tight bundle, (**g**) FM image of bundle L1M1 from sample CR_41%, (**h**) Pixel classification of the FM image, (**i**) Contour of the large bundle, (**j**) Contour of the narrow bundle, (**k**) Synthetic image based on segmented glass fibers of SEM image and classified pixels of the FM image, (**l**) Quantitative analysis of the area percentage of GF, PP, and porosity.

4. Results

4.1. Quantification of GF Single Filaments

Following the pixel classifications of SEM images, an automated object classification procedure was applied to the population of pixels representing the central core of single fibers. An additional constraint was imposed on the size of isolated pixel clusters to exclude fragmented cross sections, separate adjacent cross sections of single fibers, and identify non-separated single fibers exceeding an area threshold of 650 pixels. The conducted flow chart was based on Ilastik and did not require the implementation of any Supplementary Materials. Based on the separated objects, we extracted the coordinates of the center of each single glass fiber and quantified the number of single fibers per bundle. A summary of the collected data is provided in Figure 13. The global mean (μ) of single filaments based on all fiber bundles was 4038, with a standard deviation (σ) of 63. Most of the counted filaments fell within the 95% confidence interval, corresponding to ($\mu \pm 2 \times \sigma$), with approximately 5% of data points falling outside this range. The filaments falling outside the ($\mu + 2 \times \sigma$) interval were due to the presence of stitching fibers imbricated in the 0°-oriented bundle, such as in bundle L1M4 from plate Cr_41%, or the significant presence of fragments from glass fibers imbricated between the circular cross-sections of the 0° bundle filaments, such as in bundles L1M3, L2M3, and L3M3 from the same composite plate. These observations highlight the limits of the defined surface preparation procedure, where mechanical polishing cannot completely eliminate the breakage of GF tips. The maximum relative error computed from these identified bundles with a high number of single filaments was 3.87%. Conversely, a few fiber bundles presented a low number of single filaments, falling below the lower bound of ($\mu - 2 \times \sigma$), such as bundle L5M3 from plate CR_0%, which had 3900 single filaments, representing a relative error of 3.42%. Beyond experimental bias that may explain variations in the number of counted single filaments, this operation was also conducted to estimate the precision of the object identification procedure in identifying single filaments. The obtained data appear to fall within the 95% confidence interval.

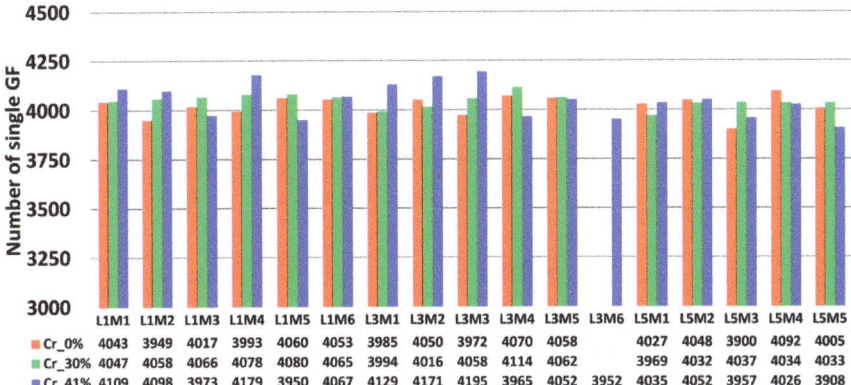

Figure 13. Overview of the total extracted single glass fiber filaments from all non-excluded 0°-oriented fiber bundles. The color indicates a change in the considered composite sample. The similarity in nomenclature designation (LiMj) of bundles does not consider any bundle order correspondence between the plates.

4.2. GF Area Fraction Quantification

Glass fiber (GF) area fractions within the bundles oriented at 0° were quantified to assess the sensitivity of the SEM image segmentation to local brightness variations in the BED-C mode. The obtained quantitative data are graphically represented in Figure 14. For the sample produced by film stacking, the average GF area percentages in layers 1, 3, and 5 were 46.6% (±3), 46.5% (±2.8), and 44.8% (±2.2), respectively. The percentages extracted from the narrow contours were 47.4% (±3.0), 47.1% (±3.0), and 45.6% (±2.0). As the fiber bundles in this configuration are expected to have a high level of impregnation, the quantified surface areas are consistent. For the plate Cr_30%, the average GF area percentages in layers 1, 3, and 5 were 53.7% (±0.6), 57.4% (±0.4), and 57.3% (±0.4), respectively. The percentages extracted from the narrow contours were 54.6% (±0.5), 57.9%

(±0.6), and 57.7% (±0.4). For the plate Cr_41%, the average GF area percentages in layers 1, 3, and 5 were 51.1% (±1.7), 57.7% (±1.1), and 57.8% (±1.1), respectively. The percentages extracted from the narrow contours were 52.7% (±1.7), 58.3% (±1.1), and 57.7% (±1.2). For the samples extracted from the plates manufactured using simplified-CRTM, the average GF area percentages were within the same ranges for both samples. However, the bundles from layer 1 for both samples contained less GF than layers 3 and 5.

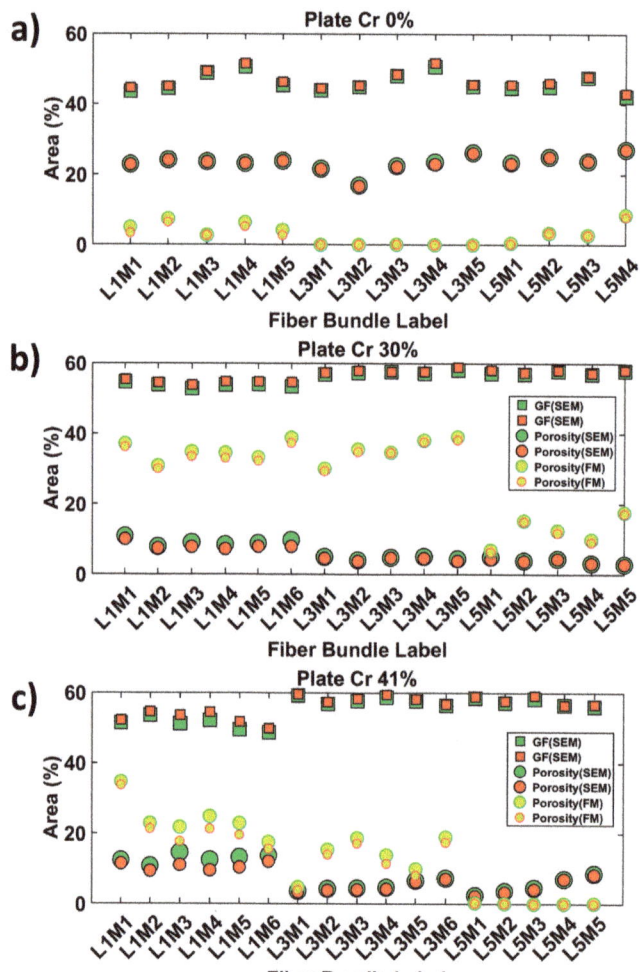

Figure 14. Graphical representation of the meso-scale quantitative output of the area fractions of porosity and glass fibers within 0°-oriented fiber bundles, considering the narrow (red bundle contour) and large (green bundle contour) bundle contours. GF and porosity area percentages in SEM and FM images: (**a**) Cr_0%, (**b**) Cr_30%, and (**c**) Cr_41% samples.

This observation can be attributed to a higher exclusion of loose GF, reducing the bundle contour and the number of single GFs included. The global observations did not show significant variations due to contrast differences in the SEM images used to reconstruct the extended field of view. This indicates that the Random Forest classifiers are less sensitive to these image biases. The variation in percentage between Cr_0% and the other plates can be explained by the morphology of the extracted bundles, which are less flattened in Cr_0% compared to the other plates.

4.3. Quantification of Porosity Area Fractions Based on Narrow and Large Bundle Contours

Porosity area fractions within the 0°-oriented bundles were quantified based on the large and narrow bundle contours and are reported in Figure 14. For the sample produced by film stacking, the average porosity area percentages in layers 1, 3, and 5 were 5.2% (±1.8), 0.02% (±0.04), and 3.8% (±3.4), respectively. The percentages extracted from the narrow contours were 4.0% (±1.7), 0.02% (±0.03), and 3.6% (±3.2). These results indicate minimal porosity within the core of the composite plate, suggesting a high degree of impregnation. For the plate Cr_30%, the average porosity area percentages in layers 1, 3, and 5 were 34.9% (±2.8), 35.5% (±3.5), and 12.4% (±4.24), respectively. The percentages extracted from the narrow contours were 33.7% (±2.6), 34.9% (±3.5), and 11.9% (±4.4). The average porosity levels in layers 1 and 3 of plate Cr_30% were similar, whereas layer 5, which was closest to the PP-rich bottom zone, had about 11.9% porosity. The bundles in layer 5 showed significant impregnation but were not fully saturated. For the plate Cr_41%, the average porosity area percentages in layers 1, 3, and 5 were 24.1% (±5.7), 13.6% (±5.5), and 0.2% (±0.1), respectively. The percentages extracted from the narrow contours were 21.5% (±6.3), 12.0% (±5.3), and 0.1% (±0.1). Plate Cr_41% displayed a clear trend of a through thickness impregnation gradient from the bottom to the top layers compared to plate Cr_30%. Layer 5 had a negligible amount of porosity, while a significant increase was observed from layer 5 to layer 3 to layer 1. This observation was confirmed by the qualitative inspection of the extended field images, where the bundles of plate Cr_41% showed loose fibers towards the outer surface, indicating the presence of PP dry zones.

4.4. Statistical Significance of the Generated Data: One-Way Anova Test

The purpose of this section is to assess the statistical significance of the obtained data in detecting the effect of manufacturing configurations on the variation of the degree of impregnation of 0°-oriented bundles between layers 1, 3, and 5. For this purpose, a one-way analysis of variance (ANOVA) at a 95% confidence interval was conducted. The obtained data, illustrated in Figure 14, were converted into the degree of impregnation of a bundle, defined by Equation (2), which allows the estimation the saturation of the unfilled spaces within the fiber bundle cross-section with the PP matrix after the exclusion of the area filled by single GF filaments.

$$Degree\ of\ impregnation = 100 \times \frac{Total\ Bundle\ area - GF\ area - Porosity\ area}{Total\ Bundle\ area - GF\ area} \quad (2)$$

4.4.1. Hypothesis 1: Consideration of All Fiber Bundles without Any Distinction between Composite Layers

The null hypothesis for ANOVA considers that no differences among the mean values of impregnation degrees between the layers from plates Cr_0%, Cr_31%, and Cr_40%. The decision rule based on ANOVA considers the rejection of the null hypothesis if the p-value is lower than or equal to the significance level. The significance level is commonly defined at 0.05 as set in another study by the authors (i.e., meaning that there is a 5% risk of concluding that there is an effect) [35]. Since the current study is exploratory, a less stringent level of risk defined at 10% was considered to increase the test's sensitivity. Table 3 represents the average degree of impregnation of bundles layer per layer as well as the manufacturing configurations. The obtained p-values from the one-way ANOVA were evaluated based on data evaluated from the large and narrow contours. The computed "p-values" that were obtained were, respectively, 0.053 and 0.0463 for the large and narrow contours. With consideration of a risk level of 5%, the p-value for the narrow contour of fiber bundles successfully rejects the null hypothesis. For the case of the large bundle contour, the p-value, despite being close to the 5% limit, technically fails to reject the null hypothesis, indicating no conclusive significance of the manufacturing conditions on the impregnation of fiber bundles from layers 1, 3, and 5. However, with the increase in the sensitivity of the test, as due to the limited number of bundles per layer and due to the variability during the manufacturing of composite materials, the null hypothesis can be rejected, meaning that the three manufacturing methods have different impregnation effects on the bundles from layers 1, 3, and 5. This less restricted test limit of 10% can be supported by the qualitative observations from the bundles, but, on the other hand, it indicates that there is a need for the testing of the image-based approach with samples manufactured with a higher level of confidence of control of the manufacturing process, which is an ongoing target by the authors.

Table 3. One-way ANOVA results for the degree of impregnation of 0°-oriented fiber bundles across different manufacturing configurations. The table presents the average degree of impregnation for each layer, the manufacturing configurations, and the corresponding p-values for both large and narrow contour data.

Manufacturing Conditions	Degree of Impregnation (%)					
	Narrow Contour			Large Contour		
	Layer 1 (%)	Layer 3 (%)	Layer 5 (%)	Layer 1 (%)	Layer 3 (%)	Layer 5 (%)
Cr_0%	90.34 ± 3.46	99.95 ± 0.07	93.23 ± 5.86	92.28 ±3.23	99.95 ±0.06	93.39 ± 5.58
Cr_30%	24.44 ± 6.14	16.66 ± 8.91	70.88 ± 10.10	25.60 ± 6.23	17.08 ± 9.2	71.77 ± 10.38
Cr_41%	50.51 ± 12.28	67.95± 12.50	99.67 ± 0.31	54.32 ± 13.49	71.41 ± 12.31	99.67 ± 0.26
	p-value					
Hypothesis 1	0.0530			0.0463		
Hypothesis 2	0	0	0	0	0	0

4.4.2. Hypothesis 2: Considering the Distinction between Composite Layers

With consideration of a separation between manufacturing conditions and the layers from which bundles are extracted, the null hypothesis corresponding to the ANOVA analysis performed separately for each layer posits that there is no significant difference in the means of the porosity values across the different manufacturing methods (A, B, and C) for each layer. Specifically, for each layer, the null hypothesis asserts that the average porosity is the same regardless of the manufacturing method used: For layer 1, the null hypothesis (H0) states that the mean porosity for the manufacturing conditions of plate Cr_0% (Method A) equals the mean porosity for the manufacturing conditions of plate Cr_30% (Method B) and the manufacturing conditions of plate Cr_41% (Method C); similarly, for layer 2 and layer 3, H0 maintains that the mean porosity for Method A equals that of Methods B and C. Conversely, the alternative hypothesis for each layer suggests that at least one of the manufacturing methods has a different mean porosity compared to the others, indicating a significant effect of the manufacturing method on porosity within each layer. As indicated from results from the ANOVA analysis in Table 3, all p-values in this context are equal to zero, meaning a strong rejection of the null hypothesis and confirming a high variability index on the porosity levels of each layer while changing the process.

5. Assessing Uncertainty in Porosity Quantifications from FM- and SEM-Based Approaches

To further consolidate the previous results, an evaluation of the degree of uncertainty was conducted based on a quantitative assessment of the surface area percentage (SAP) of dry zones (including porosity and unfilled space within the preform) obtained from the multimodal approach, including (i) FM images from Figure 9, concentration levels 2, 3, and 4 of the fluorescent agent and (ii) SEM images from Figure 8, pixels corresponding to GF. In the case of SEM only-based data, only pixels corresponding to unfilled zones and epoxy mount, as indicated in the segmentation in Figure 8, were used. Continuing the previous sections, the analysis was limited to the 0°-oriented cross-section of the fiber bundles. The data obtained were added to Figure 14, with the legend modified to include additional surface area percentages of porosities based on SEM images without affecting the data. For the Cr_0% sample, the SEM-based quantification of the surface area percentage (SAP) of porosity (dry zones), as shown in Figure 14a, is higher than dry zones infiltrated by the fluorescent agent-enriched mount epoxy, according to the multimodal-based quantification method. Specifically, the average SAP of porosity based on SEM is approximately 23.4% ($\pm$2.3%), while the multimodal-based measurement relying on FM is about 2.9% ($\pm$3.0%). When using a tighter contour of the meshes (in red color in Figure 12f), the average values for both cases are not significantly different. As observed from Figure 9a and its corresponding segmented FM image in Figure 9d, there is no significant presence of high concentrations (levels 2 to 4, from Figure 9d) of the fluorescent agent in layers 1 and especially layer 3 of the bundle cross-sections oriented at 0°. This qualitative verification suggests that the SEM-based quantification may overestimate the surface area percentage of porosity (dry zones) due to the limited grayscale distinction between PP and epoxy mount using the BED-C mode, as indicated by the histogram peaks in Figure S4 (for grayscale levels ranging between 25 and 150). Indeed, the signal drift during SEM acquisitions, while collecting the 850 tiles in the case of the Cr_0% sample, is associated with local changes in the gray levels of PP and mounting epoxy between the tiles, making it challenging for user-assisted Random Forest classifiers to accurately distinguish between PP and epoxy-rich pixels (check segmentation details B and C in Figure 8d). On the other hand, as the Cr_0% sample was extracted from a plate manufactured based on an alternated film stacking according to configuration 1 (Figure 1c), it is expected that PP impregnates easily the adjacent GF plies. With consideration of all previous observations related to the Cr_0% sample, the uncertainty level is higher based on the SEM images compared to the multimodal approach integrating FM-based quantification of the surface area percentage of porosity based on the trace of infiltrated fluorescent agent under UV light.

For the 0°-oriented bundles in the plate Cr_30%, the SEM-based surface area percentage (SAP) of porosity (dry zones) is approximately of 6.1% ($\pm$2.6%). The SAP of porosity (dry zones) based on the FM quantification approach is around 28.1% ($\pm$11.4%), using the green contour of the GF bundles. In addition, the histogram of grayscale levels for the macro-scale SEM image of the Cr_30% sample in Figure S4 shows at least four distinctive peaks for gray levels ranging from 25 to nearly 125. These peaks indicate a significant gray level drift during the acquisition of the corresponding 348 tiles, making it challenging to accurately differentiate between PP and epoxy, similar to the Cr_0% case. In contrast, the FM image in Figure 9b and its corresponding segmented image in Figure 9e show high concentrations of the fluorescent agent. The FM-based quantification approach of porosity also demonstrates sensitivity to the degree of impregnation, as seen in the SAP of porosity (dry zones) in layer 5, which is closest to the PP layer in the plate manufactured according to configuration 2 (Figure 1c), compared to layers 1 and 3, which are expected to be less impregnated as they are more distant than layer 5 from the pool of melt PP during manufacturing.

The SEM-based quantification of SAP of dry zones seems to underestimate the porosity level in the 0°-oriented bundles. The Random Forest classifier appears to overestimate the pixels corresponding to PP compared to those corresponding to the epoxy used as a resin mount, especially in layers 1 and 3, with no significant change in SAP between bundles extracted from layers 3 and 5, as shown in Figure 14b. For bundles corresponding to layer 5 from the Cr_30% sample, the average SAP of porosity from SEM-based quantification is about 3.9% ($\pm$0.9%), while the SAP based on the FM approach is approximately of 12.4% ($\pm$4.24%). This suggests that, for 0°-oriented bundles with a relatively high impregnation level, notably layer 5, the FM quantification approach provides roughly a one order of magnitude greater accuracy compared to solely using SEM. For 0°-oriented bundles with limited impregnation levels, such as in layers 3 and 1, the risk of underestimation of the SAP of porosity when relying only on SEM images is around 20%. This estimation considers the mean SAP for layers 1 and 3 in the Cr_30% case, which are of 7.1% and 35.2%, respectively. When compared to relatively high impregnation bundles extracted from the Cr_0% sample, SEM seems to have about 807% overestimation of porosity levels compared to the FM-based approach. These estimations highlight the sensitivity of SEM acquisitions and the difficulties related to pixel classifications, especially for PP and epoxy based on the BED-C mode. This issue requires further investigation into correcting the signal brightness of collected SEM tiles and quantifying the error propagation due to acquisition artefacts to assess the uncertainty level of segmenting PP and epoxy in the context of epoxy-based resin mounts frequently used in microstructure characterizations of thermoplastic composite materials. These suggested investigations were considered to be outside the scope of the current study.

In the case of fiber bundles in sample Cr_41%, Figure 14c indicates less disparity in the SAP of porosity (dry zones) compared to samples Cr_0% and Cr_30%. These observations align with the assumptions made about layer 5 of the Cr_30% sample. Specifically, for the Cr_41% sample, the SAP of porosity based on SEM images for layers 5, 3, and 1 were 5.3% ($\pm$2.5%), 5.2% ($\pm$1.5%), and 12.9% ($\pm$1.3%), respectively, compared to 0.2% ($\pm$0.13%), 13.6% ($\pm$5.5%), and 24.1% ($\pm$5.7%) based on the FM-based approach. These values indicate an underestimation of porosity levels of 38.2% for layer 3 and 53.5% for layer 1 based only on the SEM image. The uncertainty levels for layers 1 and 3 seem consistent with layer 5 of the Cr_30% sample. However, for layer 5 of the Cr_41% sample, there is an overestimation level of 2650% of SEM-based SAP of porosity compared to the FM-based approach. This observation is in the context of the high impregnation level of layer 5 for the sample Cr_41%; thus, FM can also be considered to underestimate the SAP of porosity (dry zones) due to the limited infiltration of the fluorescent agent-enriched epoxy mount and the consideration of fluorescence concentration levels higher that level 2, as indicated by the scale of concentration levels provided in Figure 9. This overestimation of porosity by SEM compared to the FM-based approach aligns with the Cr_0% sample, where most layers are expected to be well impregnated. The case of layer 5 in the Cr_41% sample indicates that the multimodal approach associating FM and SEM techniques can be reliable for quantifying the SAP of porosity in partially impregnated bundles; however, more precautions are required, especially in the case of fully impregnated bundles.

6. Conclusions

This study highlighted the potential of combining multimodal and extended-field imaging, principally integrating PLM, SEM, and FM techniques, for inspecting the surface quality after mechanical polishing and for assessing the degree of impregnation at the meso-scale of 0°-oriented bundles of glass fibers. The multimodal approach focused on FM and SEM extended-field images, providing a robust characterization of fiber bundles based on SEM images quantifying single fibers and area fractions of GF, and porosity based on the FM extended-field analysis of the concentration levels of the fluorescence-enriched epoxy resin mount. The methodology involved detailed experimental workflows for surface preparation to minimize defects during the mechanical polishing of partially impregnated polymer matrix composites. Two-scale post-processing workflows of multimodal images were developed, including image alignment and pixel classification techniques at the macro-scale, ensuring precise quantitative analyses at the meso-scale of fiber bundles oriented perpendicularly to the polishing plane (0°-oriented bundles). Bundle contours at the meso-scale were objectively identified based on normalized distance maps following the object classification of single glass fibers. Large and narrow contours were defined, with the large contour encompassing all single glass fibers and the narrow contour including only single fibers with a normalized distance to their nearest neighbors lower than 0.4 to exclude isolated single fibers detached from the bundles. The image post-processing workflow at the meso-scale was comprehensively evaluated to assess the number of single GF filaments per bundle, the area fractions of glass fibers (GFs), and porosity in 0°-oriented fiber bundles across three manufacturing configurations. Analyses utilizing pixel classifications of SEM images, followed by an automated object classification procedure, revealed that full-scale extended-field images at the considered resolutions indicated an average of 4038 single filaments per bundle, with a standard deviation of 63, indicating that most data points fell within the 95% confidence interval. The GF area fractions indicated consistent impregnation levels for the film stacking sample, with percentages ranging between 44.8% and 47.4%, indicating fewer GF fractures. Higher GF area fractions were observed in samples Cr_30% and Cr_41%, suggesting less significant impregnation levels by the simplified-CRTM manufacturing process with changing compaction ratios, but the levels were, respectively, around 53.7% and 57.3%, indicating relatively controlled single fiber damage following the four-step surface preparation protocol. Porosity area fractions were minimal in the film stacking sample, whereas higher levels were detected in Cr_30% and Cr_41%, with distinctive impregnation gradients noticed between layers 1, 3, and 5 in Cr_41%, reflecting more representative through thickness flow impregnation. The statistical analysis using a one-way ANOVA confirmed significant differences in impregnation degrees across different manufacturing configurations, particularly at a 10% risk level. Furthermore, the separate layer analysis demonstrated significant variability in porosity levels due to manufacturing conditions, with all p-values being zero. Overall, the findings underscore the significant impact of multimodal imaging for the quantitative analysis of the degree of impregnation and porosity within fiber bundles. Regarding manufacturing configurations, the statistical results from the current study and the datasets of fiber bundles indicate the need for more precise control in the manufacturing process to achieve more consistent process-related correlations. Additionally, the efficacy of multimodal imaging as a powerful tool for detailed inspection and quality control in composite material production was demonstrated. Its implementation could lead to a better understanding and optimization of compression molding and CRTM manufacturing processes, which constitute an active research area for the authors. The current study also generated a significant dataset of multimodal images of $0°$-oriented fiber bundles, which opens the door to interdisciplinary topics related to machine learning-based image post-processing approaches in the context of thermoplastic matrix composites.

Supplementary Materials: The following supporting information can be downloaded at: https://www.mdpi.com/article/10.3390/polym16152171/s1, Appendix S1: Bibliometric analysis: Verification of literature gap assumptions. Appendix S2: Full-Scale, Extended-Field Acquisitions and Image Analyses. Appendix S3: Individual bundles extracted from Cr_0%, Cr_30% and Cr_41% samples based on SEM raw data, normalized individual distances between single GF, bundle contours, raw FM images and classified pixels of FM images. Appendix S4: Histogram analysis of gray levels of macro-scale SEM images of Cr_0%, Cr_30% and Cr_41%.

Author Contributions: Conceptualization and supervision by A.A.; methodology, investigation, software, validation, and data curation by S.S. and A.A.; original draft by S.S.; review and editing by A.A.; project administration by M.L.D. All authors have read and agreed to the published version of the manuscript.

Funding: This research received no external funding.

Institutional Review Board Statement: Not applicable. Our study does not involve humans or animals.

Data Availability Statement: The datasets for this manuscript are the property of IMT Nord Europe and are not publicly available. Requests to access the primary data should be addressed to the corresponding author. The authors are willing to share the data in the form of an open access article.

Acknowledgments: The authors acknowledge the European Regional Development Fund FEDER, the French state and the Hauts-de-France Region council for co-funding the PhD grant of S. Sildipura. The authors acknowledge also Dr. Vincent THIERY for his advice/training to use the 2D microscopy equipment (IMT Nord Europe, France).

Conflicts of Interest: The authors declare no conflicts of interest.

References

1. Merotte, J.; Simacek, P.; Advani, S.G. Flow analysis during compression of partially impregnated fiber preform under controlled force. *Compos. Sci. Technol.* **2010**, *70*, 725–733. [CrossRef]
2. Sidlipura, S.; Ayadi, A.; Lagardère-Delèglise, M. Multi-modal Imaging for Porosity Quantification in Partially-impregnated UD Woven Glass Fiber/Polypropylene Composites. In Proceedings of the 23rd International Conference on Composite Materials, Belfast, UK, 30 July–4 August 2023.
3. Purslow, D. On the optical assessment of the void content in composite materials. *Composites* **1984**, *15*, 207–210. [CrossRef]
4. Liu, L.; Zhang, B.M.; Wang, D.F.; Wu, Z.J. Effects of cure cycles on void content and mechanical properties of composite laminates. *Compos. Struct.* **2006**, *73*, 303–309. [CrossRef]
5. Abdelal, N.; Donaldson, S.L. Comparison of methods for the characterization of voids in glass fiber composites. *J. Compos. Mater.* **2018**, *52*, 487–501. [CrossRef]
6. Gagani, A.; Fan, Y.; Muliana, A.H.; Echtermeyer, A.T. Micromechanical modeling of anisotropic water diffusion in glass fiber epoxy reinforced composites. *J. Compos. Mater.* **2018**, *52*, 2321–2335. [CrossRef]
7. Ekoi, E.J.; Dickson, A.N.; Dowling, D.P. Investigating the fatigue and mechanical behaviour of 3D printed woven and nonwoven continuous carbon fibre reinforced polymer (CFRP) composites. *Compos. Part B Eng.* **2021**, *212*, 108704. [CrossRef]
8. Zou, A.; Shan, Z.; Wang, S.; Liu, X.; Ma, X.; Zou, D.; Jiang, X. Study on porosity of aramid fiber reinforced composites prepared by additive manufacturing. *Compos. Adv. Mater.* **2022**, *31*, 263498332211218. [CrossRef]
9. Kabachi, M.A.; Danzi, M.; Arreguin, S.; Ermanni, P. Experimental study on the influence of cyclic compaction on the fiber-bed permeability, quasi-static and dynamic compaction responses. *Compos. Part A Appl. Sci. Manuf.* **2019**, *125*, 105559. [CrossRef]
10. Breister, A.M.; Imam, M.A.; Zhou, Z.; Anantharaman, K.; Prabhakar, P. Microbial dark matter driven degradation of carbon fiber polymer composites. *bioRxiv* **2020**. [CrossRef]
11. Liu, B.; Xu, A.; Bao, L. Preparation of carbon fiber-reinforced thermoplastics with high fiber volume fraction and high heat-resistant properties. *J. Thermoplast. Compos. Mater.* **2017**, *30*, 724–737. [CrossRef]
12. Ishida, O.; Kitada, J.; Nunotani, K.; Uzawa, K. Impregnation and resin flow analysis during compression process for thermoplastic composite production. *Adv. Compos. Mater.* **2020**, *30* (Suppl. S1), 39–58. [CrossRef]
13. Little, J.E.; Yuan, X.; Jones, M.I. Characterisation of voids in fibre reinforced composite materials. *NDT E Int.* **2012**, *46*, 122–127. [CrossRef]
14. Eliasson, S.; Hagnell, M.K.; Wennhage, P.; Barsoum, Z. A Statistical Porosity Characterization Approach of Carbon-Fiber-Reinforced Polymer Material Using Optical Microscopy and Neural Network. *Materials* **2022**, *15*, 6540. [CrossRef] [PubMed]
15. Zhang, P. Correlative Cryo-electron Tomography and Optical Microscopy of Cells. *Curr. Opin. Struct. Biol.* **2013**, *23*, 763–770. [CrossRef] [PubMed]
16. Perkovic, M.; Kunz, M.; Endesfelder, U.; Bunse, S.; Wigge, C.; Yu, Z.; Frangakis, A.S. Correlative Light- and Electron Microscopy with chemical tags. *J. Struct. Biol.* **2014**, *186*, 205–213. [CrossRef] [PubMed]
17. Howes, S.C.; Koning, R.I.; Koster, A.J. Correlative microscopy for structural microbiology. *Curr. Opin. Microbiol.* **2018**, *43*, 132–138. [CrossRef] [PubMed]
18. Mitchell, R.L.; Davies, P.; Kenrick, P.; Volkenandt, T.; Pleydell-Pearce, C.; Johnston, R. Correlative Microscopy: A tool for understanding soil weathering in modern analogues of early terrestrial biospheres. *Sci. Rep.* **2021**, *11*, 12736. [CrossRef] [PubMed]
19. Su, Y.; Nykanen, M.; Jahn, K.A.; Whan, R.; Cantrill, L.; Soon, L.L.; Braet, F. Multi-dimensional correlative imaging of subcellular events: Combining the strengths of light and electron microscopy. *Biophys. Rev.* **2010**, *2*, 121. [CrossRef]
20. Arif, M.; Mahmoud, M.; Zhang, Y.; Iglauer, S. X-ray tomography imaging of shale microstructures: A review in the context of multiscale correlative imaging. *Int. J. Coal Geol.* **2021**, *233*, 103641. [CrossRef]
21. Amedewovo, L.; Levy, A.; Du Plessix BD, P.; Aubril, J.; Arrive, A.; Orgéas, L.; Le Corre, S. A methodology for online characterization of the deconsolidation of fiber-reinforced thermoplastic composite laminates. *Compos. Part A Appl. Sci. Manuf.* **2023**, *167*, 107412. [CrossRef]
22. Kirby, A. Exploratory Bibliometrics: Using VOSviewer as a Preliminary Research Tool. *Publications* **2023**, *11*, 10. [CrossRef]
23. van Eck, N.J.; Waltman, L. Software survey: VOSviewer, a computer program for bibliometric mapping. *Scientometrics* **2010**, *84*, 523–538. [CrossRef] [PubMed]
24. Ayadi, A.; Deléglise-Lagardère, M.; Park, C.H.; Krawczak, P. Analysis of Impregnation Mechanism of Weft-Knitted Commingled Yarn Composites by Staged Consolidation and Laboratory X-ray Computed Tomography. *Front. Mater.* **2019**, *6*, 255. [CrossRef]
25. Mulle, M.; Wafai, H.; Yudhanto, A.; Lubineau, G.; Yaldiz, R.; Schijve, W.; Verghese, N. Influence of process-induced shrinkage and annealing on the thermomechanical behavior of glass fiber-reinforced polypropylene. *Compos. Sci. Technol.* **2019**, *170*, 183–189. [CrossRef]
26. Liu, F.; Li, T.; Xu, F.; Li, J.; Jiang, S. Microstructure, Tensile Property, and Surface Quality of Glass Fiber-Reinforced Polypropylene Parts Molded by Rapid Heat Cycle Molding. *Adv. Polym. Technol.* **2020**, *2020*, 3161068. [CrossRef]
27. Tanimoto, Y.; Nagakura, M. Effects of polishing on surface roughness and hardness of glass-fiber-reinforced polypropylene. *Dent. Mater. J.* **2018**, *37*, 1017–1022. [CrossRef] [PubMed]
28. Zukić, D.; Jackson, M.; Dimiduk, D.; Donegan, S.; Groeber, M.; McCormick, M. ITKMontage: A Software Module for Image Stitching. *Integr. Mater. Manuf. Innov.* **2021**, *10*, 115–124. [CrossRef]

29. Preibisch, S.; Saalfeld, S.; Tomancak, P. Globally optimal stitching of tiled 3D microscopic image acquisitions. *Bioinformatics* **2009**, *25*, 1463–1465. [CrossRef] [PubMed]
30. Cardona, A.; Saalfeld, S.; Schindelin, J.; Arganda-Carreras, I.; Preibisch, S.; Longair, M.; Douglas, R.J. TrakEM2 Software for Neural Circuit Reconstruction. *PLoS ONE* **2012**, *7*, e38011. [CrossRef]
31. Kreshuk, A.; Zhang, C. Machine Learning: Advanced Image Segmentation Using ilastik. *Methods Mol. Biol.* **2019**, *2040*, 449–463. [CrossRef]
32. Berg, S.; Kutra, D.; Kroeger, T.; Straehle, C.N.; Kausler, B.X.; Haubold, C.; Kreshuk, A. ilastik: Interactive machine learning for (bio)image analysis. *Nat. Methods* **2019**, *16*, 1226–1232. [CrossRef] [PubMed]
33. Sommer, C.; Straehle, C.; Kothe, U.; Hamprecht, F.A. Ilastik: Interactive learning and segmentation toolkit. In Proceedings of the 2011 IEEE International Symposium on Biomedical Imaging: From Nano to Macro, Chicago, IL, USA, 30 March–2 April 2011; pp. 230–233. [CrossRef]
34. Breiman, L. Random forests. *Mach. Learn.* **2001**, *45*, 5–32. [CrossRef]
35. Pisupati, A.; Ayadi, A.; Deléglise-Lagardère, M.; Park, C.H. Influence of resin curing cycle on the characterization of the tensile properties of flax fibers by impregnated fiber bundle test. *Compos. Part A Appl. Sci. Manuf.* **2019**, *126*, 105572. [CrossRef]

Disclaimer/Publisher's Note: The statements, opinions and data contained in all publications are solely those of the individual author(s) and contributor(s) and not of MDPI and/or the editor(s). MDPI and/or the editor(s) disclaim responsibility for any injury to people or property resulting from any ideas, methods, instructions or products referred to in the content.

MDPI AG
Grosspeteranlage 5
4052 Basel
Switzerland
Tel.: +41 61 683 77 34

Polymers Editorial Office
E-mail: polymers@mdpi.com
www.mdpi.com/journal/polymers

Disclaimer/Publisher's Note: The title and front matter of this reprint are at the discretion of the . The publisher is not responsible for their content or any associated concerns. The statements, opinions and data contained in all individual articles are solely those of the individual Editors and contributors and not of MDPI. MDPI disclaims responsibility for any injury to people or property resulting from any ideas, methods, instructions or products referred to in the content.

www.ingramcontent.com/pod-product-compliance
Lightning Source LLC
LaVergne TN
LVHW070404100526
838202LV00014B/1385